An Introduction to Mesoscale Meteorology

Shou Shaowen   Li Shenshen   Shou Yixuan   Yao Xiuping

China Meteorological Press

An Introduction to Mesoscale Meteorology

  Mesoscale meteorology is one of the important branches of meteorology, studying majorly mesoscale atmospheric systems, which are responsible for numeral severe disasters, such as damaging winds, flash flooding, blizzards, thunderstorms, hailstorms and tornadoes etc. It is important to investigate the development mechanisms of the mesoscale systems and to develop their diagnosing and forecasting methods and therefore mesoscale meteorology is one of the most valued areas in atmospheric science researches. This book is an introduction to the basic knowledge about mesoscale meteorology. The contents of the book include the features and equation set of the mesoscale atmospheric motions, the topographically forced mesoscale systems, the gravity waves in free atmosphere, the front and jet stream, the mesoscale convective systems (MCSs), the atmospheric instabilities, the factors effecting the development of MCSs, and the diagnoses and forecasting of the mesoscale weather. This book may be used as the textbook for the students majoring in meteorology and as a reference material for the meteorological specialists and relative personnel.

图书在版编目(CIP)数据

中尺度气象学引论 ＝ An Introduction to Mesoscale Meteorology：英文 / 寿绍文等编著. — 北京：气象出版社，2020.9
  ISBN 978-7-5029-7289-9

Ⅰ.①中… Ⅱ.①寿… Ⅲ.①中尺度-气象学-研究-英文 Ⅳ.①P4

中国版本图书馆 CIP 数据核字(2020)第 186009 号

An Introduction to Mesoscale Meteorology
By Shou Shaowen, Li Shenshen, Shou Yixuan, and Yao Xiuping
Responsible editor: Huang Hongli

Copyright © 2020 by China Meteorological Press
Published by China Meteorological Press
No. 46, Zhongguancun Nandajie, Haidian District, Beijing 100081, China

http://www.qxcbs.com
E-mail: qxcbs@cma.gov.cn
Tel: 86-10-68407112

Printed in Beijing
First published in Sept. 2020

Preface

Mesoscale meteorology is an important branch of meteorology, studying majorly mesoscale atmospheric systems, which are responsible for numeral severe disasters, such as damaging winds, flash flooding, blizzards, thunderstorms, hailstorms and tornadoes etc. So that it is one of the most attractive research areas in meteorology for meteorologists. This book is an introduction to the knowledge of mesoscale meteorology.

The major contents in this book include the features and equation set of the mesoscale atmospheric motions, the topographically forced mesoscale systems, the gravity waves in free atmosphere, the front and jet stream, the mesoscale convective systems (MCSs), the instabilities of atmosphere, the factors influencing the development of MCSs, and the mesoscale weather diagnosis and forecasting etc.

This book is based on the textbooks for the courses " mesoscale meteorology " and "mesoscale dynamics of atmosphere", that were published in Chinese by Meteorological Press and High Educational Press of China respectively.

Authors of this book gratefully acknowledge all the encouragement and supports from the National characteristic specialty construction project, Jiangsu university brand specialty construction project subsidy and Jiangsu high education research project (PPZY2015A016), the China Meteorological Administration (CMA) and Nanjing University of Information Science & Technology (NUIST) joint textbook construction project, and the National Natural Science Foundation of China (NSFC) (Grant Nos.

41575048, 41965001 and 41475041). The authors also gratefully acknowledge the generous help of numerous colleagues and technical experts.

Authors

Sept. 1st, 2020

Contents

Preface

Chapter 1　Features and Equation Set of Mesoscale Atmospheric Motions ……(1)
§ 1.1　The scale division of atmospheric motion systems …………………(1)
§ 1.2　Basic features of mesoscale atmospheric motions …………………(7)
§ 1.3　Equations set for describing mesoscale motions …………………(10)
　References ………………………………………………………………………(18)

Chapter 2　Topographically Forced Mesoscale Circulations ……………(20)
§ 2.1　Topographic waves ……………………………………………………(20)
§ 2.2　Circulations in wake …………………………………………………(31)
§ 2.3　Urban heat island circulation ………………………………………(35)
§ 2.4　Sea-land breeze ………………………………………………………(39)
§ 2.5　Mountain and valley breeze …………………………………………(51)
　References ………………………………………………………………………(57)

Chapter 3　Gravity Waves in Free Atmosphere ……………………………(61)
§ 3.1　Basic features of gravity waves ……………………………………(61)
§ 3.2　Dynamic features of gravity waves …………………………………(64)
§ 3.3　Structure and influences of gravity waves …………………………(70)
§ 3.4　Development of gravity waves ………………………………………(73)
§ 3.5　Development of the gravity waves near the upper level jet streak ………(79)
　References ………………………………………………………………………(82)

Chapter 4　Front and Jet Stream ……………………………………………(84)
§ 4.1　Structure of the front …………………………………………………(84)
§ 4.2　Kinematic and thermodynamic frontogenesis ………………………(88)
§ 4.3　Dynamic frontogenesis ………………………………………………(90)
§ 4.4　The factors influencing frontogenesis ………………………………(97)
§ 4.5　Frontal lateral secondary circulation ………………………………(99)
§ 4.6　Jet stream ………………………………………………………………(104)
§ 4.7　Mesoscale fronts in boundary layer …………………………………(110)

References ·· (126)

Chapter 5 Mesoscale Convective Systems ································ (128)

§ 5.1 Isolated convective systems ······································ (128)

§ 5.2 Belt-shaped convective systems ································ (158)

§ 5.3 Mesoscale convective complex (MCC) ························ (191)

References ·· (197)

Chapter 6 Atmospheric Instabilities ······································ (202)

§ 6.1 Conditional instability ··· (203)

§ 6.2 Conditional instability of second kind (CISK) ············· (210)

§ 6.3 Wave-CISK ·· (216)

§ 6.4 Inertial instability ··· (220)

§ 6.5 Conditional symmetric instability ······························ (222)

§ 6.6 Kelvin-Helmholtz instability ···································· (237)

References ·· (240)

Chapter 7 Factors Effecting the Development of MCSs ················ (242)

§ 7.1 The relationship between the atmospheric potential instability and convection ··· (242)

§ 7.2 The factors influencing on convective clouds ·············· (243)

§ 7.3 The effect of the vertical wind shear on propagation of convective storm ·· (246)

§ 7.4 The comprehensive effect of the environmental thermal and dynamic conditions on the intensity and types of the convective storms ······ (249)

§ 7.5 The effect of the vertical wind shear on the organization and splitting of the storms ·· (252)

§ 7.6 The effect of the vertical wind shear on formation of tornado storms ··· (259)

References ·· (265)

Chapter 8 Mesoscale Weather Diagnosis Analysis ························ (269)

§ 8.1 ω equation ··· (269)

§ 8.2 Analysis of Q vector ··· (274)

§ 8.3 Analysis of potential vorticity ··································· (285)

§ 8.4 Analysis of the helicity ·· (295)

§ 8.5 Analysis of the atmospheric instability ························ (298)

§ 8.6　Some case studies of the severe weather ·············· (310)

References ············· (319)

Chapter 9　Mesoscale Weather Forecasting ············· (323)

§ 9.1　Methodology of mesoscale weather forecasting ············· (323)

§ 9.2　Diagnosis and forecasting of heavy rain ············· (331)

§ 9.3　Analysis and forecasting of the severe convective weather ············· (338)

§ 9.4　Data and tools applied in nowcasting and VSRF ············· (346)

References ············· (350)

Chapter 1
Features and Equation Set of Mesoscale Atomospheric Motions

Atmospheric circulation systems have various scales and different features. The scale divisions of the atmospheric circulation systems and the governing equation set of the mesoscale motion systems will be discussed in this chapter.

§ 1.1 The scale division of atmospheric motion systems

The atmospheric circulations include various scale of motion systems ranged from turbulence eddy to ultra-long waves etc. They are normally divided into three categories, i. e. the large scale, the mesoscale and the small scale systems roughly with horizontal scale orders of 10^3 km, 10^2 km and 10^1 km respectively. The divisions are usually based on the empirical, theoretical and operational methods.

According to the synoptic map analysis and conventional observational experiences, people understood the large scale systems such as cyclones and fronts etc. as well as the small systems such as cumulus and tornadoes etc. quite early, while the concept of mesoscale was relatively new and originally based on the meteorological radar observational experiences. According to his radar observational experiences, Ligda (1951) pointed out that there was an important kind of weather systems for precipitation, that is too large to be observed integrally by in-situ sensors at a single station, but too small to be resolved by the conventional upper-air network. He suggested to call these systems as mesoscale systems. Based on this concept, the mesoscale may be defined as a time-space scale which is smaller than that of the conventional sounding network but much larger than that of a single thunderstorm cell. In other words, the mesoscale system is implied the system with horizontal spatial span about several ten kilometers to several hundred kilometers and the temporal span about several hours to ten hours.

Recently, people attempted to classify the weather systems objectively according to the theoretical methods based on statistics and dynamics theories. For example, Vinnichenko (1970) did the power spectrum analyses for the mean kinetic energy of the west-east wind components at the surface and the free atmosphere (3—20 km above ground) respectively. The results as shown in Fig. 1.1 show that there are the energy density peaks at the time period of minute, day and month, which indicated the influences of the turbulence in the planetary boundary layer (PBL), the diurnal change and the seasonal variation respectively, while there is a valley of the energy density at the time period of several ten minutes to several hours, which is called "mesoscale gap" and reflected the mesoscale variation. The above results reflected the existence of different scales objectively.

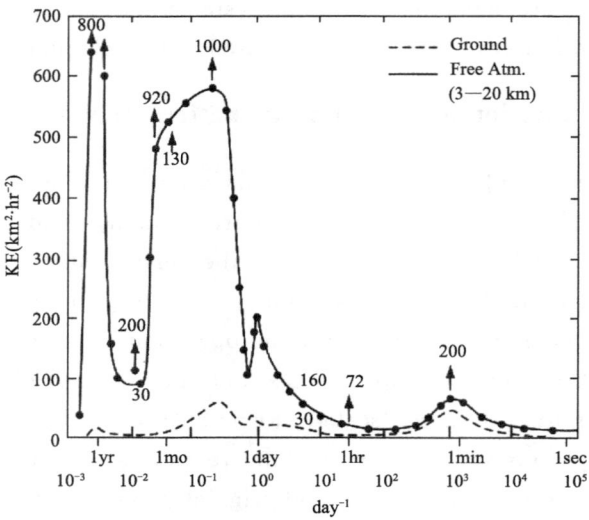

Fig. 1.1 Average kinetic energy of west-east wind component in the free atmosphere (solid line) and near the ground (dashed line). (Adapted after Vinnichenko, 1970)

More people attempt to classify the weather systems according to their atmospheric dynamic natures. One of the ways to classify the scale of the systems is based on the frequencies of the atmospheric oscillations. It is well known that the Earth's atmosphere has three basic frequencies: the buoyancy (Brunt-Väisälä) frequency N^2, the inertial frequency f, and the planetary frequency P. Where

$$N^2 = \left(\frac{g}{\Theta}\right)\frac{\partial \Theta}{\partial z}, \quad f = 2\Omega\sin\Phi, \quad P = (U\beta)^{1/2}$$

and Θ is potential temperature, Ω is the earth rotational angle velocity, Φ is latitude, U is wind velocity, and β is the change with latitude of the Coriolis parameter f. The magnitude orders of N^2, f and P are 10^{-2} s^{-1}, 10^{-4} s^{-1} and 10^{-6} s^{-1} respectively, and the corresponding time periods are minutes, hours and days respectively. Taking the above three basic frequencies as the standards, a four-fold division of atmospheric circulations is defined as follows:

$$F>N, \quad f<F<N, \quad P<F<f, \quad F<P$$
Small scale, Mesoscale, Synoptic scale, Planetary scale

In the above scale division, F is the frequency of the arbitrary atmospheric oscillation. According to the division, mesoscale may be defined as the scale of air flow with a frequency larger than f but smaller than N. So that the buoyancy and the earth rotation are the two fundamental variables for describing the mesoscale circulation.

The Richardson number, $Ri=N^2/(\frac{\partial U}{\partial z})^2$, including the effects of buoyancy and vertical wind shear, which reflects the baroclinic property of the atmosphere. The Richardson number is related with the scale of motions. For the large scale Ri is big; for the small scale Ri is small; while for the mesoscale Ri is medium as shown in Table 1.3.

The Rossby number, $Ro=2\pi/fT$ or $Ro=U/fL$, where f is Coriolis parameter and T is Lagrangian time scale, U and L are the wind velocity and horizontal scale respectively, reflected the effect of the proportion of the ageostrophic wind and the geostrophic wind respectively. The large scale motion is nearly geostrophic wind, so that Ro is small, while the small scale motions are the ageostrophic motions with strong accelerations, so that their Ro should be big.

For the mesoscale, both Ri and Ro are in medium comparing with that of the large scale and small scale. Emanuel (1983) listed various atmospheric phenomena according to the values of Ro and T to get a series of the scale, with the relationship that the bigger the system scale the smaller the Ro; Vice versa, the smaller the system scale, the bigger the Ro number (Table 1.1, Table 1.3 and Fig. 1.2).

Table 1.1 Lagrangian time scale (T) and Rossby number (Ro) for typical atmospheric systems. (Adapted from Emanuel and Raymond, 1984)

Phenomena	T	$Ro=2\pi/(fT)$
Zonal mean circulation	$2\pi a/U$	U/af
Planetary (stationary) Rossby waves	$2\pi/\sqrt{u\beta}$	$\sqrt{u\beta}/f$
Cyclone, anticyclone	$2\pi\sqrt{Ri}/f$	$1/\sqrt{Ri}$
Classical front	$2a\sqrt{Ri}/f$	$1/\sqrt{Ri}$
Sea-land breeze	1 day	$1/(2\sin\phi)$

continued

Slantwise convection	$1/f$	2π
Tropical cyclone	$2\pi R/V_T$	V_T/Rf
Inertial gravity wave	$2\pi f^{-1} - 2\pi N^{-1}$	$1 - N/f$
Thunderstorm and cumulus	N_w^{-1}	$2\pi N_w/f$
Kelvin-Helmholtz(K-H) wave	$2\pi N^{-1}$	$2\pi N/f$
PBL turbulence	$2\pi h/U^*$	U^*/fh
Tornado	$2\pi R/V_T$	V_T/fR

Where, a=earth radius, f=Coriolis parameter, β=meridional gradient of f, U=mean zonal velocity scale, U^*=frictional velocity scale, h=depth of PBL, N=mean buoyancy frequency scale, N_w=moist buoyancy frequency scale, Ri=Richardson number=N^2/U_z^2, R=radius of maximum wind scale, V_T=scale of maximum tangential wind scale, ϕ=latitude, D=depth of instability layer, U_z=scale of zonal wind vertical shear.

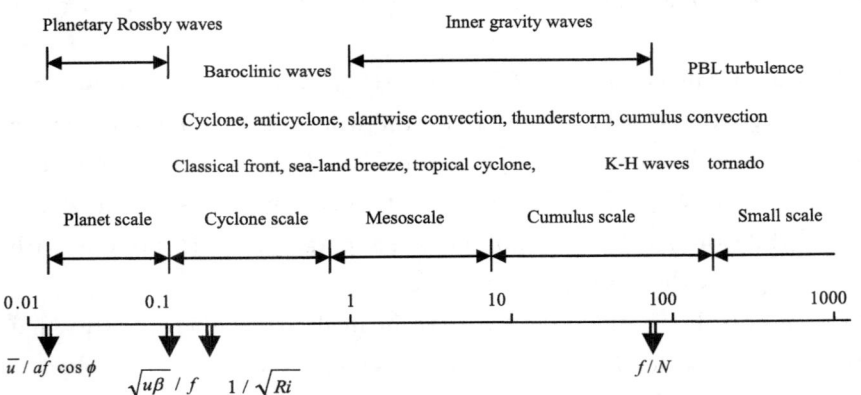

Fig. 1.2 The classification of scales according to the Lagrangian Rossby number.
(After Emanuel, 1986)

Beside the above three frequencies of the earth atmosphere, another important parameter has to be considered in the scale division of the atmospheric motions is the vertical wind shear ($U_z = \partial U/\partial z$) which reflected the non-uniform horizontal distribution of temperature on the earth surface, i.e., the baroclinic property of the earth atmosphere. Emanuel (1986) divided the air flows into four basic types, i.e., the baroclinic waves, the slantwise convection, the cumulus convection, and the PBL turbulence. They have different vertical scale, horizontal scale, time scale and Rossby number respectively as shown in the Table 1.2. Among them, the slantwise convection has the vertical scale D and the horizontal scale $U_z D/f$, so that the state ratio, the ratio between horizontal scale and vertical scale (L/D), is U_z/f. Emanuel defined the motion systems with the state ratio $L/D=U_z/f$, and the time scale $T=f^{-1}$ as the mesoscale motion systems. In the 6th chapter of this book we will discuss the atmospheric instabilities and we will understand that different atmospheric motions are related with different atmospheric instabilities and the motions with the above features (i. e., $L/D=U_z/f$, $T=f^{-1}$) are related with symmetric instability.

Table 1.2 Types of airflows and their features. (After K. A. Emanuel, 1986)

Type of motion	Vertical scale	Horizontal scale	Time scale	Rossby number
Baroclinic waves	$f^2Uz/(N^2)$	$fUz/(N)$	$2\pi N/(fUz)$	$1/Ri^{1/2}$
Slantwise convection	D	UzD/f	f^{-1}	2π
Cumulus convection	D	D	N^{-1}	$2N/f$
PBL turbulence	h	h	h/U^*	$2U^*/(hf)$

The symbols are same with that in Table 1.1.

According to the above definition of mesoscale motions given by Emanuel, we can get $L=U_z D/f \sim (U/D)(D/f) \sim U/f$, $T \sim 1/f$. If we put the typical values of wind velocity ($U \sim 10$ m·sec^{-1}) and Coriolis parameter ($f \sim 10^{-4}$ sec^{-1}) into the formulas, then we can get $L \sim 10^2$ km and $T \sim 10^4$ sec, in other words, the horizontal scale of the mesoscale motions is about several ten to several hundred kilometers, and the temporal scale of the mesoscale motions is several hours to ten hours. So the spatial and temporal scale derived from theoretical classification has good agreement with that from the experiential classification.

While in the real atmosphere the spectrum of the weather system scales is more complicated. For operational purpose it is necessary to divide the motion scales more detail. Orlanski (1975) suggested a scale division scheme based on the comprehensive analysis of the observational experiences and theoretical knowledge and has been widely adopted. According to Orlanski scheme, the weather systems can be roughly divided into three categories, i. e., the large scale (macro scale), mesoscale and small scale (micro scale) at first. And then the large scale may be further divided into Macro-α (>10000 km) and Macro-β (2000—10000 km); the mesoscale may be further divided into Meso-α (200—2000 km), Meso-β (20—200 km) and Meso-γ (2—20 km); meanwhile the small scale may be further divided into Micro-α (200—2000 m), Micro-β (20—200 m) and Micro-γ (2—20 m) respectively (Fig. 1.3).

Horizontal Scale	Lifetime	Stull (1988)	Pielke (2002)	Orlanski (1975)	Thunis and Bornstein (1996)	Atmospheric Phenomena
10000 km	1 month	Macro	Synoptic Regional	Macro-α	Macro-α	General circulation, long waves
				Macro-β	Macro-β	Synoptic cyclones
2000 km	1 week			Meso-α	Macro-γ	Fronts hurricanes, tropical storms, short cyclone waves, mesoscale convective complexes
200 km	1 day		Meso	Meso-β	Meso-β	Mesocyclones, mesohighs, supercells, squall lines, inertia-gravity waves, cloud clusters, low-level jets thunderstorm groups, mountain waves, sea breezes
20 km	1 h	Meso		Meso-γ	Meso-γ	Thunderstorms, cumulonimbi, clear-air turbulence, heat island, macrobursts
2 km				Micro-α	Meso-δ	Cumulus, tornadoes, microbursts, hydraulic jumps
200 m	30 min	Micro	Micro	Micro-β	Micro-β	Plumes, wakes, waterspouts, dust devils
20 m	1 min			Micro-γ	Micro-γ	Turbulence, sound waves
2 m	1 s	Micro-δ			Micro-δ	

Fig. 1.3 Scale divisions by I. Orlanski, Thunis and others.
(Adapted after Thunis and Bornstein, 1996)

According to the division of Orlanski, the mesoscale has a wide spatial span, which may be ranged from 2 km to 2000 km. So that the systems which may be as smaller as a thunderstorm or as larger as a hurricane or typhoon, are all belong to the category of mesoscale. While the "core" of the mesoscale is the meso-β, which has the typical features of the mesoscale motions, while the meso-α and meso-γ are the transition between mesoscale and large scale or between mesoscale and small scale respectively and they may have both the features of large scale and mesoscale or mesoscale and small scale respectively.

According to Orlanski's scale divisions, the motions with the time scale of month to day are governed by the parameter $(\beta L_R)^{-1}$, which is the inversion of the product of β, the Coriolis parameter changed with latitude, multiplied by L_R, the Rossby deformation radius, where, $L_R = (H/f)[(g/\theta)(d\theta/dz)]^{1/2}$ and H is the thickness of the homogeneous atmosphere. The motion with time scale one day to several hours, they are governed by the parameter f^{-1}, the inversion of the Coriolis parameter. While the motions with life cycle of several hours are governed by the inversion of the buoyancy frequency, $N^{-1}=[(g/\theta)(d\theta/dz)]^{-1/2}$. As to the outer gravity waves and the turbulence motions with the time scale of minutes and seconds are governed by the parameter $(g/H)^{-1/2}$ and L/u respectively.

The scale division of the atmosphere motion system by Orlanski is widely applied in the world although the standards of the scale divisions is still not totally unitary. A compare of the major divisions in the world is listed in the Fig. 1.3. From the figure we can see that the Macro-β is same with "synoptic scale", while the Meso-α is same with "sub-synoptic scale" or "medium scale". Some people call the Meso-α, β and γ as the large mesoscale, the typical mesoscale and the small mesoscale respectively.

Beside the above scale divisions, there are still various divisions for application purpose, for example, Fujita divided the atmospheric systems into five scales (Maso, Meso, Miso, Moso and Muso) and further divided the first four scales into α, β respectively. Meanwhile he also divided the small systems into the sub scales α, β, $\gamma \cdot \delta$ and ε for studying the special phenomena such as downburst, micro downburst, tornadoes and suction vortex etc.

§ 1.2 Basic features of mesoscale atmospheric motions

Based on the above discussions, it is clear that the mesoscale is a special scale different from the large scale and the small scale. Their basic features can be summarized as follows:

(1) According to the scale division given by I. Orlanski, the horizontal scale of

mesoscale is a wide extent ranged from 2×10^0 km to 2×10^3 km and the time scale ranges from several ten minutes to days. In general speaking, meso-β scale is the typical mesoscale, while Meso-α and Meso-γ systems have the features of both mesoscale and large scale or both mesoscale and small scales concurrently.

(2) The divergence D, the relative vorticity ζ and the vertical velocity W of the mesoscale motions are much stronger than that of the large scale. Since $D\sim V/L$, $\zeta\sim V/L$, and $W\leqslant HV/L$, where W and V are vertical velocity and horizontal wind, H and L are the vertical scale and horizontal scale respectively. Therefore the values of D and W are proportion to $1/L$, in other words, the smaller the horizontal scale, the greater the values of D and W. It is well known that the precipitation rate is proportional to the vertical velocity and the total rainfall is related with the precipitation rate and the life cycle of the precipitation system, therefore it is easy to understand why the heavy rain events are majorly caused by the mesoscale systems because they have much stronger vertical motion than the large scale systems although their life cycles are relatively shorter, while they have much longer life cycle than small scale systems although their vertical velocity maybe relatively weaker than small scale systems.

(3) The Coriolis force and buoyancy force have their different importance for different motion scale. In the large scale motions, the Coriolis force is relative important, while the buoyancy force can be ignored. In contrary for the small scale motions, the buoyancy force is relative important, while the Coriolis force can be ignored. However, in the mesoscale motions, both the Coriolis force and the buoyancy force are necessary to be considered. This nature can be expressed by Rossby number (Ro) and Richardson number (Ri). For the three basic motion scales, the typical values of Ro and Ri are listed in Table 1.3.

Table 1.3 Typical values of *Ro* and *Ri* for different scale systems.

	Large scale	Mesoscale	Small scale
$Ro=V/fL$	0.1	1.0	>1.0
$Ri=N^2/U_z^2$	100.0	1.0	<1

From Table 1.3, it can been seen that for large scale Ro is small and Ri is big; oppositely for small scale Ro is big and Ri is small. Since Ro is inversely proportion to the Coriolis force and Ri is also inversely proportion to the buoyancy force, therefore for larger scale the effect of Coriolis force is big and the effect of buoyancy force is small, and inversely for the small scale the effect of Coriolis force is small and the effect of buoyancy force is big. For smaller mesoscale systems the Coriolis force term is relatively small, the motion is ageostrophic motion, while for the larger mesoscale systems the Coriolis force term is relatively big so that the motion is quasi-geostrophic motion. Phillips (1963) introduced a parameter called Burger number B ($B=Ro^2Ri$), and defined two types of geostrophic motions: when $B\approx 10^{-2}$, the motion is called the second type geostrophic motion; when $B\approx 1$ the motion is called the first type of geostrophic motion. The larger mesoscale motion showed its geostrophic nature since it has the relation of $B\approx 1$. However, although both the large scale system and mesoscale systems possibly are satisfied the relation $B\approx 1$, they still can be distinguished obviously since they have different Ro and Ri. The above analysis showed that large scale motions are geostrophic and static balanced motion; the small scale motion is ageostrophic and non-static motion. The larger mesoscale motions can be the quasi-geostrophic and quasi static motion; the smaller mesoscale motions may be the ageostrophic and non-static motions; while typical mesoscale motions are possibly ageostrophic and quasi-static balanced motions. Therefore Pielke (1984) defined the mesoscale as such a special scale, which is big enough in horizontal scale so that the static balance relation is satisfied, while it is small enough so that the Coriolis force term is small relative to the advection term and pressure gradient force term. Under this situation the stream fields are different from the gradient wind and the geostrophic wind relations even under no frictional force situation above the planet boundary layer (PBL). Therefore it is not suitable to use the relations of geostrophic wind and gradient wind as the approximation of real wind fields for mesoscale analysis although the fluid static approximation still may be used to express vertical pressure distribution. But it should note that the veracity of the static relation is also related with the atmospheric stability and wind velocity, when atmosphere is stable and wind velocity is weak the static balance assumption is also suitable even for smaller systems, while the static relation will be less accuracy or even not suitable to use when the atmosphere stability decreased and the wind velocity increased.

(4) The adjustment process between pressure and wind fields is different for large scale motions and for mesoscale motions. For large scale motions, it is normally the wind field adjusts the pressure field, while for the mesoscale motions, it is inversely, the pressure field adjust the wind field. A simply explanation is as follows.

Considering an initial non limited oceanic water column, of which the horizontal extent is $2A$ and depth is H (Fig. 1.4). When the water flow is under geostrophic balance, the kinetic energy of the geostrophic flow for the unit length is E_{geo}, and the potential energy is P_{geo}. The ratio of kinetic energy and potential energy is

$$(E/P)_{geo} = L_R^2/L^2 \tag{1.2.1}$$

where $L_R = \sqrt{gH}/f$ is called as Rossby deformation radius, which is the spatial scale determined by the gravity force, the Earth rotation and the depth of fluid; L is horizontal scale of the motion system.

It can be seen from equation (1.2.1) that when $L \gg L_R$, then $E_{geo} \ll P_{geo}$. From this result we can know that if kinetic energy increase, it will need a great potential energy to restore the geostrophic balance, but if the potential energy increase, then just need a small amount of kinetic energy to restore the geostrophic balance. Obviously the later will be easy to be realized. Therefore for large systems the wind field will adjust pressure field, while for the mesoscale systems it is oppositely, the pressure field will adjust the wind field.

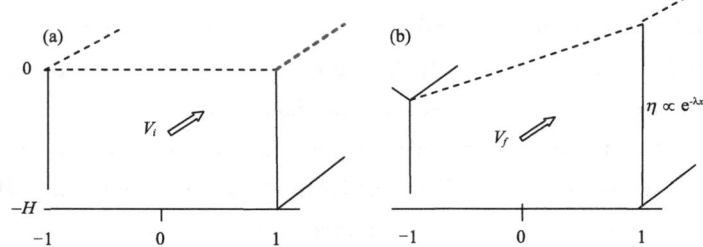

Fig. 1.4 (a) The state at initial time; (b) The state at final.

§ 1.3 Equations set for describing mesoscale motions

1.3.1 The original basic atmospheric dynamic and thermodynamic equations

In the Cartesian coordinates system (x, y, z), after neglected the turbulence diffuse, the atmospheric basic dynamic and thermodynamic equations are as following:

The motion equation:

$$\begin{cases} \dfrac{du}{dt} = -\dfrac{1}{\rho}\dfrac{\partial p}{\partial x} + fv \\[4pt] \dfrac{dv}{dt} = -\dfrac{1}{\rho}\dfrac{\partial p}{\partial y} - fu \\[4pt] \dfrac{dw}{dt} = -\dfrac{1}{\rho}\dfrac{\partial p}{\partial z} - g \end{cases} \tag{1.3.1}$$

The continuity equation:

$$\frac{d\rho}{dt} + \rho\left(\frac{\partial u}{\partial x} + \frac{\partial v}{\partial y} + \frac{\partial w}{\partial z}\right) = 0 \tag{1.3.2}$$

The state equation:
$$p = \rho RT \tag{1.3.3}$$

The potential temperature equation:
$$\theta = T\left(\frac{1000}{p}\right)^{AR/C_p} \tag{1.3.4}$$

The thermodynamic equation:
$$\frac{d\theta}{dt} = \frac{\theta}{C_p T}\frac{dQ}{dt} \tag{1.3.5}$$

where,
$$\frac{d}{dt} = \frac{\partial}{\partial t} + u\frac{\partial}{\partial x} + v\frac{\partial}{\partial y} + w\frac{\partial}{\partial z}$$

The above equations are normally not used for discussing mesoscale weather problems directly, since firstly the equation contains the acoustic waves, which is the noise for atmospheric motions; secondly the orders of the terms in the equations are different and can be simplified; and thirdly, some of the terms in the equations are non-linear terms, which represent the interactions between different meteorological fields and are important for mesoscale problems, while some of them, such as the pressure gradient force terms etc. may be linearized under some assumptions and let the problem simplified for mathematic processing.

1.3.2 The assumptions for simplifying the equations

Now let us discuss the mainstays for simplifying the equations. Firstly, to divide arbitrary atmospheric thermodynamic variable f into two parts, and let
$$f = \bar{f} + f' \tag{1.3.6}$$
where $\bar{f}$ is the synoptic scale background and f' is the mesoscale disturbance. Based on the observational experiences the following assumptions are made:

(1) The change with time of the synoptic scale state is much slow than that of the mesoscale disturbance, i. e.
$$\left|\frac{\partial \bar{f}}{\partial t}\right| \ll \left|\frac{\partial f'}{\partial t}\right| \tag{1.3.7}$$

(2) The horizontal gradient of the synoptic scale background is much smaller than that of mesoscale disturbance, i. e.
$$\left|\frac{\partial \bar{f}}{\partial x}\right| \ll \left|\frac{\partial f'}{\partial x}\right|, \quad \left|\frac{\partial \bar{f}}{\partial y}\right| \ll \left|\frac{\partial f'}{\partial y}\right| \tag{1.3.8}$$

(3) The mesoscale disturbance is much smaller than the synoptic scale background, i. e.

$$|f'/\bar{f}| \ll 1 \quad \text{or} \quad |f'| \ll |\bar{f}| \qquad (1.3.9)$$

1.3.3 The derivation of the Boussinesq and anelastic approach equations

(1) The motion equations

Based on the above assumptions, we can get the simplified horizontal motion equation easily as follows:

$$\begin{cases} \dfrac{du}{dt} = -\dfrac{1}{\bar{\rho}+\rho'}\dfrac{\partial(\bar{p}+p')}{\partial x} + fv \approx -\dfrac{1}{\bar{\rho}}\dfrac{\partial p'}{\partial x} + fv \\ \dfrac{dv}{dt} = -\dfrac{1}{\bar{\rho}+\rho'}\dfrac{\partial(\bar{p}+p')}{\partial y} - fu \approx -\dfrac{1}{\bar{\rho}}\dfrac{\partial p'}{\partial y} - fu \end{cases} \qquad (1.3.10)$$

Since in the vertical direction the pressure gradient force can be written as:

$$-\frac{1}{\rho}\frac{\partial p}{\partial z} = -\frac{1}{\bar{\rho}+\rho'}\left(\frac{\partial \bar{P}}{\partial z} + \frac{\partial p'}{\partial z}\right) \qquad (1.3.11)$$

Substituting the above equation (1.3.11) into the vertical motion equation and combining with the static equation ($\dfrac{d\bar{p}}{dz} = -\bar{\rho}g$), then we can get the vertical motion equation as following:

$$\begin{aligned}
\frac{dw}{dt} &= -\frac{1}{\bar{\rho}+\rho'}\left(\frac{\partial \bar{P}}{\partial z} + \frac{\partial p'}{\partial z}\right) - g \\
&= -\frac{1}{\bar{\rho}+\rho'}\frac{\partial p'}{\partial z} + \left(\frac{1}{\bar{\rho}+\rho'} - 1\right)g \\
&= -\frac{1}{\bar{\rho}+\rho'}\frac{\partial p'}{\partial z} - \frac{\rho'}{\bar{\rho}+\rho'}g \\
&\approx -\frac{1}{\bar{\rho}}\frac{\partial p'}{\partial z} - \frac{\rho'}{\bar{\rho}}g
\end{aligned} \qquad (1.3.12)$$

(2) The continuity equation

Changing the original continuity equation (1.3.2) into the following form:

$$\frac{\partial \rho}{\partial t} = -\left(u\frac{\partial \rho}{\partial x} + v\frac{\partial \rho}{\partial y} + w\frac{\partial \rho}{\partial z}\right) - \rho\left(\frac{\partial u}{\partial x} + \frac{\partial v}{\partial y} + \frac{\partial w}{\partial z}\right)$$

or

$$\frac{\partial \alpha}{\partial t} = -\left(u\frac{\partial \alpha}{\partial x} + v\frac{\partial \alpha}{\partial y} + w\frac{\partial \alpha}{\partial z}\right) + \alpha\left(\frac{\partial u}{\partial x} + \frac{\partial v}{\partial y} + \frac{\partial w}{\partial z}\right) \qquad (1.3.13)$$

where $\alpha = \dfrac{1}{\rho}$ is specific volume. Set $\alpha = \bar{\alpha} + \alpha'$, and satisfying the following in-equations:

Chapter 1 Features and Equation Set of Mesoscale Atomospheric Motions • 13 •

$$\left|\frac{\partial \bar{\alpha}}{\partial t}\right| \ll \left|\frac{\partial \alpha'}{\partial t}\right|, \quad \left|\frac{\partial \bar{\alpha}}{\partial x}\right| \ll \left|\frac{\partial \alpha'}{\partial x}\right|, \quad \left|\frac{\partial \bar{\alpha}}{\partial y}\right| \ll \left|\frac{\partial \alpha'}{\partial y}\right|, \quad |\alpha'/\bar{\alpha}| \ll 1$$

Substituting the above in-equations into the equation (1.3.13), we get

$$\frac{\partial \alpha'}{\partial t} = -\left(u \frac{\partial \alpha'}{\partial x} + v \frac{\partial \alpha'}{\partial y} + w \frac{\partial \alpha'}{\partial z}\right) - w \frac{\partial \bar{\alpha}}{\partial z} + \bar{\alpha}\left(1 + \frac{\alpha'}{\bar{\alpha}}\right)\left(\frac{\partial u}{\partial x} + \frac{\partial v}{\partial y} + \frac{\partial w}{\partial z}\right)$$

$$\approx -\left(u \frac{\partial \alpha'}{\partial x} + v \frac{\partial \alpha'}{\partial y} + w \frac{\partial \alpha'}{\partial z}\right) - w \frac{\partial \bar{\alpha}}{\partial z} + \bar{\alpha}\left(\frac{\partial u}{\partial x} + \frac{\partial v}{\partial y} + \frac{\partial w}{\partial z}\right)$$

(1.3.14)

By using the scale analysis method given by Dutton and Fichtl (1969) (adapted from Pielke, 1984) to estimate the magnitude of terms in the equation (1.3.14), we get

$$\left|\frac{\partial \alpha'}{\partial t}\right| \sim \frac{\alpha'}{t_a}, \quad \left|u \frac{\partial \alpha'}{\partial x}\right| \sim U \frac{\alpha'}{L_x}, \quad \left|v \frac{\partial \alpha'}{\partial y}\right| \sim V \frac{\alpha'}{L_y},$$

$$\left|w \frac{\partial \alpha'}{\partial z}\right| \sim W \frac{\alpha'}{L_z}, \quad \left|w \frac{\partial \bar{\alpha}}{\partial z}\right| \sim W \frac{\alpha'}{H_a}$$

$$\left|\bar{\alpha} \frac{\partial u}{\partial x}\right| \sim \bar{\alpha} \frac{U}{L_x}, \quad \left|\bar{\alpha} \frac{\partial v}{\partial y}\right| \sim \bar{\alpha} \frac{V}{L_y}, \quad \left|\bar{\alpha} \frac{\partial w}{\partial z}\right| \sim \bar{\alpha} \frac{W}{L_z}$$

(1.3.15)

where t_a^{-1} represents the characteristic frequency value of the variation of the mesoscale specific volume; U, V and W are the characteristic value of the horizontal and vertical components of the wind velocity respectively; L_x, L_y and L_z are the horizontal and vertical scales of the mesoscale disturbance respectively; H_a is the atmospheric density elevation, which is defined as follows:

$$H_a = \left|\frac{1}{\bar{\alpha}} \frac{\partial \bar{\alpha}}{\partial z}\right|^{-1} \quad \text{or} \quad H_a^{-1} = \left|\frac{1}{\bar{\alpha}} \frac{\partial \bar{\alpha}}{\partial z}\right|$$

(1.3.16)

It presents that the scale of the variation of the synoptic scale reference of the specific volume in vertical direction has the same magnitude order with the synoptic scale reference of the specific volume itself, i. e.

$$\frac{1}{\bar{\alpha}} \frac{\partial \bar{\alpha}}{\partial z} \sim \frac{1}{A} \frac{\Delta A}{H_a} \sim \frac{1}{H_a}$$

(1.3.17)

where A and ΔA are the scales of $\bar{\alpha}$ and its variation in vertical direction respectively; H_a is the length scale in vertical direction, i. e. the density elevation in vertical direction. In the earth atmosphere H_a is nearly 8 km.

Dividing the terms in equation (1.3.14) by $\bar{\alpha}(\partial w/\partial z)$, then the magnitudes of the terms may be estimated as follows:

$$\left|\frac{\partial \alpha'}{\partial t}\right| / \left|\bar{\alpha} \frac{\partial w}{\partial z}\right| \sim \frac{\alpha'}{\bar{\alpha}} \frac{L_z}{W t_a}$$

$$\left|u\frac{\partial \alpha'}{\partial x}\right| \Big/ \left|\bar{\alpha}\frac{\partial w}{\partial z}\right| \sim \frac{\alpha'}{\bar{\alpha}}\frac{UL_z}{WL_x}$$

$$\left|u\frac{\partial \alpha'}{\partial y}\right| \Big/ \left|\bar{\alpha}\frac{\partial w}{\partial z}\right| \sim \frac{\alpha'}{\bar{\alpha}}\frac{VL_z}{WL_y}$$

$$\left|w\frac{\partial \alpha'}{\partial y}\right| \Big/ \left|\bar{\alpha}\frac{\partial w}{\partial z}\right| \sim \frac{\alpha'}{\bar{\alpha}}$$

$$\left|w\frac{\partial \bar{\alpha}}{\partial z}\right| \Big/ \left|\bar{\alpha}\frac{\partial w}{\partial z}\right| \sim \frac{L_z}{H_a}$$

$$\left|\bar{\alpha}\frac{\partial u}{\partial z}\right| \Big/ \left|\bar{\alpha}\frac{\partial w}{\partial z}\right| \sim \frac{UL_z}{WL_x}$$

$$\left|\bar{\alpha}\frac{\partial v}{\partial z}\right| \Big/ \left|\bar{\alpha}\frac{\partial w}{\partial z}\right| \sim \frac{VL_z}{WL_y} \qquad (1.3.18)$$

It can be seen from the equation (1.3.18) that, since $|\alpha'/\bar{\alpha}| \ll 1$, so long as

$$L_z/(Wt_a) \sim 1, L_z/H_a \ll 1 \qquad (1.3.19)$$

Then the terms such as $\frac{\partial \alpha'}{\partial t}, u\frac{\partial \alpha'}{\partial x}, v\frac{\partial \alpha'}{\partial y}, w\frac{\partial \alpha'}{\partial z}, w\frac{\partial \bar{\alpha}}{\partial z}$ etc. in the equation (1.3.14) may be neglected. So that the equation (1.3.14) may be written as

$$\bar{\alpha}\left(\frac{\partial u}{\partial x}+\frac{\partial v}{\partial y}+\frac{\partial w}{\partial z}\right)=0 \quad \text{or} \quad \frac{\partial u}{\partial x}+\frac{\partial v}{\partial y}+\frac{\partial w}{\partial z}=0 \qquad (1.3.20)$$

The above equation is set up under the condition $L_z/H_a \ll 1$. This condition means that the vertical scale of the motion, i.e. the depth of the circulation is much smaller than the density elevation of the atmosphere. In other words this is the shallow convection situation. Therefore the equation (1.3.20) is called as the shallow convection continuity equation, which is also usually called as the assumption of incompressibility. Under the assumption not only the acoustic waves are removed, but also the spatial variation of density is ignored. Under the homogeneity fluid field (i.e., the density is constant) situation, this is the accurate form of the mass conservation, which is adapted by many mesoscale models to express the mass conservation.

(3) State equation

Substituting the relations $p=P+p'$, $T=\bar{T}+T'$, $\rho=\bar{\rho}+\rho'$ into the state equation $p=\rho RT$, we can get

$$P\left(1+\frac{p'}{P}\right)=\bar{\rho}\left(1+\frac{\rho'}{\bar{\rho}}\right)R\bar{T}\left(1+\frac{T'}{\bar{T}}\right)$$

Taking logarithmic operation for two sides of the above equation, we can get

$$\ln P\left(1+\frac{p'}{P}\right)=\ln \bar{\rho}\left(1+\frac{\rho'}{\bar{\rho}}\right)+\ln R\bar{T}\left(1+\frac{T'}{\bar{T}}\right)$$

Since $\ln P = \ln \bar{\rho} + \ln R \bar{T}$, so that

$$\ln\left(1+\frac{p'}{P}\right) = \ln\left(1+\frac{\rho'}{\bar{\rho}}\right) + \ln\left(1+\frac{T'}{\bar{T}}\right)$$

Since $\ln\left(1+\frac{p'}{P}\right) \approx \frac{p'}{P}$, $\ln\left(1+\frac{\rho'}{\bar{\rho}}\right) \approx \frac{\rho'}{\bar{\rho}}$, $\ln\left(1+\frac{T'}{\bar{T}}\right) \approx \frac{T'}{\bar{T}}$

therefore

$$\frac{p'}{P} \approx \frac{\rho'}{\bar{\rho}} + \frac{T'}{\bar{T}} \quad (1.3.21)$$

According to the observational data, the following in-equation is settled for most mesoscale weather processes:

$$\frac{p'}{P} \ll \frac{T'}{\bar{T}} \quad (1.3.22)$$

So that we have also the following approximate expression:

$$-\frac{\rho'}{\bar{\rho}} \approx \frac{T'}{\bar{T}} \quad (1.3.23)$$

It shows that in the vertical motion equation the buoyancy term is majorly caused by the temperature disturbance.

(4) The potential temperature equation

$$\theta = T\left(\frac{1000}{p}\right)^{AR/C_p} = \frac{p}{\rho R}\left(\frac{1000}{p}\right)^{AR/C_p}$$

Substituting the relations $\theta = \Theta + \theta'$, $T = \bar{T} + T'$, $\rho = \bar{\rho} + \rho'$ and $p = P + p'$ into the above equation and taking the logarithmic operation and then we can get

$$\ln\Theta\left(1+\frac{\theta'}{\Theta}\right) = \ln\frac{P(1+p'/P)}{R\bar{\rho}(1+\rho'/\bar{\rho})} + \kappa \ln\frac{1000}{P(1+p'/P)}$$

where $k = AR/C_p$, since $\ln\Theta = \ln\frac{P}{R\bar{\rho}} + \kappa \ln\frac{1000}{P}$, so that

$$\ln\left(1+\frac{\theta'}{\Theta}\right) = \ln\left(1+\frac{p'}{P}\right) - \ln\left(1+\frac{\rho'}{\bar{\rho}}\right) - \kappa \ln\left(1+\frac{p'}{P}\right)$$

$$= \frac{1}{\chi}\ln\left(1+\frac{p'}{P}\right) - \ln\left(1+\frac{\rho'}{\bar{\rho}}\right)$$

where $\chi = 1 - k = C_P/C_V$

Due to $\left(1+\frac{\theta'}{\Theta}\right) \approx \frac{\theta'}{\Theta}$, $\ln\left(1+\frac{p'}{P}\right) \approx \frac{p'}{P}$, $\ln\left(1+\frac{\rho'}{\bar{\rho}}\right) \approx \frac{\rho'}{\bar{\rho}}$

hence

$$\frac{\theta'}{\Theta} \approx \frac{1}{\chi}\frac{p'}{P} - \frac{\rho'}{\bar{\rho}} \quad (1.3.24)$$

By using the equation (1.3.21), the above equation can be changed as follows

$$\frac{\theta'}{\Theta} \approx \frac{T'}{\bar{T}} - k\frac{p'}{P} \quad (1.3.25)$$

According to the equation (1.3.22), we also have

$$\frac{\theta'}{\Theta} \approx \frac{T'}{T} \qquad (1.3.26)$$

(5) Thermodynamic equation

From the equation, in which the term $\Delta_a \theta$ has been omitted already, we can get

$$\frac{1}{T}\frac{dQ}{dt} = C_P \frac{d\ln\theta}{dt} = C_P \frac{d\ln T}{dt} - R \frac{d\ln P}{dt}$$

By using the state equation the thermodynamic equation may be changed into the form expressed by the pressure and density as following

$$\chi p \frac{d\ln\theta}{dt} = \frac{dp}{dt} - C_s^2 \frac{d\rho}{dt} \qquad (1.3.27)$$

where $C_s^2 = \chi RT$ is the adiabatic acoustic velocity

$$\frac{d\ln\theta}{dt} = 0 \quad \text{or} \quad \frac{dp}{dt} - C_s^2 \frac{d\rho}{dt} = 0 \qquad (1.3.28)$$

Expressing the adiabatic equation in the potential temperature form and by using the relation (1.3.6), we get

$$\frac{1}{\Theta}\frac{d\theta'}{dt} + SW = 0 \qquad (1.3.29)$$

Where $S = \frac{1}{\Theta}\frac{d\Theta}{dz}$. The above equation may be also expressed as follows

$$\frac{d\theta'}{dt} = -W \frac{d\Theta}{dz} = -\alpha w \qquad (1.3.30)$$

where

$$\alpha = \begin{cases} \frac{\partial \Theta}{\partial z} = \gamma_d - \gamma, & \text{when} \quad q < q_s \\ \frac{\partial \Theta}{\partial z} = \gamma_m - \gamma, & \text{when} \quad q \geq q_s \end{cases} \qquad (1.3.31)$$

γ_d and γ_m are the dry adiabatic lapse rate and the moist adiabatic lapse rate respectively; q and q_s are the specific humidity and saturation specific humidity respectively.

To sum up, we can get the mesoscale weather equation as follows:

$$\begin{cases} \frac{du}{dt} = -\frac{1}{\rho}\frac{\partial p'}{\partial x} + fv \\ \frac{dv}{dt} = -\frac{1}{\rho}\frac{\partial p'}{\partial y} - fu \\ \frac{dw}{dt} = -\frac{1}{\rho}\frac{\partial p'}{\partial z} - \frac{\rho'}{\rho}g \\ \frac{\partial u}{\partial x} + \frac{\partial v}{\partial y} + \frac{\partial w}{\partial z} = 0 \end{cases} \qquad (1.3.32)$$

Chapter 1 Features and Equation Set of Mesoscale Atomospheric Motions • 17 •

$$\begin{cases} \dfrac{p'}{P} = \dfrac{\rho'}{\rho} + \dfrac{T'}{T} \quad \text{or} \quad \dfrac{\rho'}{\rho} \approx -\dfrac{T'}{T} \\ \dfrac{\theta'}{\Theta} = \dfrac{T'}{T} - \kappa \dfrac{p'}{P} \quad \text{or} \quad \dfrac{\theta'}{\Theta} \approx \dfrac{T'}{T} \\ \dfrac{d\theta'}{dt} = -\alpha w \end{cases} \qquad (1.3.32)$$

In the derivation process of the above motion equations, the following approaches were made: ① in the horizontal motion equation, since the changes of density in horizontal direction is very small, so that $\dfrac{1}{\rho}$ is replaced by $\dfrac{1}{\bar{\rho}}$, and let the horizontal pressure gradient force term to be linearized; ② in the vertical motion equation, the buoyancy force caused by density disturbance was taken into account; ③ in the continuity equation, the atmosphere was assumed as incompressible, so that the acoustic waves are removed. The above approximate treatment is called Boussinesq approach. In the derivation of the above equation set, the fluid motion is confined to in a shallow layer, so that the simplified equations can just be applied for studying the mesoscale motions in a shallow layer such as cumulus convection, sea-land breeze, gravity waves in PBL etc.

While due to the big vertical span, the vertical change of the density, i. e., $\dfrac{d\bar{\rho}}{dz}$, has to be considered for the motions in a deep layer and can not allow to be ignored. For this circumstance, the continuity equation will be in following form:

$$w \dfrac{\partial \bar{\alpha}}{\partial z} - \bar{\alpha} \left(\dfrac{\partial u}{\partial x} + \dfrac{\partial v}{\partial y} + \dfrac{\partial w}{\partial z} \right) = 0 \quad \text{or}$$

$$w \dfrac{\partial \bar{\rho}}{\partial z} + \bar{\rho} \left(\dfrac{\partial u}{\partial x} + \dfrac{\partial v}{\partial y} + \dfrac{\partial w}{\partial z} \right) = 0 \qquad (1.3.33)$$

To set $\bar{\rho} = \bar{\rho}(z)$, then the formula (1.3.33) can be written as

$$\partial \bar{\rho} u / \partial x + \partial \bar{\rho} v / \partial y + \partial \bar{\rho} w / \partial z = 0 \qquad (1.3.34)$$

When Ogura and Phillips (1962) treated the deep convection, they neglected $\partial \rho / \partial t$ term in the continuity equation and meanwhile kept the term in the adiabatic equation. They call this approximate treatment as anelastic approximation.

Now assuming the atmosphere is stationary, the disturbance is three dimensional motion in x, y, z directions, i. e., the motion state is $u = u'$, $v = v'$, $w = w'$. Regardless of the earth rotation effect and taking anelastic approximation, then we have the equation set as following:

$$\begin{cases} \dfrac{\partial u'}{\partial t} = -\dfrac{1}{\bar{\rho}}\dfrac{\partial p'}{\partial x} \\ \dfrac{\partial v'}{\partial t} = -\dfrac{1}{\bar{\rho}}\dfrac{\partial p'}{\partial y} \\ \dfrac{\partial w'}{\partial t} = -\dfrac{1}{\bar{\rho}}\dfrac{\partial p'}{\partial z} - \dfrac{\rho'}{\bar{\rho}}g \\ \dfrac{\partial \bar{\rho} u'}{\partial x} + \dfrac{\partial \bar{\rho} v'}{\partial y} + \dfrac{\partial \bar{\rho} w'}{\partial z} = 0 \\ \dfrac{\partial p'}{\partial t} - C_s^2 \left(\dfrac{\partial \rho'}{\partial t} - \bar{\rho} w's \right) = 0 \end{cases} \quad (1.3.35)$$

Compared with the Boussinesq equations, one of the major differences of anelastic equations is that the effect of the term $\dfrac{\partial \bar{\rho}}{\partial z}$ is taken into consideration. Thus the anelastic equations can be regarded as the generalized Boussinesq equations.

The three types of motion equation set: the original equation, the Boussinesq equation and the anelastic equation, treated the density in different ways, they are therefore called as elastic, non-elastic and anelastic approximate equations respectively. They are used for the treatment of the mesoscale problems in different circumstances.

References

COTTON W R, ANTHES R A, 1989. Storm and Cloud Dynamics[M]. Academic Press.
EMANUEL K A, 1983. On the Dynamical Definition of "Mesoscale". Mesoscale Meteorology—Theories, Observations and Models[M]. Lilly D K, Gal-Chen T (ed.). Boston: Reidel Publishing Co.
EMANUEL K A, 1984. Dynamics of Mesoscale Weather Systems[M]. Am Meteor Soc.
EMANUEL K A, 1986. Overview and Definition of Mesoscale Meteorology[M]. Am Meteor Soc.
FUJITA T T, 1986. Mesoscale Classifications: Their History and Their Applications to Forecasting. In Mesoscale Meteorology and Forecasting[M]. Ray P S (ed.). Am Meteor Soc:18-35.
KESSLER E, 1986. Thunderstorm Morphology and Dynamics[M]. University of Oklahoma Press.
LIGDA M G H, 1951. Radar storm observation, compendium of meteorology[J]. Bull Am Meteor Soc, 1265-1282.
LORENZ E N, 1969. The predictability of a flow which possesses many scales of motions[J]. Tellus, 21:289-307.
OGURA Y, PHILLIPS N A, 1962. Scale analysis of deep and shallow convection in the atmosphere [J]. J Atmos Sci: 173-179.

ORLANSKI I, 1975. A rational subdivision of scales for atmospheric processes[J]. Bull Am Meteor Soc, 56:527-530.

PIELKE R A, 1984. Mesoscale Meteorological Modeling[M]. Academic Press.

RAY P S, 1986. Mesoscale Meteorology and Forecasting[M]. Am Meteor Soc.

SHOU S W, 1993. Mesoscale Atmospheric Dynamics[M]. Beijing: China Meteorological Press.

THUNIS P, BORNSTEIN R, 1996. Hierarchy of mesoscale flow assumptions and equations[J]. J Atmos Sci, 53:380-397.

VINNICHENKO N K, 1970. The kinetic energy spectrum in the free atmosphere-1 second to 5 years [J]. Tellus, 22:158-166.

Chapter 2
Topographically Forced Mesoscale Circulations

  Mesoscale circulation systems include the topographically forced circulations and the circulations in free atmosphere. Furthermore, the generalized topographically forced circulations may be further divided into the topographic mechanical forced circulations such as the mountain waves, the downslope wind and the wake circulations etc. and the topographic thermal forced circulations such as the heat island circulation, the sea-land breeze and the mountain-valley breeze circulations etc. due to the inhomogeneity of the topographic condition and the surface heat distributions. In the free atmosphere, the mesoscale circulations can be further divided into the non-convective circulations such as the gravity waves, the front-jet streak etc., and the convective circulations. While the later can be further divided as shallow convections and deep convections, which will be left for discussing in the subsequent chapters. In current chapter we will discuss various topographically forced circulations. Although the topographically forced circulations themselves are normally not very strong, they may possibly play important roles for triggering severe weather such as heavy rain or severe convective weather etc.

§ 2.1 Topographic waves

2.1.1 Basic type of topographic waves

  Mountains normally can influence the air flows notably. Many experienced soaring airplane or parasail drivers understand how to control promotion and demotion of the aircraft by taking advantage of the air flow waves near the mountain area. Early in 1913, Von Ficker used the "free balloon", which has zero buoyancy at the discharged place, for observing the trajectory of the air flows, and some of the results are showed in Fig. 2.1.

  The air flow waves caused by the air flows run over the mountain is called topographic waves. Forchgott(1949) divided the common topographic waves into four basic types: the sheet flows, the standing eddy, the lee wave flows and the rotor streaming as shown in Fig. 2.2.

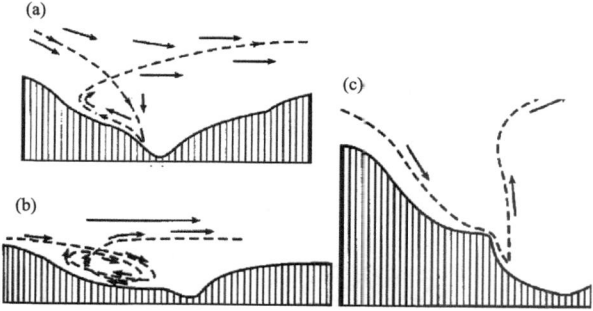

Fig. 2.1 Some air flow trajectories (dashed lines) observed on the lee side of the mountains. (After Von Ficker, 1913)

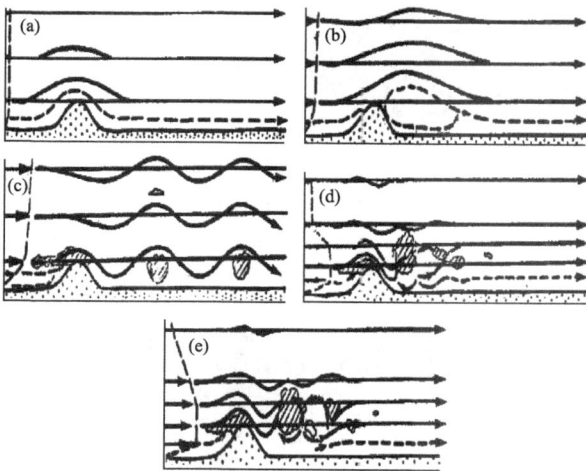

Fig. 2.2 Four types of the airflow over the mountains. (a) the sheet flows; (b) the standing eddy; (c) the wave flows; (d) and (e) the rotor streaming. (Adapted from Forchgott, 1949)

Different types of topographic waves are majorly depending on the types of wind profiles. The sheet flows normally occur under the weak wind condition. The waves existed in a shallow layer over the mountain. It is a smooth shallow wave and is usually called as "Mountain wave". When the wind velocity near the top of the mountain is stronger, it is possibly to form the semi-stationary waves and there is a shallow wave over it. This semi-stationary wave is called as standing eddy. When the wind velocity increase with height, the wave flow can be formed at the lee side of the mountain and it is called "lee wave", which can extend to upper troposphere and lower stratosphere. By the ground based observations and the satellite and aircraft observation as shown in Fig. 2.3 and Fig. 2.4, it can be found that the wavelike cloud systems existed in the downwind direction of the mountains. The clouds are usually caused by lee waves and when the wind velocity existed maximum in the vertical direction might occur the rotor streaming, which is a special type of the lee wave.

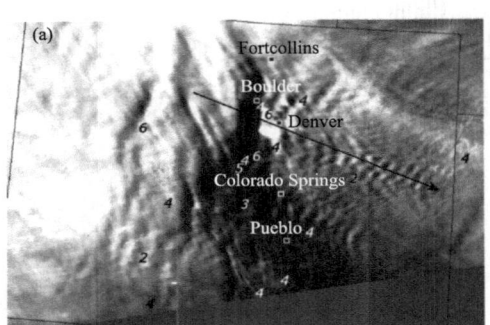 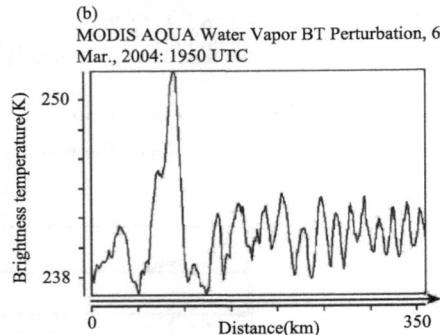

Fig. 2.3 (a) MODIS water image at 1950 UTC 6 Mar. 2004., Plotted numbers represent the turbulence intensity at the pilot report (PIREP) locations within ±2 h of the MODIS image. (b) The 6.7 mm brightness temperature image across the transect marked in the image to the left. (Adapted from Uhlenbrock et al., 2006)

Fig. 2.4 A Blue Ridge lee wave visible on a layer of stratus clouds at 5000 ft* above the mountain (0800 16 Apr., 1974). (Adapted from Smith, 1976)

---

* 1 ft=30.48 cm.

2.1.2 Features and atmospheric conditions of lee waves

Lee waves are induced by vertical oscillation of atmosphere due to the mechanical forcing of the barrier. When air flow is forced by mountain to cause ascent motion, in the stable stratification the recover force-the gravity force will cause the air to recover to initial place and therefore to induce vertical oscillation and spread the wave along the airflow toward downstream. The features of the lee waves are as follows.

(1) Wave length: According to the observations the wave lengths of lee waves are ranged from 1.8 km to 70 km, normally about 5—20 km. The wave length changes with height. It is longer at the higher level, shorter at the lower level.

(2) Wave amplitude: The amplitude of lee wave ranges from several hundred meters to 3 kilometers, normally 0.3—0.5 km. The amplitude has no certain relation with the wave length. Some lee wave amplitude can be as greater as more than 6 km, they are called the hydraulic jump type of lee waves.

(3) Vertical velocity: The magnitude of the vertical velocity of lee waves is about 2—6 m·s^{-1}, the maximum can be 15 m·s^{-1}.

The lee waves usually occurred under the certainly atmospheric conditions. For given barrier, the lee wave formation depends on two atmospheric conditions, i. e., the static stability and the wind shear. In general speaking, when lee wave forms, the height of the most stable layer is just located at the height of the top of the mountain, in other words, there is a stable stratification at the height near the top of mountain, and it is true at least for strong lee wave. Furthermore, the maximum amplitude of the lee waves could usually be found in the layer with highest static stability. Fig. 2.5 shows the mean vertical wind profiles of the non lee wave days, the lee wave days and the strong lee wave days respectively. It is clear that the vertical wind shear is favorable to form lee waves and the stronger wind shear will be more favorable to form the stronger lee waves. As to the stationary waves on lee side of the mountain are usually occurred under the following conditions: ① The mountain ridge is a long mountain ridge or a belt of mountain but not a isolated of mountain peak; ② At the windward side of the mountain, low level atmosphere is stable and the stability decrease with height; ③ The wind direction is close to the direction perpendicular to the mountain ridge with the angle less than 30° and it is mainly not changed with height; ④ The wind velocity at the height of mountain ridge is over a critical wind velocity (about 10 m·s^{-1}), and the wind velocity increase with height in the atmosphere layer from top of the mountain to the tropopause.

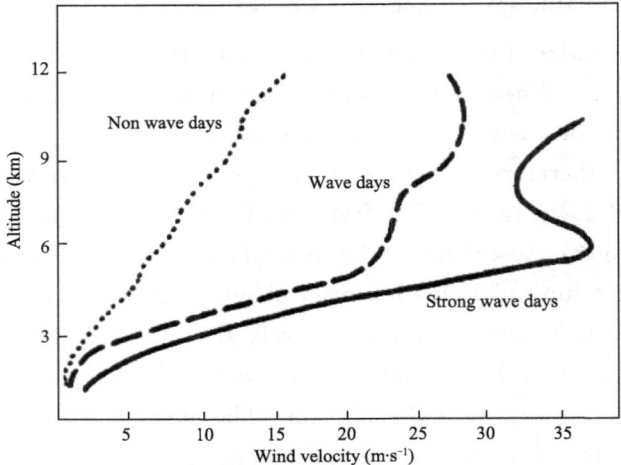

Fig. 2.5 The mean vertical profile of wind velocity at the non lee wave days, lee wave days and the strong lee wave days. (After Colson, 1954)

Since 40's of the 20 century, many people did the dynamic studies on lee waves. One of the representative works was the two layer model given by Scorer, who defined a parameter

$$l^2 = \frac{\beta g}{U^2} - \frac{1}{U}\frac{\partial^2 U}{\partial z^2} \qquad (2.1.1)$$

where l^2 is called Scorer parameter, β is stability parameter, $\beta = \frac{1}{\theta}\frac{\partial \theta}{\partial z}$, θ is potential temperature, g is gravity acceleration, U is wind velocity. In generally speaking, the second term in the right hand of the above formula is much less than the first term, thus

$$l^2 \approx \frac{\beta g}{U^2} \qquad (2.1.2)$$

According to the theoretical analysis by Scorer, the lee waves just can be formed under certain atmospheric condition, i. e. , when $\frac{\partial l^2}{\partial z} < 0$, the lee waves can be formed. Actually, the condition $\frac{\partial l^2}{\partial z} < 0$ means that there is a inversion layer and the wind velocity increases with height as shown in Fig. 2.6.

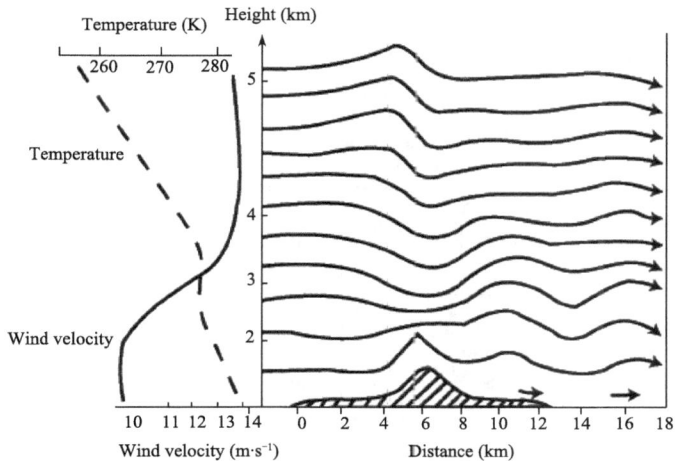

Fig. 2.6 The inversion and vertical wind shear are favorable to lee wave formation. (After Scorer, 1949)

Scorer parameter l^2 usually can be used as a criteria for forecasting the lee wave formation. As the above criterion, lee waves occurred under the condition when the l^2 decreases with height, so that the high value of l^2 at low level will be favorable to form lee waves. Fig. 2.7 showed the sounding analysis and vertical distribution of l^2 in a lee wave event occurred over northwest England. It can be seen clearly from the diagram that in the layer 900—800 hPa the stratification is stable, wind direction changes only little, wind velocity increases with height, and the maximum of l^2 occurred on the height of 1—2 km. Many cases have similar results. That means the temperature inversion (or at least a stable layer) below 600 hPa is almost the necessary condition for lee wave formation.

2.1.3 The influences of lee wave on precipitation

The formation of lee waves has distinguish influences on precipitation. Li et al. (1978) made a numerical simulation and experiment for a lee wave process occurred near the area of Taihang Mountain in northern China. Fig. 2.8 showed the stream lines in the vertical section along the direction perpendicular to the mountains, the shape of which as shown by shadow area in the left of the low part of the diagram is steep in east slope and gentle in west side. The lee wave was located over the foot of the mountain. Comparing the vertical velocity ω and the 24 hours rainfall, it can be found that the precipitation area is coincided with the ascending area of the lee wave.

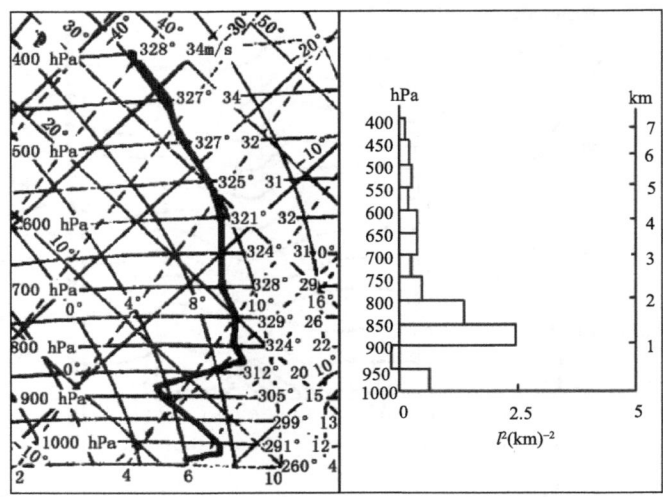

Fig. 2.7 The tephigram and vertical profile of l^2 before the lee wave event occurred on Mar., 11, in Scotland. (After Corby, 1957)

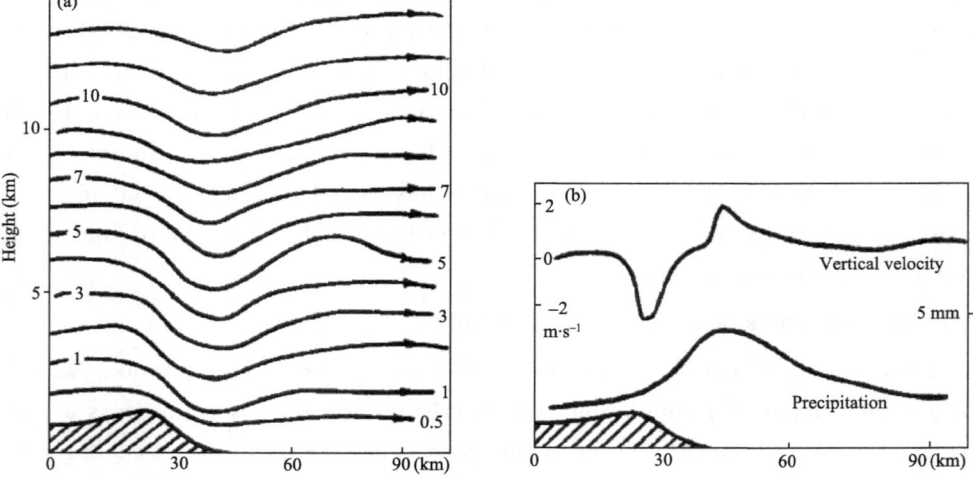

Fig. 2.8 Numerical simulation and experiment of the lee wave process in east side of the Taihang Mountain on Oct. 14, 1975. (a) stream lines (10^4 m$^2 \cdot$ s^{-1}); (b) vertical velocity and precipitation. (After Li et al., 1978)

This case shows that the large scale air flow interacted with the mountain can cause lee waves, which then might cause precipitation and convective weather phenomena under suitable atmospheric condition.

2.1.4 Downslope wind

In the lee side of the mountain, wind velocity is normally weak due to the obstruct effect of the mountain, while sometimes at the foot of mountain of the lee side it will occur local strong winds, the phenomena so called "downslope winds", which can cause severe damage. For instance in the "30 miles* wind area" to the north of Turfan Depression of Xinjiang, China, the local severe wind events occurred quite often. The strong wind can even blow down the train.

The velocity of the downslope wind is commonly 5 m·s^{-1}—45 m·s^{-1}, some can be as high as 54 m·s^{-1} even over 64 m·s^{-1}. The wind speed usually increase suddenly, the strong wind can last 4—8 h, even 16 h or longer. Fig. 2.9 shows the distribution of the mean surface wind velocity of the 20 downslope windstorm events occurred to the east side of the Rocky mountain. In the figure the thick dot line represented the horizontal distribution of gust wind speed, the thin dot line represented the mean wind velocity and the solid line is westerly component. The wind velocity is strongest near the foot of the hill in the lee side of the mountain. The observation showed that the lee waves are usually related with the hydraulic jump type lee waves, which are characterized by a strong gliding downward air flow along the lee side slope of the mountain and a strong ascending flow in the downwind side. The downslope wind is directly related with the hydraulic jump type lee waves. The descending of the lee wave brings the air in middle level (700—500 hPa) with great momentum downward to the ground to cause strong wind on the ground. Therefore the maximum surface instant wind velocity is close to the prophase wind velocity on 500 hPa. But when the mid level air is colder than the ambient air, the surface wind velocity will be greater than mid level wind velocity because the potential energy will be changed into kinetic energy during the gliding process. For instance, a downslope wind speed was near a hurricane, while the wind velocity on 500 hPa was only 27 m·s^{-1}.

The development of hydraulic jump type lee waves needs an inversion layer existed in mid or low level of troposphere. Fig. 2.10 is a vertical section of the isolines of

---

* 1 mile=1609.344 m.

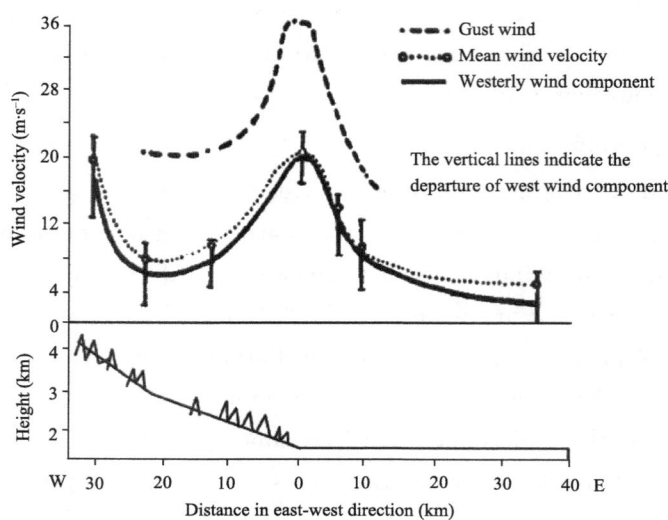

Fig. 2.9 Mean surface wind velocity of 20 downslope wind events occurred in the area near Boulder in Colorado state at the east side of the Rocky Mountain of U. S. (After Brinkmann, 1974)

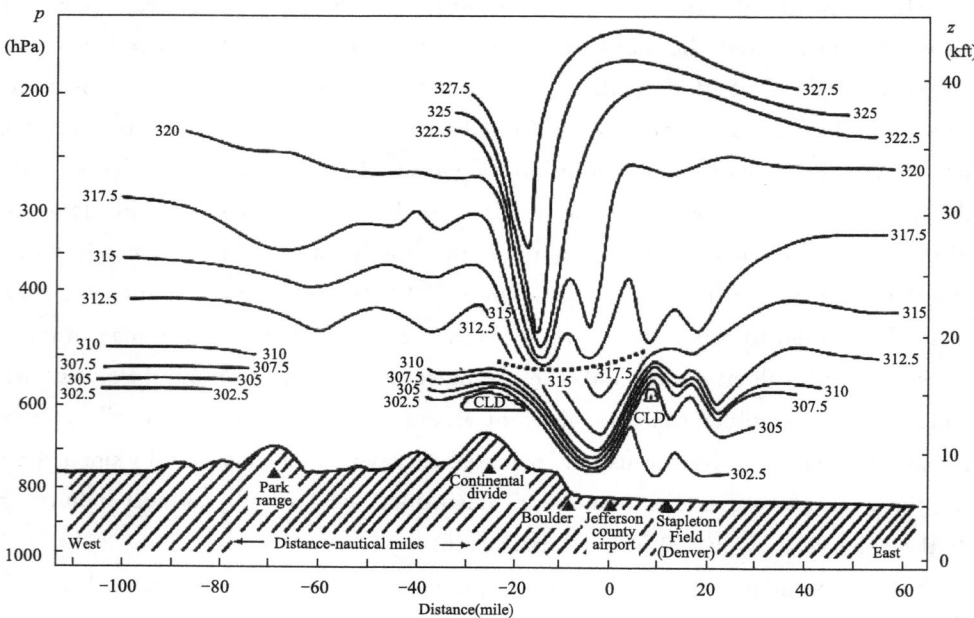

Fig. 2.10 The west-east direction vertical section of the potential temperature (unit: K) across Boulder Colorado of U. S. on Jan. 11, 1972. (After Klemp and Lilly, 1975)

potential temperature during a downslope wind process occurred in east of Rocky Mountain according to the observation of aircraft. In the diagram there is a large amplitude wave (the amplitude is greater than 6 km), this is a hydraulic jump type lee wave. The layer with dense isolines of potential temperature on 500 hPa in the upwind direction indicates the existence of the inversion layer. The inversion layer normally is corresponding to discontinuity surface. When the strong and thick cold air move over mountains, the cold front inversion can be found in the upwind side of the mountains.

In general speaking, the formation of the downslope wind requires stable low level stratification, and the greater difference of Scorer parameter between high and low level atmosphere, as well as the slope topography. The situation favorable to downslope wind is usually a deep cold trough at high level and stronger cold advection at the front of the trough as shown in Fig. 2.11, and there is a greater temperature and pressure differences between two sides of the mountain on the surface synoptic map. A quite similar situation on the 500 hPa and surface weather maps of the downslope wind process occurred in the east side of Rocky Mountain are shown in Fig. 2.12 and Fig. 2.13 respectively.

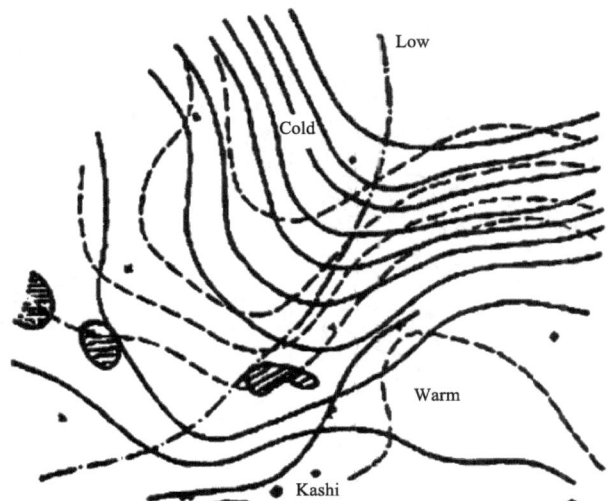

Fig. 2.11　The synoptic situation on 500 hPa constant pressure surface at 0000 GMT Jun. 13, 1963.

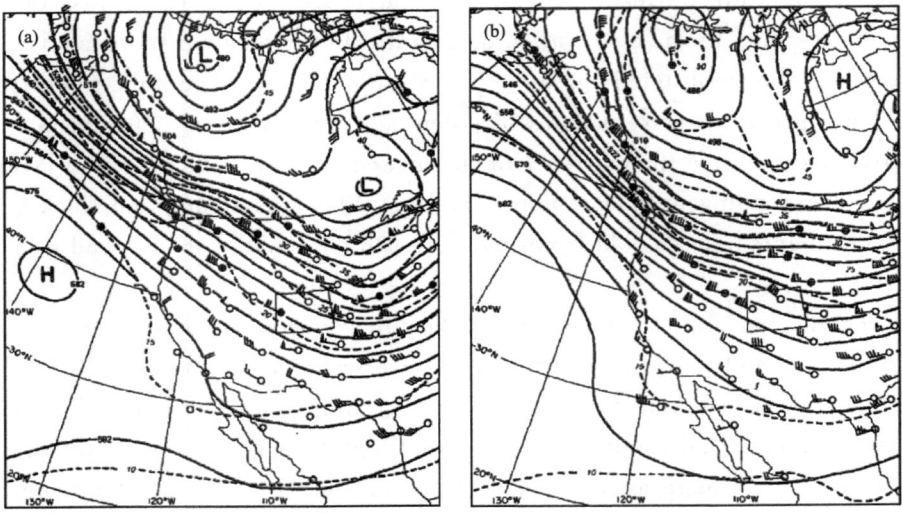

Fig. 2.12   500 hPa weather maps for 1200 GMT 11 Jan. (a) and 0000 GMT 12 Jan., 1972 (b). The solid lines are height contours, and dashed lines are temperatures. The state of Colorado is outlined by a light solid line. (After Lilly, 1978)

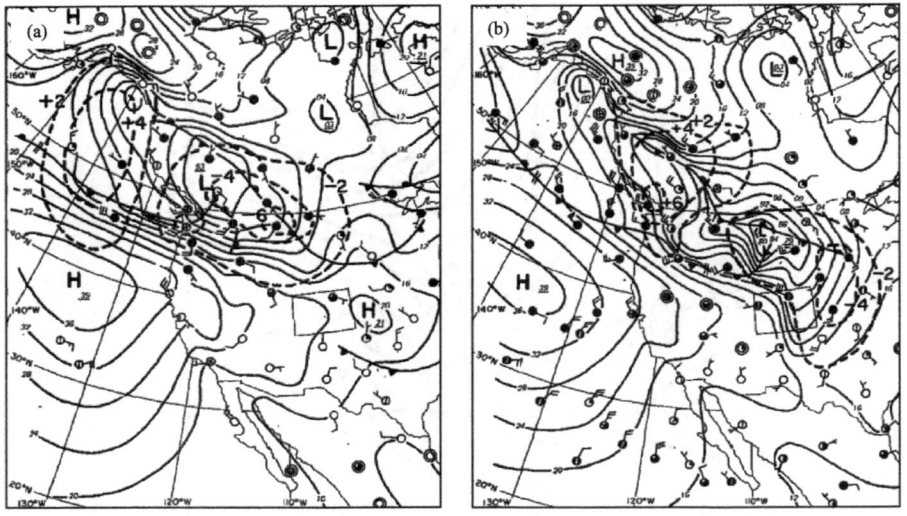

Fig. 2.13   The surface weather maps for 1200 GMT 11 Jan. (a) and 0000 GMT 12 Jan., 1972 (b). The state of Colorado is outlined by a light solid line. (After Lilly, 1978)

§ 2.2 Circulations in wake

The turbulence area formed behind the substance moving relative to the airflow is commonly called as wake, which can be found in the lee side of island or mountain quite often. In the lee side of the mountain sometimes the mesoscale low pressure or troughs, which are called lee lows or lee troughs, can be also formed. The physical formation about lee lows or troughs can be explained by using the principle of potential vorticity conservation. For a air column with the depth D, the potential vorticity conservation equation can be expressed as:

$$\frac{d}{dt}\left(\frac{f+\zeta}{D}\right)=0 \qquad (2.2.1)$$

where ζ is the vertical component of relative vorticity, f is the earth rotation vorticity, $(f+\zeta)$ is the absolute vorticity, and $(f+\zeta)/D$ is potential vorticity. When the air flow goes up the mountain, the cyclonic vorticity will decrease since D decrease. In the lee side of the mountain, when the air flow goes downhill, the cyclonic vorticity will increase since D increase. This explanation is suitable to the large scale low pressure in lee side and high pressure in windward side of the mountain. But for the formation of mesoscale high or low pressure, it is necessary to consider the convergence and divergence of wind velocity and air mass. In the lee side, the mesoscale low pressure will be caused due to the air flow acceleration will cause the low level divergence of the wind velocity and air mass. Beside of the low pressure in lee side, under some conditions, series of vortexes can be formed in the wake propagating toward downstream. This series of the vortexes formed in wake is called as vortex street. Some of them are cyclonic circulations and some are anti-cyclonic circulations.

A Von Kármán vortex street, which is derived from laboratory by Von Kármán comprised by two rows of vortexes which are nearly parallel to each other (Fig. 2.14). The vortex in one row is located at the midpoint of the ligature of two vortexes on another row. The vortexes on the same row have same circulations, but just opposite one on the another row. The diameter of the initial discharged vortex is corresponded to the scale of diameter of the island or mountain. For example, the diameter of the vortex initial discharged from Cheju Island is about 40 km. When the vortexes move toward downstream, their longitudinal interval (a) is about 50—100 km and the crosswise interval (h) is about 30—50 km. The diameter of the vortex will increase when it moves toward downstream. An island with diameter of 40 km can cause

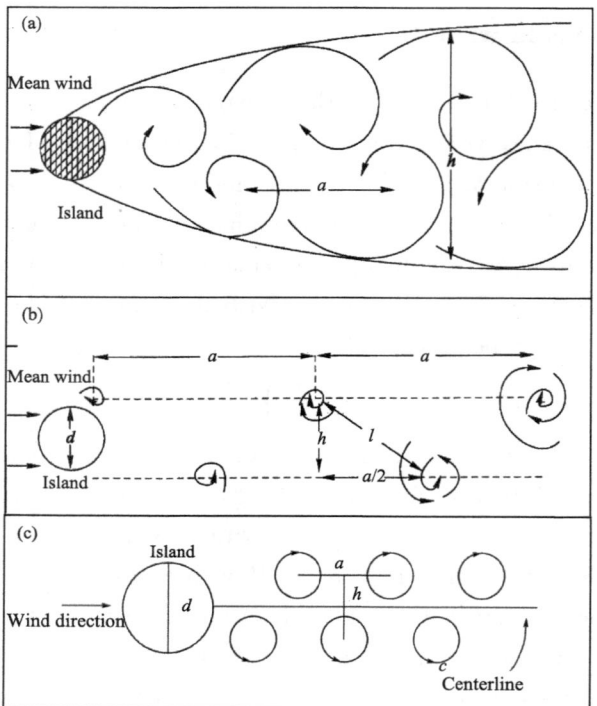

Fig. 2.14 (a) An illustration of the atmospheric vortex street in the lee side of an isolated island; (b) The Kármán vortex street in a 2 dimensional fluid caused by an obstacle with diameter d (After Chopra and Hubert, 1964); and (c) A schematic diagram of a mountainous island of cross-wind. Width d and its Von Kármán vortex street showing the cross-street distance h and the along-street distance a between two like-rotating vortices. (After Young and Zawislak, 2006)

a wake with width of 100 km and length of 400—600 km. Normally, the time interval to discharge a couple of vortexes is about 8 hours and the process can last more than 30 hours. From his research of the atmospheric vortex street near Cheju Island, Tsuchiya (1969) found that their moving velocity is about 3/4 of the basic flow speed. The vortex tangent speed is about 1/3 of the speed of the undisturbed flow.

The typical weather conditions for the vortex street are that there is a low level temperature inversion layer existed on the height of 450—2000 m above the ocean surface and the top of the island is just over the inversion layer; meanwhile, the basic flow is stable, with the wind velocity about 10 m · s^{-1}. The low level inversion layer is important, because that means the movement of the airflow bypassing the barrier is faster than that of overpassing the barrier.

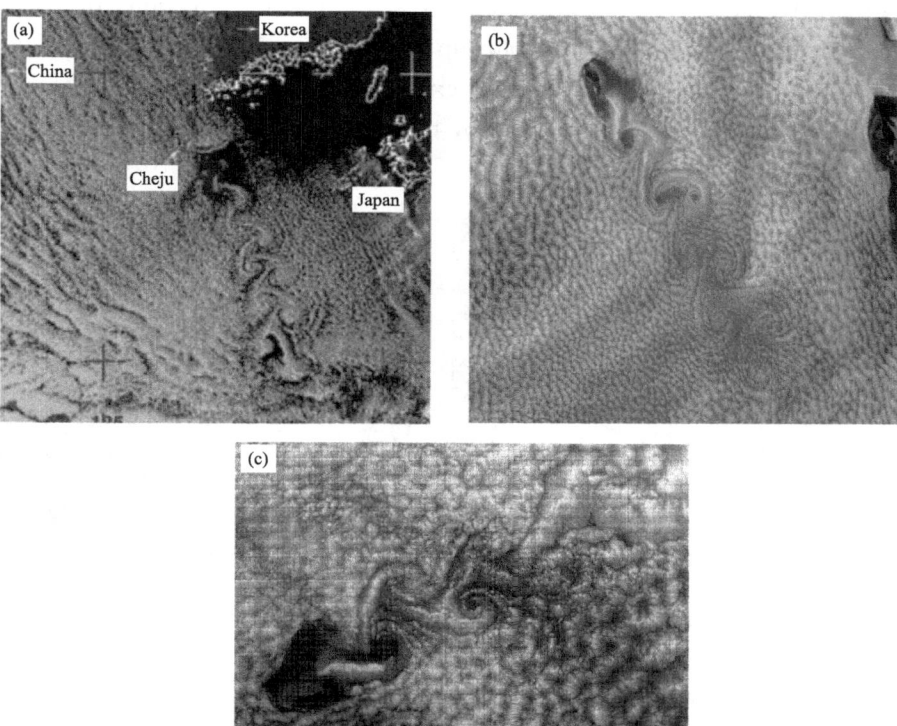

Fig. 2.15 Some classic examples of lee disturbances in mesoscale. (a) A composite imagery of NOAA-14 observed at 0743 UT, 31 Dec., 2000. A clear area is visible in the lee side of Cheju Island, while six vortices are produced farther (After Chunga and Kim, 2008); (b) A Von Kármán vortex street in the stratocumulus off the island of Guacalope at 1845 UT 13 Jun., 2004. The image was taken by the Terra MODIS satellite. Image courtesy of MODIS Rapid Response Project at NASAGSFC. (After Young and Zawislak, 2006); (c) A Von Kármán vortex street that formed to the lee of the Guadalupe Island, off the coast of Mexico's Baja Peninsula, revealed by MISR images from 11 Jun., 2000 detected by NASA satellite Terra. (After Visible Earth, NASA)

Chunga and Kim (2008) and Young and Zawislak (2006) discussed the mountain-generated vortex streets over the Korea South Sea. They pointed out that the Von Kármán vortices are quite often found in the lee side of Cheju Island. In this area an isolated mountain often disturbs airflow over the Korea South Sea. The vortex often occurs with a strong northerly flow with persistent patterns of cumulus and stratiform clouds. The generation of vortices and cloud streaks including enhancement of precipitation in the lee side observed with NOAA satellites. Observations of vortex and cloud formation by the Halla Mountain up to 800 km are recorded. Some cases as shown in Fig. 2. 15a, b and c are classic examples of lee disturbances in mesoscale. The typical 500 hPa synoptic situation of the atmospheric vortex street phenomena near Cheju Island is shown in Fig. 2. 16.

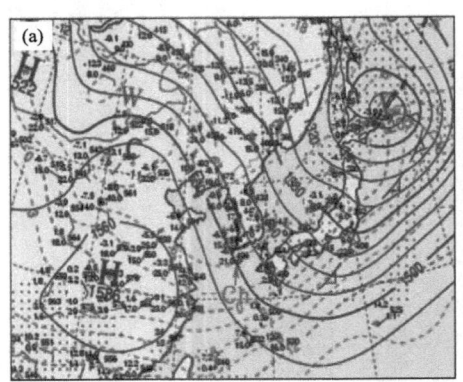 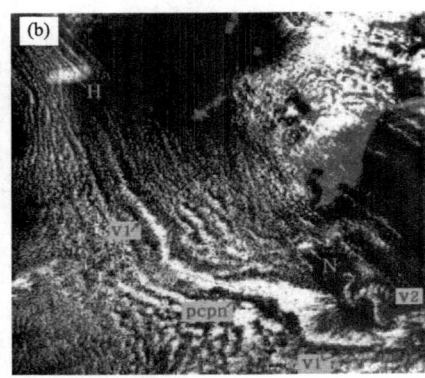

Fig. 2. 16 The 500 hPa synoptic situation at 0000 UT 18 Mar. , 2005 and the atmospheric vortex street phenomena. (a) A meteorological map of 850 hPa, 0000 UT, 18 Mar. , 2005, produced by the Korea Meteorological Administration. Ch refers to Cheju Island and Halla Mountain. The letters L, H, W refer to low, high pressure and warm centers. The contours over Korea and SW Japan resemble the shape of the long cloud streak from Halla Mountain. shown in Fig. 2. 16b; (b) Same as Fig. 2. 15a but of NOAA-17 at 0203 UT, 18 Mar. , 2005. A very long "linear/sticky" cloud (v1) was formed from Halla Mountain for 800 km far downstream. Also, rainy (precipitable) clouds were in the centre of the cloud streak. A much smaller cloud streak (v2) was generated from Niyanomura Mountain (N). (After Chunga and Kim, 2008)

Smith et al. (1993) studied the Hawaii wake. They pointed out that under the influence of the east-northeasterly trade winds, the island of Hawaii generates a wake that extends about 200 km to the west-southwest. The patterns of wind and aerosol concentration, which are based on the data obtained from five wake surveys, suggest that Hawaii's wake consists of two large quasi-steady counter rotating eddies. The southern clockwise-rotating eddy carries a heavy aerosol load due to input from the Kilauea volcano. At the eastern end of the wake, the eddies are potentially warmer and more humid than the surrounding trade wind air. Near the northern and southern tips of the island, there are sharp shear lines. Along the shear lines there are dry and warm air bands. Behind the Kohala peninsula, there is a small embedded wake. Furthermore, there are the wake centerline clouds, the hydraulic jumps to the north and south of the island, a descending inversion connected with accelerating trade winds and evidence for side-to-side movement as shown in Fig. 2.17.

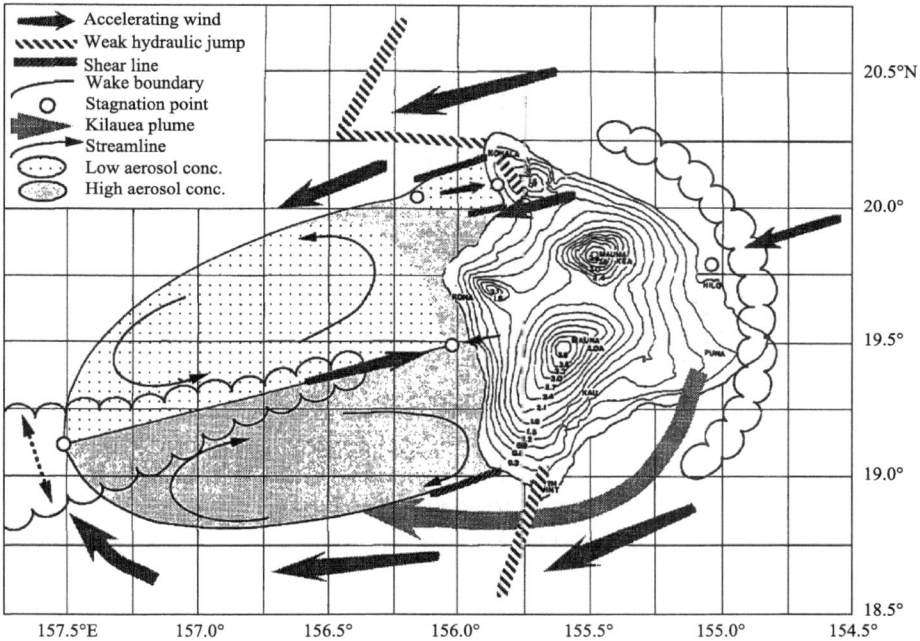

Fig. 2.17 Summary diagram depicting features observed in Hawaii's wake. Dashed two-way arrow at the downstream end of the wake is suggesting the existence of a north to south drift. The upstream rain band and "centerline" cloud are also outlined. (After Smith et al., 1993)

§ 2.3 Urban heat island circulation

The local circulation occurred on the island caused by the temperature difference between land and ocean is called as heat island circulation. For instance, such heat island circulation phenomena can be found near Puerto Rico Islands in Atlantic Ocean. During the daytime the convective clouds formed on the island due to the temperature on island increased. When the convective clouds became mature, it produced downdraft, which converged with the flows from out of the island and to cause a cumulus cloud ring around the island as shown in Fig. 2.18. Similarly, after the thick fog area, which existed in the morning, disappeared a convective cloud ring around the area can be formed in the afternoon. These effects to cause the local circulations due to the local temperature difference are called heat island effects.

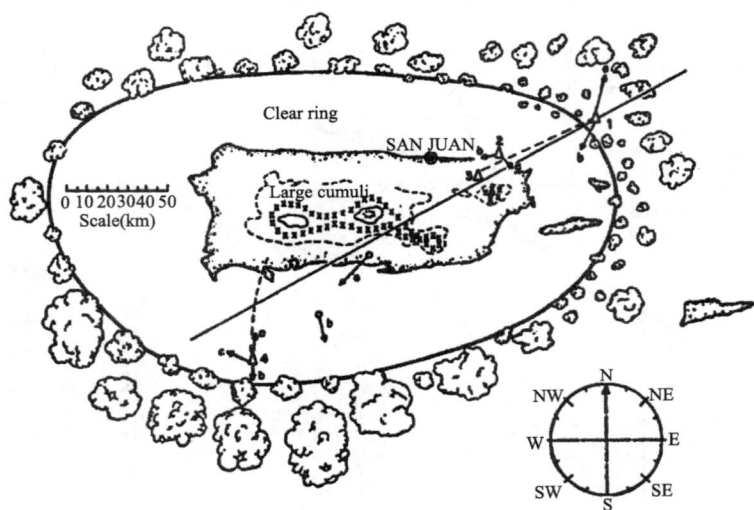

Fig. 2.18 The heat island phenomena near Puerto Rico Islands in Atlantic Ocean. (After Fujita, 1963)

In urban area, pyro-conductivity and heat capacity are much higher than the rural area and easy to form the temperature difference between urban and rural. This phenomena are similar to the heat island effect in ocean and are commonly called as urban heat-island effects.

The formation of urban heat-island is closely related with prevailing wind and sky situation. When the urban heat-island occurs, if prevailing wind is weaker, the heat-island will move toward downwind direction of the prevailing wind. When the prevailing wind increases to certain value the heat-island will decrease and even disappear since under the stronger air flow condition the heat will be quickly brought out and the dynamic exchanging effect will increase. The wind velocity value that the heat island disappeared is called as the critical wind velocity, which is proportional to the scale of the city. According to the statistics, for the large city with million of population the critical wind velocity is about 10 m·s^{-1}, for the middle city with several hundred thousand of population the critical wind velocity is about 8 m·s^{-1}, and for small city less than hundred thousand of population the critical wind velocity is about 5 m·s^{-1}. When the wind velocity is over the critical wind, the heat island phenomena tends to disappear. The intensity of the urban heat island can be expressed by the maximum difference of temperature between urban and rural. The statistical relationship between the intensity of the urban heat island and the population of the city can be written as follows:

$$(\Delta T_{u-r})_{max} = 3.06 \lg P - 6.79 \qquad (2.3.1)$$

where $(\Delta T_{u-r})_{max}$ is the maximum difference of temperature, P is the population of the city.

The urban heat island phenomena are commonly founded in many large cities in the world. The annual average temperature of urban region in many large cities is usually 0.6 ℃ to 1.1 ℃ higher than that in rural region. Under the clear air and weak wind condition, the temperature difference between urban and rural can be higher than 5 ℃ in a large city with several million population; 3—5 ℃ in a middle city with several hundred thousand population; and 2—3 ℃ in a smaller city. As the examples, Fig. 2.19a and b show the urban heat island phenomena in the areas near Beijing and Tianjin in northern China by the urban island detecting maps of satellite FY-1C at 0831 (BJT) Sept. 1, 2000 and 1332, Sept. 1, 2001 respectively. In general speaking, the larger the city size the stronger the urban heat island phenomena. According to the investigation on the urban heat island phenomena in the areas near Nanjing City is also obviously like many other large cities. The temperature difference between urban and rural area was greater more than 2—3 ℃. Fig. 2.20 shows the urban heat island effect near the area of Nanjing City based on the MODIS data. The characteristics of urban heat island (UHI) effect and its cause are investigated by using MODIS data in April 2004. The surface parameters from the MODIS data have surface temperature (ts), albedo (α), and normalized difference vegetation index ($NDVI$). Their heterogeneities over urban and rural area are analyzed based on land cover classification, and their relations are also presented in order to explain the UHI effect. The results show that there exists obvious the UHI effect. Ts over urban areas are by 10.83% higher than those over rural area, and $NDVI$ and α over urban area are by 62% and 18.75% less than those over rural area, respectively. Surface temperature has significantly negative correlation with $NDVI$ and their correlation coefficient is -0.73. Correlation between $NDVI$ and albedo is determined by the spectrum of light. Difference in vegetation cover is the primary cause of the UHI effect.

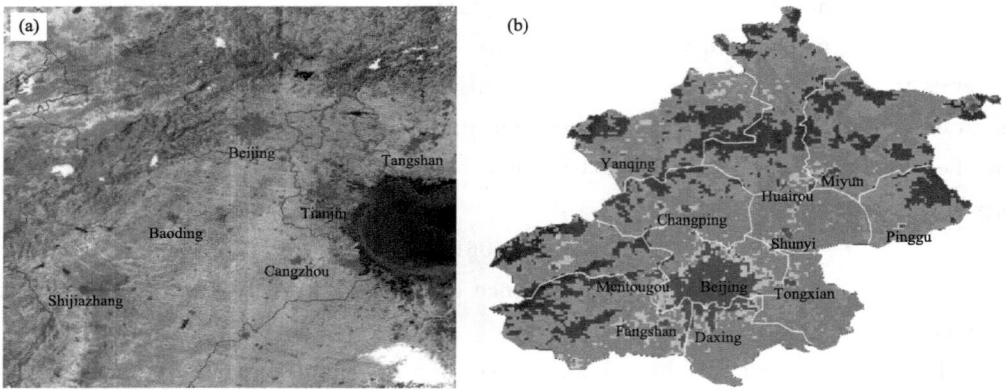

Fig. 2.19 The urban island detecting maps of satellite FY-1C of Hebei China. (a) at 0831 BJT Sept. 1, 2000; (b) at 1332 BJT Sept. 1, 2001, Beijing(see the color illustrations)

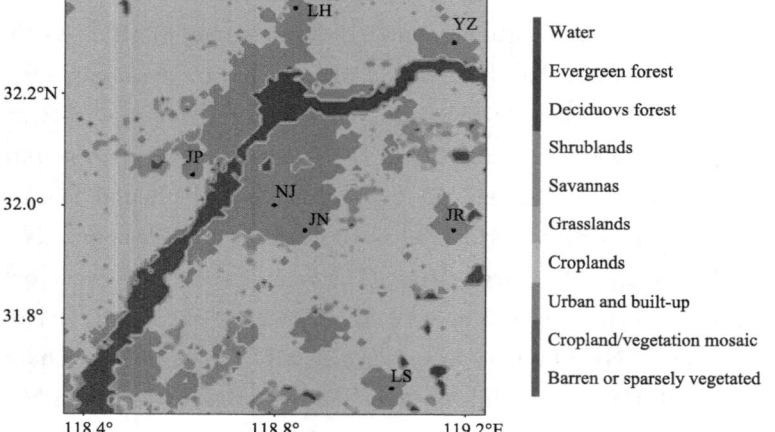

Fig. 2.20 The urban heat island effect near the area of Nanjing City based on the MODIS data. The horizontal and vertical axises of the diagram are the latitude and longitude respectively. (After Wang et al., 2008)(see the color illustrations)

    The urban island phenomena may induce the wind blowing from rural area to the urban area, and to cause cyclonic convergence ascending flow over the urban area. Its intensity can be as great as 5—10 m · s^{-1}. The depth of the vertical ascending flow of the circulation can reach to the height of several hundred meters and then returns to the rural area horizontally. The urban island circulation can cause the pollutants concentrate over the urban area, and to cause the smokescreen, and it can also cause the rainfall and meteorological elements distribution non-uniformly.

Most of the studies of the urban island circulations were based on the statistics analysis. Recent years more studies were based on the dynamic theoretical analysis and numerical simulations. Li (1990) derived an expression of critical wind velocity, based on the Boussinesq approximation equations considered the local thermodynamic disturbing caused by heat island effects. Li et al. (1990) studied the influences of the slopped terrain on the urban heat island by using Boussinesq approximation two dimensional numerical model and pointed out that even a small terrain slope angle, $\beta=0.0014$, still can induce significant influences on the boundary wind velocity distribution and the rotation of the wind direction as well as the turbulence features in the near surface layer.

Along with the establishment of the dense automatic meteorological stations network in urban region, the observation and analysis for the urban heat island phenomena is getting more delicate. Very often the local heavy rain events in the urban region of the cities are closely related with the urban heat island effects. For instance, during the heavy rain process occurred in Tianjin City in Jun. 28 to 29, 2005, the rainfall was mainly concentrated in the urban region, reflecting the influencing of the urban heat island circulation on the heavy rainfall distinctly.

Recently, there has been increased interest in the modification of atmospheric circulations by surface heterogeneities. The urban surface, in particular, affects sensible heat, latent heat and momentum fluxes, surface convergence, boundary-layer height, and other boundary-layer features. Such changes impact upon pollution dispersion, thunderstorm initiation, optical properties of the atmosphere, and many other physical and chemical processes in the urban environment.

§ 2.4 Sea-land breeze

2.4.1 Observation of sea-land breeze

In the coast area or near the lakeside it is very often to see the shoreward or offshore winds, they are called as sea (or lake) wind and land wind respectively. The sea-land breeze is caused due to the temperature difference between sea and land. In the daytime the sea surface is cooler than the land, so that the wind blows from sea to land; while in the night the land is cooler than sea surface, so that the wind blows from land to sea. In the vertical direction above the sea breeze or land breeze on the surface will form opposite flows. The depth of the vertical circulations is about 100—1000 meters, and the penetrated distance is about 50—300 km (Fig. 2.21 and Fig. 2.22).

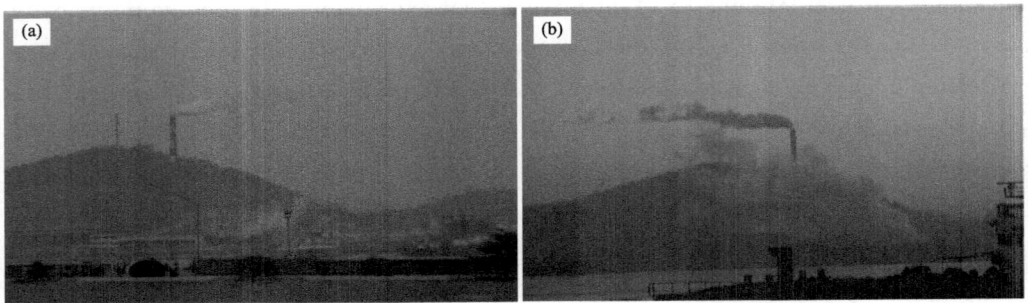

Fig. 2.21 The pictures of the smoke fluid from a power factory chimney (with height 104 m) near the coast of Beilun Port, Ningbo, China show the sea-breeze in afternoon (a) and land-breeze in the morning (b) (left side is sea, right side is land). (After Song et al., 2008)

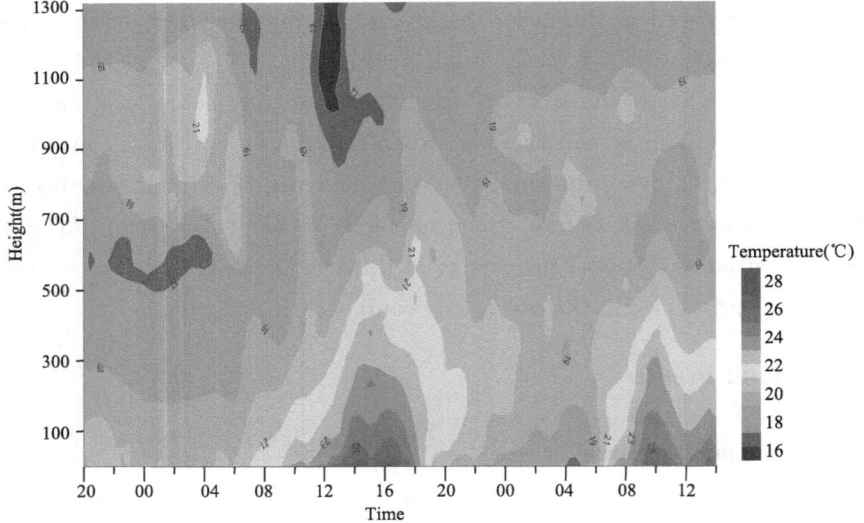

Fig. 2.22 The pictures of the vertical distribution of temperature near the coast of Beilun Port, Ningbo, China observed by kite balloon. (After Song et al., 2008)(see the color illustrations)

As early as the beginning of 20 century, people begin to observe the sea-land breeze by using instrumentation. The recent modern observational technologies include wind profiles, fixed or mobile meteorological towers, the sounding balloon, kite-airship or meteorological airplanes, acoustic radar or Doppler radar, and satellite tracing etc. The research on sea-land breeze is mainly applied in weather forecasting of the coastal cities and monitoring of the atmospheric environment in the planetary boundary layer, including the meteorological characteristics of the city boundary, city weather forecasting, atmospheric pollution etc. In general speaking, the research on sea-land breeze has distinct meaning in weather forecasting and meteorological services in the coast areas, in which the sea-land breeze is normally one of the important factors necessary to be considered in local weather forecasting. The sea-land breeze circulation has direct influences on many aspects, such as the intensity and distribution of local precipitation; the development of convective clouds; the diurnal variation of wind direction and speed; the diurnal range of atmospheric temperature; the occurring time of daily maximum temperature; the development and dissipating of fog and the daily variation of visibility and so on. Meanwhile, the sea-land breeze is also a natural source, which can be used as a wind energy source and also can be used in inshore fishing and breeding industry etc. The sea-land breeze circulation can also significantly influence the diffusion of the land pollutants. In recent years due to the people much increased their attention to the environmental and ecological problems, the research scope of the sea-land breeze had been expanded to the mechanisms of the pollutant diffusion. It is meaningful to guide the reasonable pollutant emission time and to improve the atmospheric environment quality. Besides, the research on sea-land breeze also has practical significances on the safe-guide of aviation and navigation, urban planning, forest fire dangers forecasting as well as the development of tourism resource etc.

2.4.2 The rotation of sea-land breeze

The sea-land breeze normally obeys distinct rules on its variation of occurring time and intensity. Generally, at noon time, around 12 o'clock, the sea breeze begins; around 15 o'clock the sea breeze reaches maximum; the land breeze begins around 03 o'clock at night and reaches maximum about 06 o'clock in the early morning. The wind direction of the sea-land breeze changes with time also regularly. In the northern hemisphere, the wind direction normally clockwise rotates with time, while it is in contrast in southern hemisphere, where the wind rotates anticlockwise.

Fig. 2.23 is the wind rose chart of Ashdod, a port city in Israel, in July from 1958 to 1968. In the figure every small chart indicates the frequency of different wind directions in different time. The wind poles indicate the wind coming direction, its length represents the frequency magnitude of the wind direction occurring. The longer the pole, the greater the frequency. According to the diagram it is clearly that there is maximum frequency wind direction for every time. They are distinctly clockwise rotation with time.

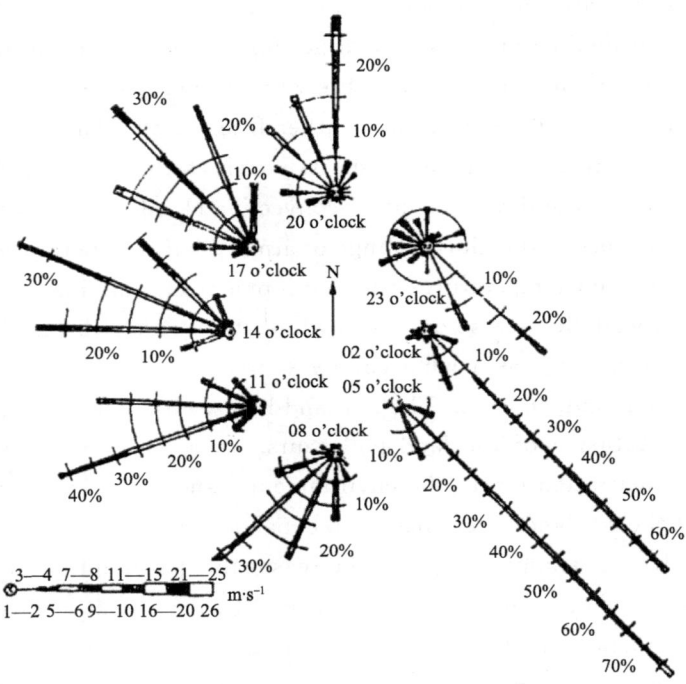

Fig. 2.23 Wind rose chart for Ashdod Port, July, 1958—1968. The lines point out into the direction from which the wind comes, the speed being denoted by the width and by the shading of the wind rose lines. Ashdod Port is located on Israel's Mediterranean coast. (Adapted from Neumann, 1977)

As an example, Fig. 2.24 showed the shifting direction of the tracer in Beilun port of Ningbo City of Zhejiang Province, China in the time from 17 o'clock Oct. 21 to 11 o'clock Oct. 22, 2006, based on the numerical simulation. Since there was no obvious influences of strong synoptic wind systems, the shifting direction of the tracers mainly depends on the influence of the local sea-land breeze circulation. Hence the variation of the shifting direction of the tracers actually represents the time history of wind direction of the sea-land breeze. From the diagram it is also clearly showed that

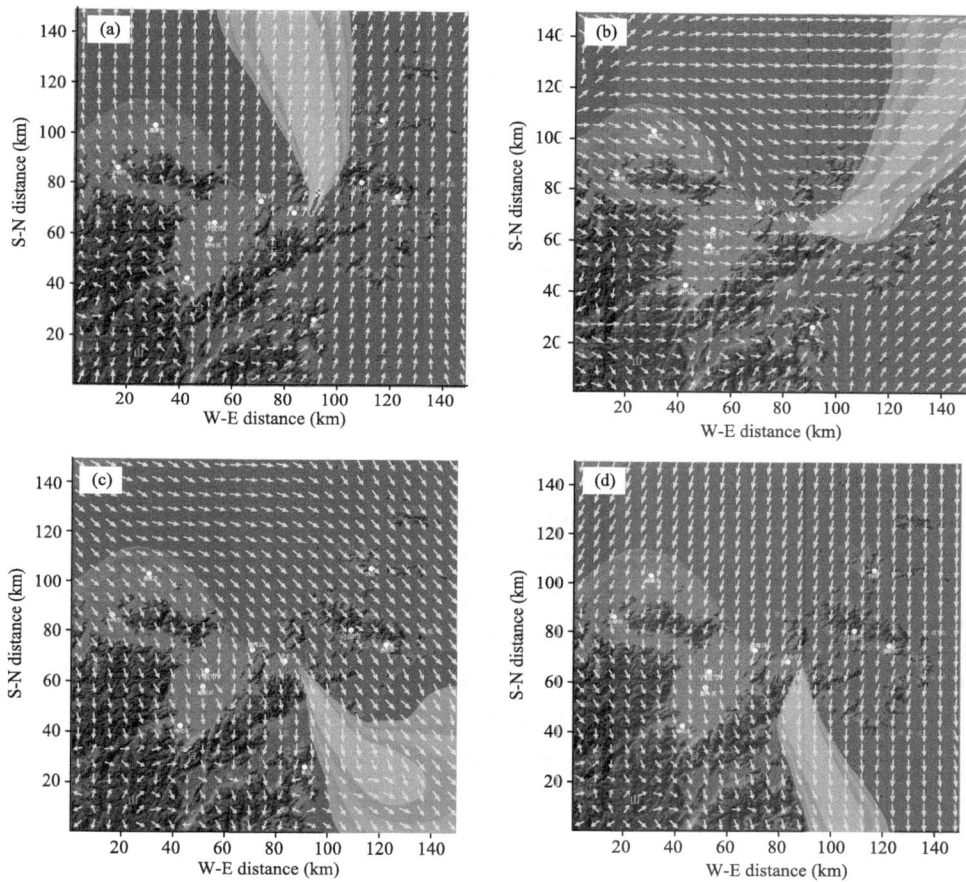

Fig. 2.24 Variation of the shifting direction of the pollutant tracers and the wind fields on (a) 1700 BJT 21 Oct.; (b) 2300 BJT 21 Oct. (c) 0500 BJT 22 Oct.; (d) 1100 BJT 22 Oct., 2005 near Beilun Port, Ningbo, China based on Mesoscale numerical simulation. (Adapted from Song et al., 2008)(see the color illustrations)

the wind direction clockwise rotates with time. While the wind direction of sea-land breeze changes with time probably can be influenced by various factors and sometimes it is complicated. Under the different situation, sometimes the wind direction rotates clockwise, while sometimes it rotates anticlockwise, sometimes it rotates quickly and sometimes slowly.

In order to investigate the rule of sea-land breeze direction changing with time,

Neumann (1977) derived an equation describing the rotating rate of the sea-land breeze direction. He divided the pressure P into two terms, $P=P_L+P_M$, where P_L is the pressure caused by large scale disturbance, P_M is the Mesoscale excess pressure, set α as the angle between horizontal wind direction and any fixed horizontal axis (say x axis). Then set $\omega=\partial\alpha/\partial t$, where $\partial\alpha/\partial t$ is the rotation rate of the horizontal wind rotating around local vertical direction. When $\partial\alpha/\partial t>0$, it denotes the wind direction rotated anticlockwise, while $\partial\alpha/\partial t<0$ denotes clockwise turning. Finally the rotating rate can be written in the following form:

$$\frac{\partial\alpha}{\partial t}=-f-\underbrace{\frac{1}{\rho U^2}\left(u\frac{\partial P_m}{\partial y}-v\frac{\partial P_m}{\partial x}\right)}_{(1)}+\underbrace{\frac{f}{U^2}(uu_g+vv_g)}_{(2)}$$

$$+\underbrace{\frac{1}{U^2}(uF_y-vF_x)}_{(3)}-\underbrace{\left(u\frac{\partial\alpha}{\partial x}+v\frac{\partial\alpha}{\partial y}+w\frac{\partial\alpha}{\partial z}\right)}_{(4)} \quad (2.4.1)$$

The equation (2.4.1) is the equation for describing the rotating rate of of the sea-land breeze direction. Where f is Coriolis parameter; u, v are the components of the horizontal wind in x and y directions respectively. u_g, v_g are the components of geostrophic wind in x and y directions respectively; F_x, F_y are the components of frictional force in x and y directions. The first term in the right hand side is the local constant term, while the others are variable terms, which are in turn called as mesoscale pressure gradient term, large scale term, frictional term and non-linear term respectively. The contribution of local term in northern hemisphere is always negative, $(\partial\alpha/\partial t=-f<0)$, that means in the northern hemisphere the term will always cause the sea-land breeze direction rotating clockwise. The actions of the variable terms are depending on the special weather situation. They can be positive or negative. When they are positive the clockwise turning of the wind direction in northern hemisphere will be speed up, otherwise the clockwise turning will become slower, or even can change to anticlockwise turning.

2.4.3 The influences of sea-land breeze on weather and environment

The sea-land breeze effect can cause distinct changes of temperature and other meteorological elements. For instance, due to the sea breeze normally reaches maximum around 15 o'clock, so the temperature is relatively low at the time as shown in Fig. 2.25.

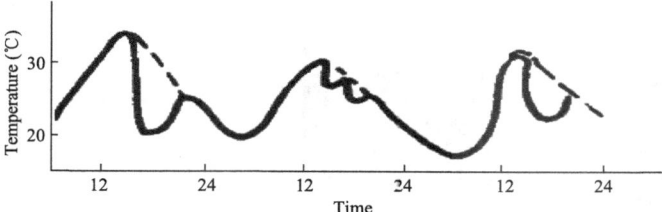

Fig. 2.25 The diurnal variation of temperature based on the observational data. The dashed line indicates the temperature variation under assumption that there is no sea breeze. (Adapted from Guteman, 1969)

When the sea-land breeze happens, a discontinuity line of the meteorological elements similar to a front, which is called as sea breeze front or land breeze front, will be formed at the front edge of the sea-land breeze. They can play a role of trigger mechanism to trigger severe convective weather. Especially when two sea (or land) breeze fronts meet each other, or a sea (or land) breeze front and another discontinuity line meet each other, possibly to form an occlusion situation and cause very severe disaster weather.

As an example of the impacting on weather of the sea-land breeze let us look at the phenomena of the coast-parallel atmospheric fronts, which may be occasionally formed off the east coast of Taiwan as shown in Fig. 2.26. Werner et al. (2007) concluded that the frontal features visible on the ERS SAR images are not sea surface manifestations of the Kuroshio front but are sea surface manifestations of a coastal atmospheric front caused by the collision of two airflows from opposite directions. The first airflow originates from a (weak) onshore synoptic-scale wind and the second one from an offshore wind. They hypothesize that the offshore wind results partially from flow reversal of the onshore blowing synoptic-scale wind and partially from thermally driven land breeze/katabatic wind (Fig. 2.27). Although the atmospheric fronts can persist over a longer period, their position from the coastline exhibits pronounced diurnal variations, which can be explained only by the presence of a downslope density current or thermally driven land breeze/katabatic wind. They conclude that this thermally driven wind plays an important role in the formation and time evolution of the atmospheric front off the east coast of Taiwan.

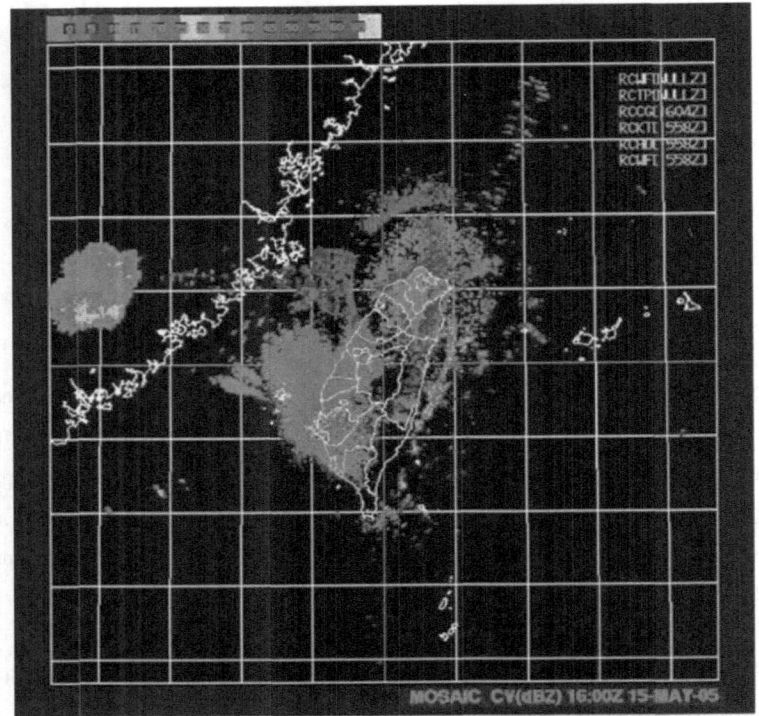

Fig. 2.26 Weather radar image acquired at 1600 UTC 15 May, 2005 (0000 LST 16 May), that is, 10 h, 24 min before the ERS-2 SAR data acquisition. (After Werner et al., 2007) (see the color illustrations)

Another example of the impacting on weather of the sea-land breeze is the regional phenomenon occasionally occurred in the west Washington State of the United States called the Puget Sound Convergence Zone (PSCZ), which is produced by topographic channeling of low-level onshore flow convergence. This phenomenon has significant impact on cloud, precipitation, and wind patterns across the Puget Sound Basin, especially in the spring season.

The Puget Sound lowlands are bound on the east by the Cascade Range and on the west by the Coast Ranges, which include the Willapa Hills, the Olympic Mountains, and the Vancouver Island Ranges. The western barrier is breached north and south of the Olympic Mountains by the Strait of Juan de Fuca and the Chehalis Gap respectively (see Fig. 2.28a for locations of all referenced geographical features and stations).

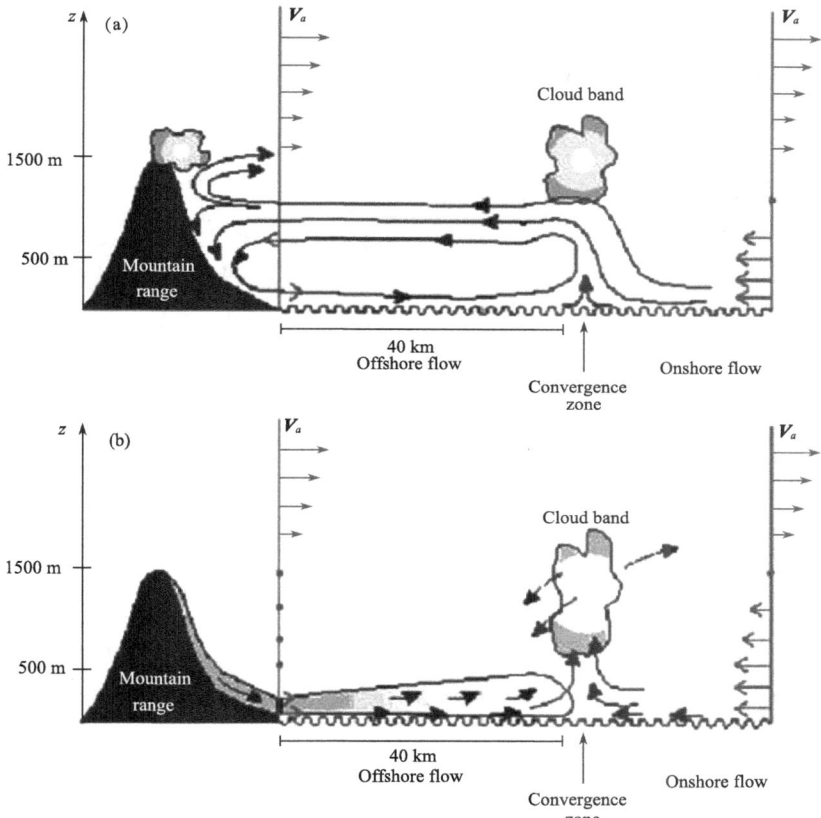

Fig. 2.27 A schematic diagram for explaining the formation of the convective belt in the east of Taiwan after (a) Schematic plot of the airflow for the case when the recirculation of air causes the formation of the coast-parallel atmospheric front. The left and right vertical lines with the horizontal arrows attached denote the horizontal components of the wind vectors close to the shoreline and east of the convergence line, respectively. (b) Same as in Fig. 2.27a, but for land breeze/katabatic wind. Note that this is a schematic plot, which does not show the circulation pattern at a fixed time. At 1025 LST (the time of the SAR observations), the flow of cold air from the mountain slope has already ended. (After Werner et al., 2007)

Under certain conditions, low-level onshore winds are deflected around the Olympic Mountains to flow through the Chehalis Gap and the Strait of Juan de Fuca. Subsequently, downstream from the Olympic Mountains, components of this flow turn north from the Chehalis Gap and south from the strait. This results in low-level convergence over Puget Sound, the general location of which is determined by the direction of the winds on the outer coast. It is this area of convergence, first discussed by Safford (1967), that has come to be known as the PSCZ (Bundy, 1969; Mass, 1981). A conceptualization of this flow is shown in Fig. 2.28b.

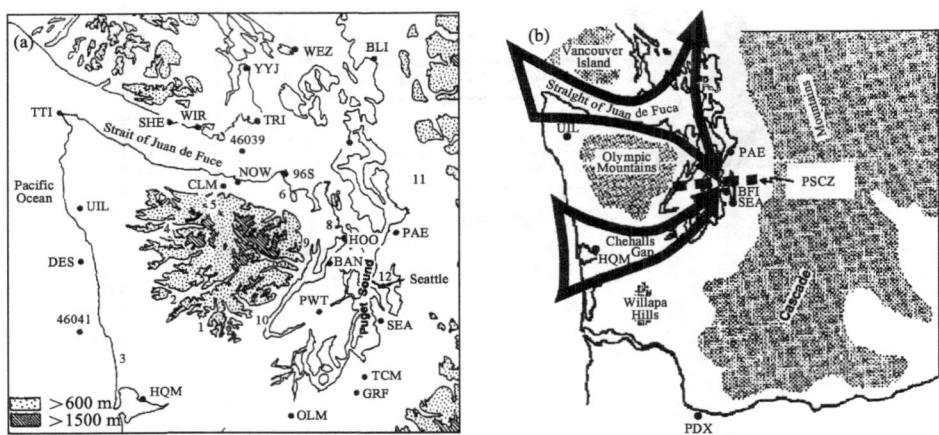

Fig. 2.28 The topographic map of Washington State (a) and the conceptualized flow associated with a PSCZ (b). Stippled area generally above 600 m (2000 ft). The microbarograph sites and other observational stations are indicated by numbers and abbreviation respectively. (After Whitney et al., 1993)

The typical PSCZ is associated with the passage across Washington State of a 500 hPa short-wavelength trough and an associated surface front. The link is the tendency for low-level winds to veer into the west-northwest in this postfrontal environment, which is the necessary direction for PSCZ development. As the west-northwest coast, several interactions occur. First, a sea level pressure trough forms on the lee side of the Coast Ranges. Second, a mesoscale surface low pressure center forms within the lee trough downstream (northeast) from the highest terrain of the Olympic Mountains. The mesoscale low is characterized by ① a small area of calm winds near its center, ② weak northerly winds along the northeast side of the Olympic Mountains, and ③ an enhancement of the southwesterly flow across the southern Puget Sound lowlands. The intensification of this mesoscale low seems to initiate the formation of the PSCZ.

The PSCZ has a seasonal cycle with maximum occurrences during the spring and early summer (from April to June) when two to four events are likely each month (Mass, 1981).

Chapter 2 Topographically Forced Mesoscale Circulations

The PSCZ also displays a significant diurnal modulation, especially during the warm season, that is due to a phasing of the synoptic-scale control (i. e., the low-level west-northwesterly winds) with western Washington sea breeze circulations. As the sun heats the western Washington interior, an ocean-to-land wind regime becomes established that increases the westerly flow through both the strait and the Chehalis Gap. There is also a diabatically produced north-to-south flow. These diurnal breezes combine to reinforce the circulation of a PSCZ and favor its southward advance. This reinforcement also combines with an afternoon minimum in static stability to favor increased shower activity. Thus there is a tendency for PSCZs to form across the northern portions of the sound early in the day and then move southward, reaching the central sound during the afternoon and early evening with a maximum in attendant weather. PSCZs may move back toward the north overnight as the sea breeze circulations dissipate. Additionally, PSCZs are typically much weaker in character at night due to the increasing static stability in the boundary layer.

Under the most favorable conditions, the PSCZ can be strong enough to produce thunderstorms with brief heavy rain, snow, ice pellets, or hail. During the coldest months, heavy snow is possible, especially if the PSCZ should become quasi-stationary; accumulations of over 4 in*. (102 mm) can occur, and winter storm warnings are occasionally necessary. In contrast, a weak PSCZ in a dry air mass may produce nothing more than a surface wind-shift boundary.

National Weather Service forecasters in Seattle have developed and used a combination of approaches to forecast the PSCZ. The approaches including medium range forecast (beyond 1 h) and those that address the "nowcasting" or very short range forecast (0—12 h). A decision tree outlines the process to make a forecasting is shown in Fig. 2.29.

For the cities located near the coast of sea, the regional weather will usually be influenced by the interaction between sea breeze and urban-heat island circulations (UHIC). For example, Yoshikado (1994) suggested that the UHIC may persist under the influence of the sea breeze and interact with it. The sea-breeze front has a tendency to remain over the city due to the effect of the UHIC, which can cause convergent flow patterns, particularly in cities close to a coast. Observations reveal that

---

* 1 in=2.54 cm.

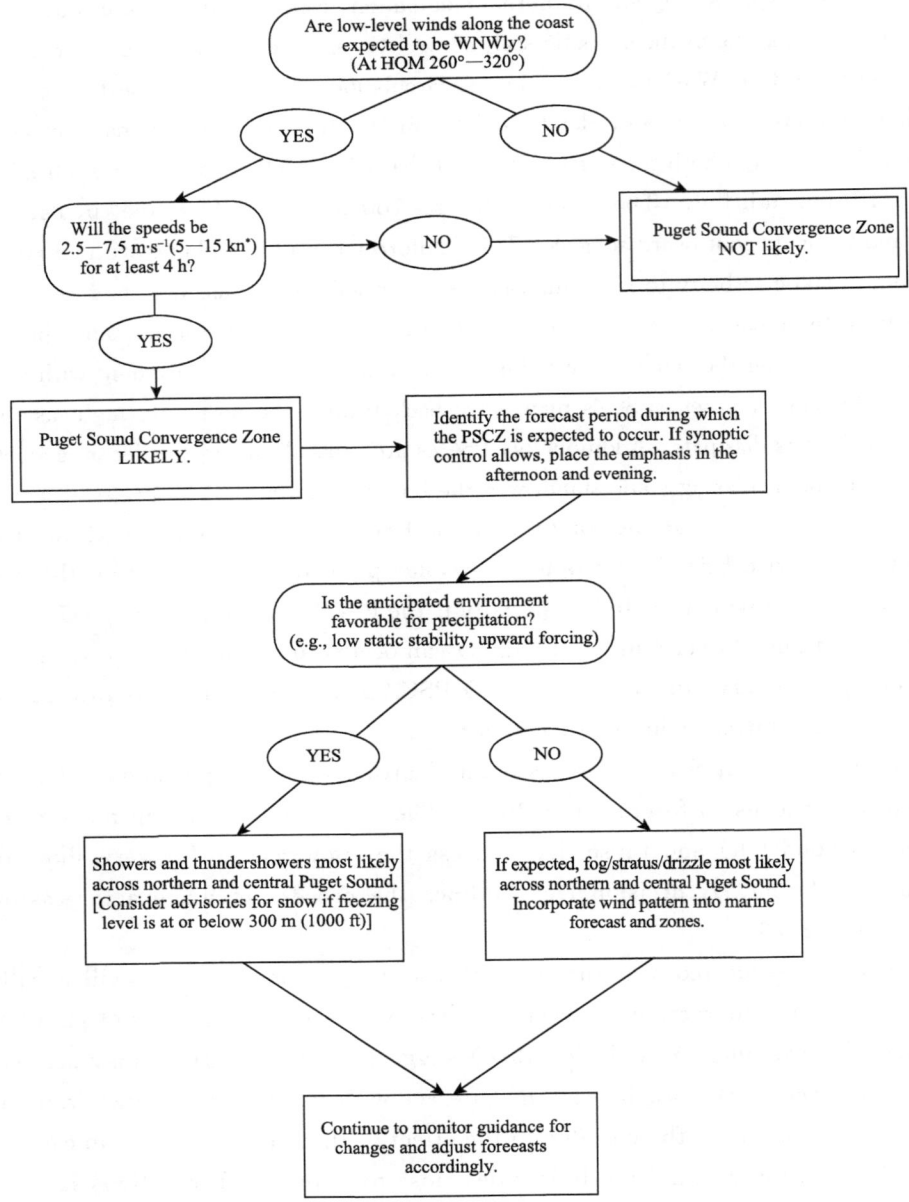

Fig. 2.29 Decision tree for forecasting of the PSCZ.
(After Whitney et al., 1993)

* 1 kn = 0.514444 m·s^{-1}.

the interaction between the UHIC and the sea breeze is more significant, when: ① larger urban regions are involved; ② the urban area is sufficiently far from the coast such that the UHIC has ample time to develop. The modeling results from Ohashi and Kida (2002), which show that sea-breeze and heat-island coupling prevents further inland movement of the sea breeze. Because of a persistent convergence zone over the urban area, significantly high concentrations of pollutants result over the city.

The sea breeze can also modify the UHIC pattern. Gedzelman et al. (2003), studying the UHIC of New York with a mesoscale network of weather stations, verified that during spring and summer sea breezes commonly reduce and delay the UHIC and displace it about 10 km inland. Thus, there is no complete agreement concerning the impact of the UHIC on the propagation speed of the sea-breeze front. Freitas et al. studied the interactions of an urban heat island and sea-breeze circulations during winter over the metropolitan area of São Paulo, Brazil. The interactions between the UHICs and the sea breeze were explored using sensitivity tests. Initially, the UHIC acts to accelerate the sea-breeze front into the metropolitan area of São Paulo (MASP) until it reaches the centre of the urban region. The presence of the urban region increases the sea-breeze-front propagation speed by about $0.32 \text{ m} \cdot \text{s}^{-1}$ when compared with the situation where no city exists. Additionally, due to the strong convergence zone in the centre of the city the circulations induced from MASP act to stall the sea breeze over the city for about two hours, carrying a large amount of humidity from the surface to upper levels of the urban atmosphere. The UHI also contributes to increased wind speed in the sea-breeze circulation cell. The prevention of sea-breeze-front propagation continues until the UHI effect disappears (see Fig. 2.30).

§ 2.5 Mountain and valley breeze

2.5.1 Features of the mountain and valley breeze

The breeze induced due to the temperature difference between a slope and the atmosphere nearby is called slope breeze. In the daytime the wind blowing from flat ground up to slope is called anabatic breeze; at night the wind blowing from slope down to the flat ground is called katabatic breeze. When the slope is composed by mountain and valley, the slope wind is called as mountain and valley breeze. In the daytime the wind blowing from valley up to the mountain is called valley breeze or up

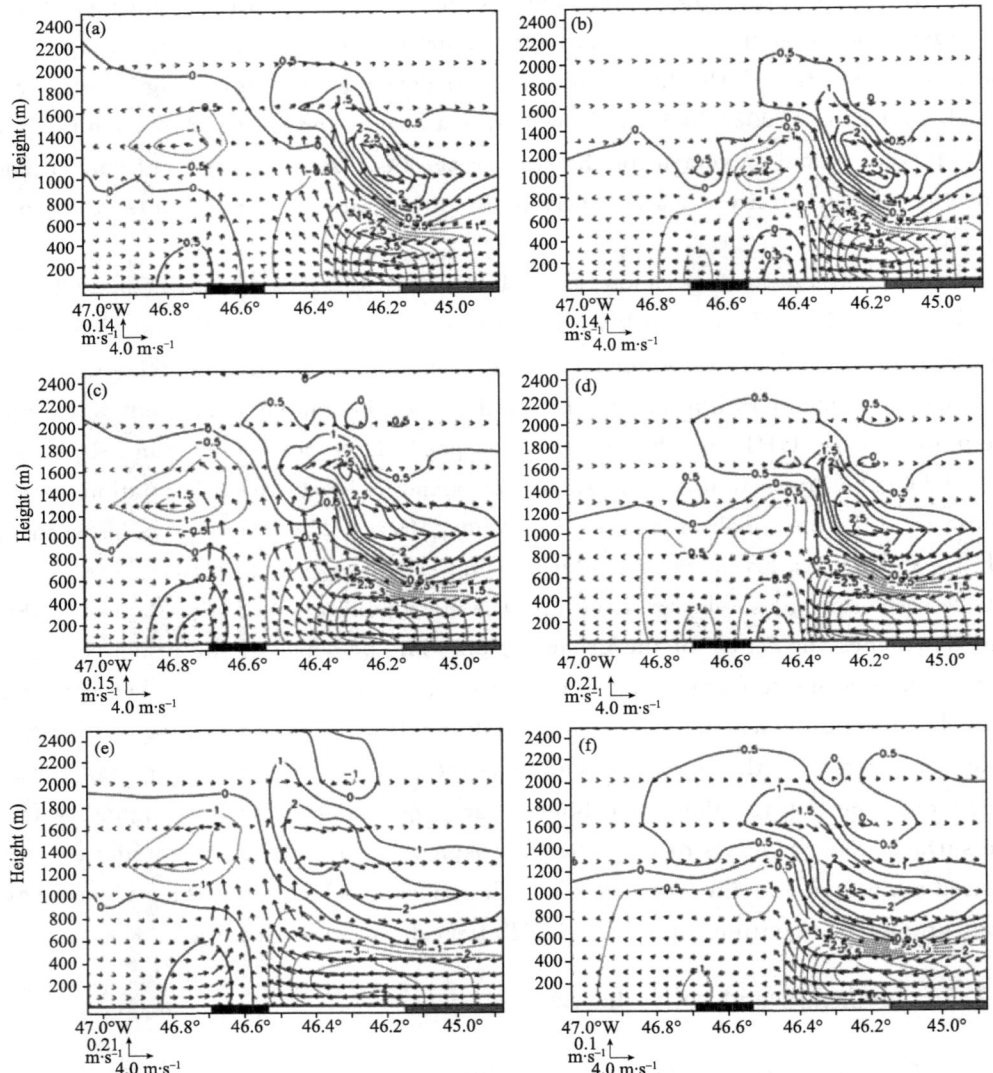

Fig. 2.30 Time evolution of the sea-breeze front for simulations with city (WC) and without city (NC). The bar in the bottom of the pictures indicates regions of ocean (dark grey), urban (black) and any type of vegetation (white). Vectors correspond to the composition between vertical (w) and sea-breeze wind (U_{sb}). The intensity of these two components is presented in the bottom left corner of each figure. Contours represent the magnitude of U_{sb} (in m·s^{-1}). (After Freitas et al., 2007) (a) 1500 LT (WC); (b) 1500 LT (WC); (c) 1600 LT (WC); (d) 1600 LT (WC); (e) 1700 LT (WC); (f) 1800 LT (WC)

valley breeze; at night the wind blowing from mountain down to the valley is called as mountain breeze or down valley breeze (Fig. 2.31).

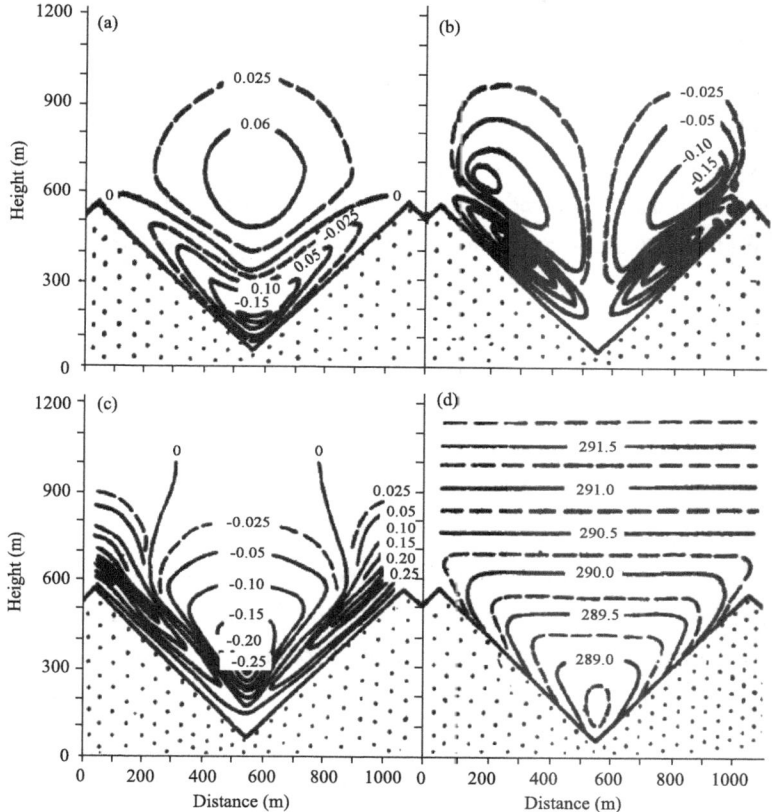

Fig. 2.31 Section of wind and temperature at the entrance of valley (a) longitudinal (length wise direction) component of wind (m · s^{-1}), positive is up valley wind or valley wind, negative is down valley wind or mountain wind (the maximum is 0.15 m · s^{-1}); (b) lateral (crosswise direction) component of wind (m · s^{-1}), at the right side of the figure, positive value is the wind points from the valley bed to right side of the mountain ridge; it is in axisymmetric between left and right sides; (c) vertical velocity (m · s^{-1}), positive is upward; (d) potential temperature (K) (After Thyer, 1966)

The formation mechanism of the slope breeze is similar to the sea-land breeze, both are induced due to the thermo forcing. Suppose under the initial condition the atmosphere is calmness and cloudless. In the daytime the slope surface absorbs radiation heat and warms the surface, hence the temperature of the air near the slope surface will higher than that of the free atmosphere at the same sea level. As a result the vertical pressure gradient in the cooler free atmosphere will be greater than that in the warmer air near the mountain slope. That means at the given influencing height of the slope heating the pressure will become higher than that at the point far away from the slope. Hence the horizontal pressure gradient force will cause the air flows from the slope to the place where the pressure is lower. While it is opposite at the low level, where the pressure at flat ground is higher than that on the slope, so that the upslope breeze and circulation are induced. At night, the surface cooling will cause a pressure gradient pointed to the slope at higher level and leads to cooler air moving down from the slope.

Only a tiny temperature difference between slope and free atmosphere, say few one tenth degree, might enough to cause slope wind. According to the calculation of Wenger (1923), the slope wind velocity can be as strong as $9\Delta T$ (m · s^{-1}) in 3 hours after initial time, where ΔT is the temperature difference between slope and free atmosphere at the initial time.

2.5.2 The influences of the mountain on weather

The influences of the mountain-valley breeze are similar to those of the sea-land breeze. When the valley breeze advances or withdraws, it can cause the variation of temperature and other meteorological elements. It can also cause convergence or shear lines on the wind field, which may act as a trigger mechanism to trigger convective weather.

Under some situation, the influences of mountain-valley breeze and sea-land breeze will exist simultaneously. For instance, the variation of the atmospheric circulation near the Taiwan Strait area is normally closely related with both the influence of sea-land breeze and mountain-valley breeze.

Beside those mentioned above, the influences of mountain ranges including many other aspects, so that the effects of the mountains is one of the most important factors, which have to be considered carefully for making the weather forecasting.

Garvert et al. (2007) studied the multiscale mountain waves influencing a major orographic precipitation event. Their study combines high-resolution mesoscale model simulations and comprehensive airborne Doppler radar observations to identify kinematic structures influencing the production and mesoscale distribution of precipitation and microphysical processes during a period of heavy prefrontal orographic rainfall over the Cascade Mountains of Oregon of United States on 13—14 Dec., 2001. A three-dimensional idealized schematic of topography and wind flow over the study area as shown in Fig. 2.32 showed that the significant vertical velocity perturbations in different scales are presented over the Cascade Mountains.

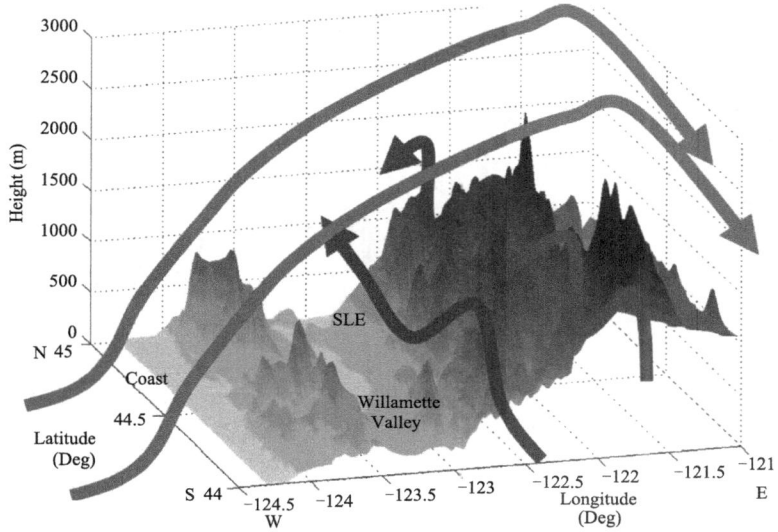

Fig. 2.32 Three-dimensional idealized schematic of topography and wind flow over the study area from 2300 to 0100 UTC 13—14 Dec., 2001. Blue arrows show strong southerly low θ_e airflow at low levels along the windward (west facing) slopes of the Cascade range which was subsequently involved in wave generation over multiple small-scale east-west-oriented ridges-valleys within the Cascade foothills. Red arrows show the high θ_e cross barrier flow that surmounted the low θ_e air and exhibited a vertically propagating mountain wave structure anchored to the mean north-south Cascade crest. (After Garvert et al., 2007)(see the color illustrations)

Hobbs et al. (1996) discussed the influences of the Rocky Mountains on the cyclones in the lee of the Mountains and developed a new conceptual model for cyclones in the central United States. Their research results show that, when a short wave trough moves eastward over the Rocky Mountains and into the central United States, the following important features may form: a dry trough (i.e., a lee trough that also has the characteristics of a dryline), an artic front, a low level jet, and two synoptic-scale rainbands (called the cold front aloft rainband and the pre-dry-trough rain band) that can produce heavy precipitation and severe weather well ahead of the dry-trough. These features are in corporate into a new conceptual model for cyclones in the central United States. A schematics illustration for the approach of the cold front aloft

(CFA) and the formation of the CFA rain band is shown in Fig. 2.33. Using this model can aid the interpretation of observational data and numerical model output, and it may also help to improve short range forecasting in the central United States.

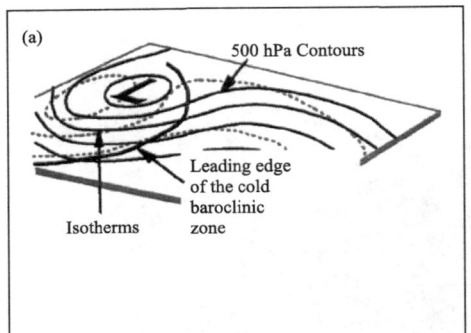
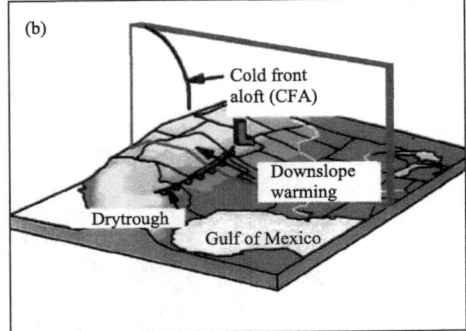
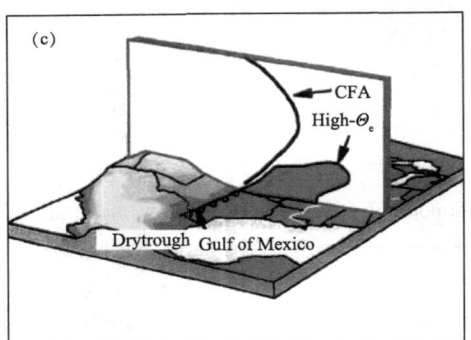
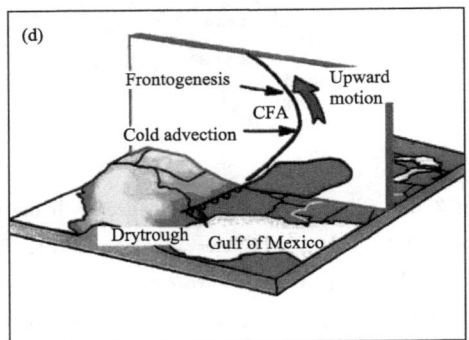
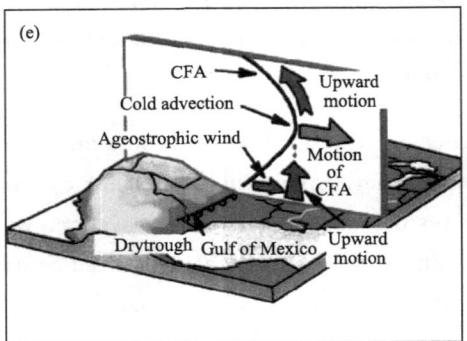
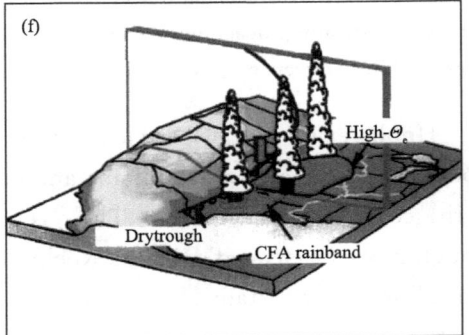

Fig. 2.33 Schematics illustrating the approach of the cold front aloft (CFA) and the formation of the CFA rain band. (After Hobbs, 1996)(see the color illustrations)

References

ANTHES R A, DANIEL K, DEARDORFF J W, 1982. Further considerations on modeling the sea breeze with a mixed-layer model[J]. Mon Wea Rev, 110:757-765.

ARAKAWA S, 1969. Climatological and dynamical studies on the local strong winds, mainly in Hokkaido[J]. Japan Geophys Magaz, 34(4):359-425.

ATKINSON B W, 1981. Meso-Scale Atmospheric Circulations[M]. Academic Press: 495.

BANTA R M, 1995. Sea breezes shallow and deep on the California coast[J]. Mon Wea Rev, 123: 3614-3622.

BANTA R M, OLIVIER L D, LEVINSON D H, 1993. Evolution of the monterey sea-breeze layer as observed by pulsed Doppler lidar[J]. J Atmos Sci, 50:3959-3982.

CASE J L, MANOBIANCO J, 2004. An objective technique for verifying sea breezes in high-resolution numerical weather prediction models[J]. Wea Forecasting, 19:690-705.

CHOPRA K F, HUBERT I F, 1964. Karman Vortex Streets in the earth's atmosphere[J]. Nature, 203:1341-1343.

CHOPRA K P, 1972. Velocity field in vortices leeward of island[J]. J Atmos Sci, 39:396-399.

CHUNG Y S, KIM H S, 2008. Mountain-generated vortex streets over the Korea South Sea[J]. International Journal of Remote Sensing, 29(3):867-877.

FEIT D M, 1969. Analysis of the Texas Coast Land Breeze[R]. Tech Rep, U S A, 52.

FINKELE K, HACKER J M, KRAUS H, et al, 1995. A complete sea-breeze circulation cell derived from aircraft observations[J]. Bound-Layer Meteor, 25:63-88.

FISHER E L, 1960. An observational study of the sea breeze[J]. Atmos Sci, 17:645-660.

FREITAS E D, ROZOFF C M, COTTON W R, et al, 2007. Interactions of an urban heat island and sea-breeze circulations during winter over the metropolitan area of São Paulo, Brazil[J]. Bound-Layer Meteor, 122:43-65.

FRIZZOLA J A, FISCHER E L, 1963. A series of sea breeze observations in the New York City area [J]. J Appl Meteor, 2:722-739.

GARVERT M F, SMULL B, MASS C, 2007. Multiscale mountain waves influencing a major orographic precipitation event[J]. J Atmos Sci, 64: 711-737.

HOBBS P V, LOCATELLI J D, MARTIN J E, 1996. A new conceptual model for cyclones generated in the lee of the Rocky Mountains[J]. Bull Am Meteor Soc, 77:1169-1178.

HOLTON J R, PYLE J, CURRY J A, 2003. Encyclopedia of Atmospheric Sciences[M]. Lee Waves and Mountain Waves. Durran D R, University of Washington, Department of Atmospheric Sciences, Seattle, USA, 1161-1169.

HUFF F A, CHANGNON S A, 1972. Climatological assesment of urban effects on precipitation at St. Louis[J]. J Appl Meteor, 11(5):823-842.

JOHNSON R H, TOTH J J, 1982. Topographic effects and weather forecasting in the Colorado PROFS meso-network area[C]. Prior Vol. Am Meteor Soc Conf on Weather Forecasting and Analysis, Seattle, Washington.

KINGSMILL D E, 1995. Convection initiation associated with a sea-breeze front, a gust front and their collision[J]. Mon Wea Rev, 123:2913-2933.

KLEMP J B, LILLY D K, 1975. The dynamics of wave induced down-slope wind[J]. J Atmos Sci, 32:320-329.

KLEMP J B, LILLY D K, 1978. Numerical simulation of hydrostatic mountain waves[J]. J Atmos Sci, 35:78-107.

KOO Y S, REIBLE D D, 1995. Flow and transport modeling in the sea-breeze. Part II: Flow model application and pollutant transport[J]. Bound-Layer Meteor, 75: 209-234.

KOZO T L, 1982. A mathematical model of sea breeze along the Alaskan Beaufort Sea Coast:Part II [J]. J Appl Meteor, 21:906-924.

KRAUS H, HACKER J M, HARTMANN J, 1990. An observational aircraft based study of sea-breeze frontogenesis[J]. Bound-Layer Meteor, 53:223-265.

LAIRD N F, KRISTOVICH D A R, RAUBER R M, et al, 1995. The cape canaveral sea and river breezes:Kinematic structure and convective initiation[J]. Mon Wea Rev, 123:2942-2956.

LANDSBERG H E, 1981. The Urban Climate[J]. International Geophysics Series, 28:275.

LI J, DU Y X, LIU W, et al, 1978. The numerical experiment on the formation of the lee waves and their influence on precipitation[J]. Atmos Sci, 2(3):210-218.

LI X F, 1990. The dynamic features of the mesoscale circulation forced by heat-island and the wind speed limitation[J]. Journal of Meteorology, 48(3):327-335.

LI X S, LI L Q, 1990. The influence of slope on the urban heat island[J]. Journal of Meteorology, 48 (3):293-302.

LILLY D K, 1978. A severe downslope windstorm and aircraft turbulence event induced by a mountain wave[J]. J Atmos Sci, 35:59-77.

LIN W S, WANG A Y, WU C S, et al, 2001. A case modeling of sea-land breeze in Macao and its neighborhood[J]. Adv Atmos Sci, 18: 1231-1240.

MAHRER Y, PIELKE R A, 1977. The effects of topography on the sea and land breezes in a two dimensional numerical model[J]. Mon Wea Rev, 105:1151-1162.

MASS C F, DEMPSEV D P, 1985. A one level mesoscale model for diagnosing surface winds in mountainous and coastal regions[J]. Mon Wea Rev, 113:1211-1227.

NEUMANN J, 1977. On the rotation rate of the direction of sea and land breezes[J]. J Atmos Sci, 34:1913-1917.

NEUMANN J, MAHRER Y, 1971. A theoretical study of the land and sea breeze circulation[J]. J Atmos Sci, 28:532-542.

OHASHI Y, KIDA H, 2002. Local circulations developed in the vicinity of both coastal and inland urban areas: A numerical study with a mesoscale atmospheric model[J]. J Appl Meteor, 41: 30-45.

OKOUCHI Y, SEGAL M, KESSLER R C, et al, 1984. Evaluation of soil moisture effects on the generation and modification of mesoscale circulation[J]. Mon Wea Rev, 112:2281-2292.

PEARSON R A, 1975. On the asymmetry of the land-breeze sea-breeze circulation[J]. Q J Roy Meteor Soc, 101:529-536.

SANG J G, 1989. The solution of downslope motion[J]. Journal of Meteorology, 47(2):191-198.

SCORER R S, 1949. Theory of waves in lee of mountains[J]. Q J Roy Meteor Soc, 75:41-56.

SCORER R S, 1954. Theory of airflow over mountains Ⅲ-Air-stream characteristics[J]. Q J Roy Meteor Soc, 80: 417-428.

SCORER R S, 1978. Environmental Aerodynamics. Chapter 6[M]. Ellis Horwood Publisher.

SHIR C C, 1973. A preliminary numerical study of atmospheric turbulent flows in the idealized planetary boundary layer[J]. J Atmos Sci, 30:1327-1339.

SHOU S W, ZHANG S Y, 1992. A numerical experiment with the effect of a complicated terrain on the mesoscale systems[J]. Acta Meteorologica Sinica, 6(4):452-461.

SMITH R B, 1976. The generation of lee waves by the Blue Ridge[J]. J Atmos Sci, 33:507-519.

SMITH R B, GRUBISIC V, 1993. Aerial observations of Hawaii wake [J]. J Atmos Sci, 50 (22):3728.

SONG J H, SHOU S W, Li Q T, 2009. Observational analysis and numerical simulation of a typical sea-land breeze process in Ningbo in summer season[J]. Tropical Meteorology, 37:336-342.

STEYN D G, 1998. Scaling the vertical structure of sea breezes [J]. Bound-Layer Meteor, 86: 505-524.

STEYN D G, 2003. Scaling the vertical structure of sea breezes revisited[J]. Bound-Layer Meteor, 107:177-188.

THYER N H, 1966. A theoretical explanation of mountain and valley winds by a numerical method [J]. Arch Met Geophys Bioklim A, 15: 318-347.

TIJM A B C, HOLTSLAG A A M, VAN DELDEN A J, 1999. Observations and modeling of the sea breezes with the return current[J]. Mon Wea Rev, 127: 625-640.

TSUCHIYA K, 1969. The clouds with the shape of Karman vortex street in the wake of Cheju Island, Korea[J]. J Met Soc Japan, 47:457-465.

UHLENBROCK N L, BEDKA K, FELTZ W F, et al, 2006. Mountain wave signatures in MODIS 6.7-μm imagery and their relation to pilot report of turbulence[J]. Wea Forecasting, 22:662-670.

VERGEINER I, LILLY D K, 1970. The dynamic structure of lee wave flow as obtained from balloon and airplane observations[J]. Mon Wea Rev, 98:220-232.

WANG G L, JIANG W M, WEI M, 2008. An assessment of urban heat island effect using remote sensing data[J]. Marine Science Bulletin, 10(2):14-26.

WERNER A, CHEN J P, LIN I-I, et al, 2007. Atmospheric fronts along the east coast of Taiwan studied by ERS synthetic aperture radar images[J]. J Atmos Sci, 64:922-937.

WHITENY W M, DOHERTY R L, COLMAN B R, 1993. A methodology for predicting the Puget sound convergence zone and its associated weather[J]. Wea Forecasting, 8:214-222.

XU Q, XUE M, DROEGEMEIER K K, 1996. Numerical simulation of density currents in sheared environments within a vertically confined channel[J]. J Atmos Sci, 53:770-786.

YOSHIKADO H, 1994. Interaction of the sea-breeze with urban heat islands of different sizes and locations[J]. J Met Soc Japan, 72:139-143.

YOSHIKADO H, TSUCHIDA M, 1996. High levels of winter air pollution under the influence of the urban heat island along the shore of Tokyo Bay[J]. J Appl Meteor, 35:1804-1814.

YOUNG G S, ZAWISLAK J, 2006. An observational study of vortex spacing in island wake vortex streets[J]. Mon Wea Rev, 134:2285-2294.

ZHANG Y X, CHEN Y L, SCHROEDER T A, 2005. Numerical simulations of sea-breeze circulations over northwest Hawaii[J]. Wea Forecasting, 20:827-846.

Chapter 3
Gravity Waves in Free Atmosphere

Gravity wave (GW) is one of the basic waves in atmosphere, and also one of the most simple and basic mesoscale motion systems. GWs can play important roles on weather variations through many ways such as triggering convections, transferring energy and momentum, and causing clear air turbulence (CAT) etc. Especially the GWs with large amplitude can cause extreme severe weather very often. In this chapter the essential features and structures as well as the development of GWs will be introduced briefly.

§ 3.1 Basic features of gravity waves

Generally, the gravity waves existed only under the action of the external conditions are called external gravity waves, while those existed inner the fluid when the external conditions are restricted, for instance the top and bottom boundaries are fixed, are called the inner gravity waves. Here is a brief introduction to the features, natures, structure of the inner gravity waves and their influences on weather.

Gravity waves are the propagation of the gravity oscillation of the static stable atmosphere. When the air parcel in static balance is disturbed and moved upward from the original balance height with adiabatic cooling, the gravity force will cause it to return to its original balance height. While as the air parcel moved downward and with adiabatic warming, the buoyancy force will cause it to return to the original balance height again. Such oscillation will propagate outward to cause the waves (see Fig. 3.1), which are therefore called as gravity waves or buoyancy waves. When the earth rotation effect is considered the wave is called inertial inner gravity wave.

Gravity wave is a kind of vertical transverse wave. The particle oscillation direction and the wave propagation direction are perpendicular to each other. When the wave propagates horizontally, the air particle moves up and down in vertical direction. In contrast, for the longitudinal wave or compression wave the oscillation direction of particle is parallel to that of wave propagation. For the horizontal transverse waves, both the air particle motion and the wave propagation are in a horizontal surface. The lee waves discussed in Chapter 2 is also a kind of gravity waves. They are

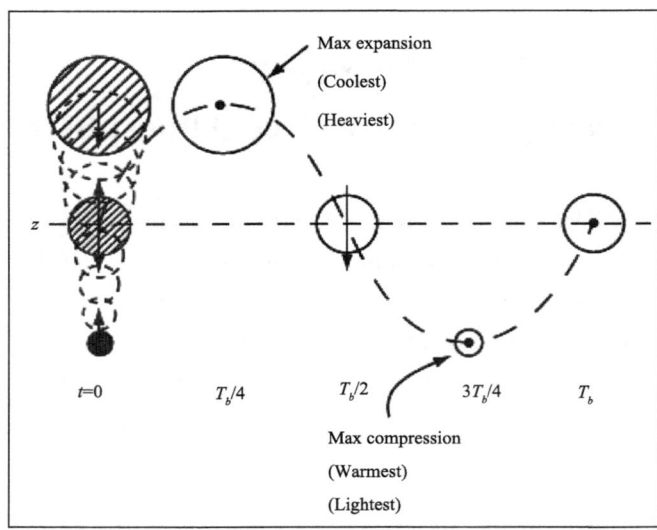

Fig. 3.1 Vertical oscillation of an air parcel in a stably stratified atmosphere when the Brunt Vaisala frequency is N. The oscillation period of the air parcel is $T_b = 2p/N$ and the volume of the air parcel is proportional to the area of the circle. (Adapted after Hooke, 1986)

related with topography, which is a fixed development source of the wave. In this chapter we will focus on the gravity waves in free atmosphere.

According to observations, the gravity waves can occur in different levels of the atmosphere from near-surface layer to low level of troposphere, and at high level up to the level higher than 75—100 km. They can be divided into three categories: the first, the GWs occurred in high altitude higher than 20 km above sea level; the second, the GWs occurred in very low level, say lower than 500 meters, which are normally Kelvin-Helmholtz (K-H) waves. Their wave lengths are normally very short and they can play role of a connector between mesoscale and micro scale systems; and the third, the GWs occurred in the main body of the atmosphere, i.e. the layer from 500 m to 20 km above sea level.

The gravity waves occurred in main part of the atmosphere include sub-synoptic scale and mesoscale. The wave length of sub-synoptic scale gravity waves is as long as thousand kilometers, while the wave length of mesoscale gravity waves has a wide range (4.4—300 km) and with the mean value about 34 km, amplitude of 853 meters, pressure amplitude of 0.1—5 hPa and with the mean value of 0.9 hPa; the period is 4—160 minutes, with mean value of 27 minutes, the phase velocity is 5.9—

60 m·s^{-1}, with the average value of 26 m·s^{-1}.

Microbarograph is one of the main instruments to measure the gravity waves. The modern microbarograph has high sensitivity. The pressure variation lower than 1 Pa can be measured. According to the observations, there are two types of typical mesoscale gravity waves: the larger amplitude irregular gravity waves as shown in Fig. 3.2a and the smaller amplitude regular one as shown in Fig. 3.2b. Beside micro-

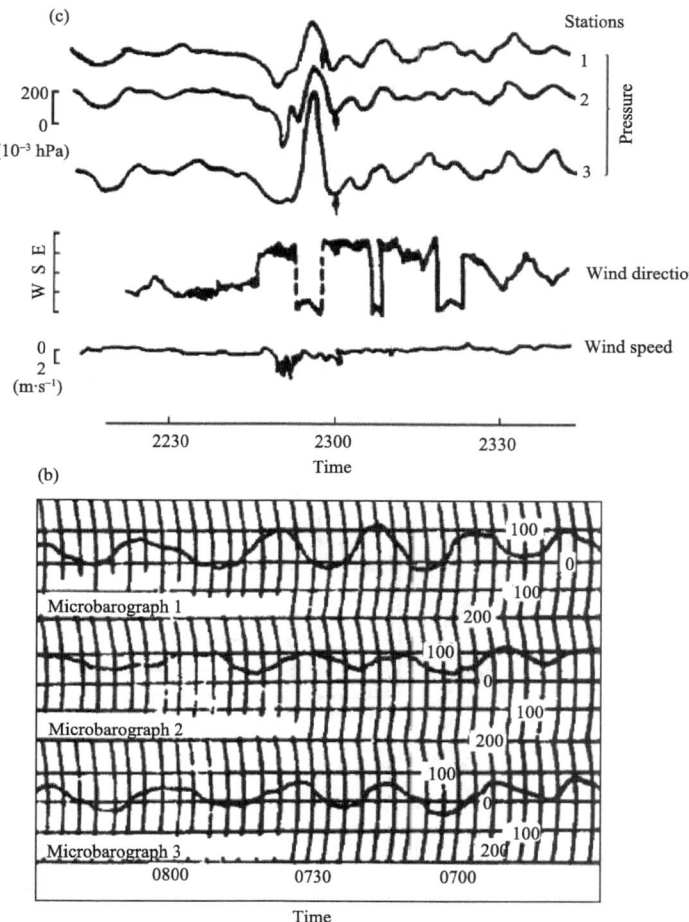

Fig. 3.2 The gravity wave recorded on surface. (a) large amplitude irregular type. The perturbation on the right side of the arrows are the GW caused by thunderstorms; (b) small amplitude regular type. The pressure unit is 10^{-3} hPa. The pressure records are from three observational stations and microbarographs. (Adapted from Curry and Murty. 1974)

barograph, satellite, radar, aircrafts and acoustic detecting etc. all can be used as the methods and tools for detection of gravity waves. On the satellite images or radar pictures the waves can be clearly presented very often.

Based on the observation, it is found that the gravity waves are very often occurred under the background existed temperature inversion or stable layer and vertical wind shear. In the wave layer the typical wind shear is 16—30 m·s^{-1}·km^{-1}. Fig. 3.3 shows the background of the gravity wave occurred in east of the United States on Mar. 18, 1969. From the picture it is clearly seen that the gravity wave occurred on 8—10 km height were clearly related with the upper level frontal zone and the maximum vertical wind shear as well as the minimum Richardson number. In general speaking, the gravity waves formed in the layer with $Ri<0.5$ (sometimes $Ri<0.25$), and normally the smaller the Richardson number value, the larger the amplitude of gravity waves.

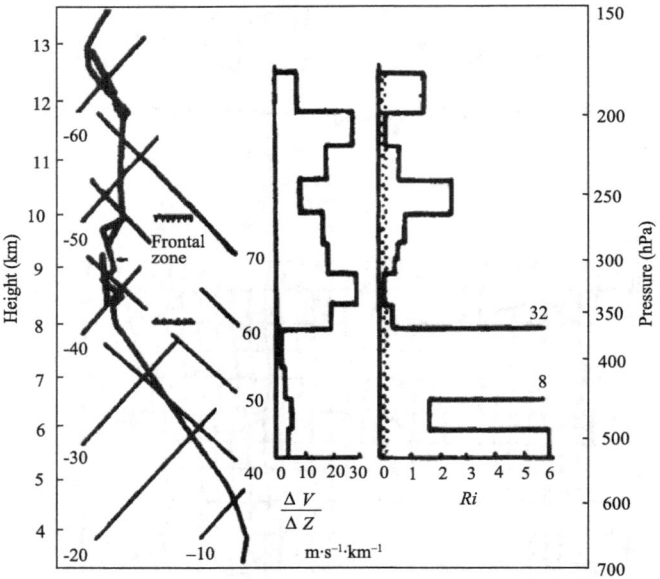

Fig. 3.3 The related temperature, wind shear and Richardson number (the thick line represents the smoothed value for calculating Richardson number on the sounding curves) on Mar. 18, 1969 over east of the United States. (Adapted from Reed and Hardy, 1972)

§ 3.2 Dynamic features of gravity waves

Gravity waves can be described by using Boussinesq equation set. Set the basic state of the atmosphere is stationary, the perturbation is three dimensional, using the anelastic approximation, then the linear Boussinesq equation is as following:

$$\begin{cases} \dfrac{\partial u'}{\partial t} = -\dfrac{1}{\bar{\rho}}\dfrac{\partial p'}{\partial x} + fv' \\ \dfrac{\partial v'}{\partial t} = -\dfrac{1}{\bar{\rho}}\dfrac{\partial p'}{\partial y} - fu' \\ \lambda \dfrac{\partial w'}{\partial t} = -\dfrac{1}{\bar{\rho}}\dfrac{\partial p'}{\partial z} - \dfrac{\rho'}{\bar{\rho}}g \\ \dfrac{\partial \bar{\rho}u'}{\partial x} + \dfrac{\partial \bar{\rho}v'}{\partial y} + \dfrac{\partial \bar{\rho}w'}{\partial z} = 0 \\ \dfrac{1}{\Theta}\dfrac{\partial \theta'}{\partial t} + Sw' = 0 \\ -\dfrac{\rho'}{\bar{\rho}} \approx \dfrac{T'}{T} \approx \dfrac{\theta'}{\Theta} \end{cases} \quad (3.2.1)$$

where u', v', w' are the wind disturbance components in x, y, and z direction respectively; p', θ' are the disturbances of pressure and potential temperature respectively; $\bar{\rho}$, T, P, Θ are the background field of density, temperature, pressure and potential temperature respectively. They are the function of height and the static balance is satisfied as follows:

$$\frac{\partial P}{\partial z} = -\bar{\rho}g \quad (3.2.2)$$

In equation (3.2.1), λ is a tracing coefficient. By taking $\lambda=1$ to indicate non-static balance, $\lambda=0$ indicates static balance. $\chi=C_p/C_v$, $S=\partial \ln\Theta/\partial z$; Set $u=\bar{\rho}u'$, $v=\bar{\rho}v'$, $w=\bar{\rho}w'$, $\theta=\bar{\rho}\theta'/\Theta$, and omitting the pressure disturbance term in the vertical equation. Then equation (3.2.1) can be written as follows:

$$\begin{cases} \dfrac{\partial u}{\partial t} - fv = -\dfrac{\partial p'}{\partial x} \\ \dfrac{\partial v}{\partial t} + fu = -\dfrac{\partial p'}{\partial y} \\ \lambda \dfrac{\partial w}{\partial t} - g\theta = -\dfrac{\partial p'}{\partial z} \\ \dfrac{\partial u}{\partial x} + \dfrac{\partial v}{\partial y} + \dfrac{\partial w}{\partial z} = 0 \\ \dfrac{\partial \theta}{\partial t} + Sw = 0 \end{cases} \quad (3.2.3)$$

Since the incompressible hypothesis has been made in the equations, the acoustic wave has been filtered out. By operating the Laplacian operator on the third equation in equation (3.2.3), and then taking derivation for t, then by using the fifth equation in equation set (3.2.3), we can obtain the following equation:

$$\lambda \nabla^2 \frac{\partial^2 w}{\partial t^2} = -\frac{\partial}{\partial z}\frac{\partial}{\partial t}\nabla^2 p' - N^2 \nabla^2 w \qquad (3.2.4)$$

where $N^2 = gS$. By acting Laplacian operator on p' in the first and second equation in equation (3.2.3), and then taking derivation for t, we get:

$$\frac{\partial}{\partial t}\nabla^2 p' = \left(\frac{\partial^2}{\partial t^2} + f^2\right)\frac{\partial w}{\partial z} \qquad (3.2.5)$$

Putting equation (3.2.5) into equation (3.2.4), then we get the equation about vertical velocity w as following:

$$\left[\left(\frac{\partial^2}{\partial t^2} + f^2\right)\frac{\partial^2}{\partial z^2} + \nabla^2\left(\lambda\frac{\partial^2}{\partial t^2} + N^2\right)\right]w = 0 \qquad (3.2.6)$$

Set a wave solution of the above equation as follows:

$$w = w_0 e^{i(k_x X + k_y Y + k_z Z - \sigma t)} \qquad (3.2.7)$$

where $k_x = \frac{2\pi}{L_X}$, $k_y = \frac{2\pi}{L_Y}$, $k_z = \frac{\pi}{H}$, they are the wave numbers in horizontal zonal, meridional and vertical directions respectively; L_X, L_Y are the horizontal wave length; H is the thickness of the disturbance; σ and w_0 are frequency and amplitude of the disturbance. Putting equation (3.2.7) into (3.2.6), we can get the frequency equation of gravity waves:

$$\sigma^2 = \frac{k_z^2 f^2 + k_H^2 N^2}{k_z^2 + \lambda k_H^2} = \frac{f^2 + N^2 m^2}{1 + m^2 \lambda} \qquad (3.2.8)$$

where $k_H^2 = k_X^2 + k_Y^2$, $m = k_H/k_z$, represented non-dimensionless number of the ratio between the horizontal scale and vertical scale.

$$\left.\begin{array}{ll}\text{When} \quad \lambda=0, & \sigma_0^2 = f^2 + N^2 m^2 \\ \text{When} \quad \lambda=1, & \sigma_1^2 = (f^2 + N^2 m^2)/(1+m^2)\end{array}\right\} \qquad (3.2.9)$$

From the above equation, it can be seen: $|\sigma_1| \leqslant |\sigma_0|$.

The horizontal wave speed of the inertial gravity wave is as follows:

$$C_H = \frac{\sigma}{k_H} \qquad (3.2.10)$$

$$\left.\begin{array}{ll}\text{When} \quad \lambda=0, & C_{H_0}^2 = \frac{1}{k_H^2}(f^2 + N^2 m^2) \\ \text{When} \quad \lambda=1, & C_{H_1}^2 = \frac{1}{k_H^2}\frac{(f^2 + N^2 m^2)}{(1+m^2)}\end{array}\right\} \qquad (3.2.11)$$

Based on above equation, to get $C_{H_0}^2/C_{H_1}^2 = \sigma_0^2/\sigma_1^2 = (1+m^2)$, i.e., $C_{H_1}^2 \leqslant C_{H_0}^2$, or

$|C_{H_1}| \leqslant |C_{H_0}|$. Since the horizontal wave pocket speed $C_g = \dfrac{d\sigma}{dk_H}$, so from the equation (3.2.8), we can get

$$C_g = \pm \frac{L}{2\pi} \frac{m^2(N^2 - \lambda f^2)}{(1+m^2\lambda)^{\frac{3}{2}}(f^2+N^2m)^{\frac{1}{2}}} \qquad (3.2.12)$$

When $\lambda = 0$, $C_{g_0} = \pm \dfrac{L}{2\pi} \dfrac{N^2 m^2}{\sqrt{f^2+N^2m^2}}$

When $\lambda = 1$, $C_{g_1} = \pm \dfrac{L}{2\pi} \dfrac{m^2(N^2-f^2)}{(1+m^2)^{\frac{3}{2}}(f^2+N^2m^2)^{\frac{1}{2}}}$

$\qquad\qquad(3.2.13)$

When stratification is stable, $N^2 > 0$, and generally $N^2 \gg f^2$, from equation (3.2.12), $|C_{g_1}| \leqslant |C_{g_0}|$ may be deduced. By comparison between equations (3.2.13) and (3.2.11), it can get $|C_{g_0}| \leqslant |C_{H_0}|$, $|C_{g_1}| \leqslant |C_{H_1}|$.

The following are the discussion about the natures of gravity waves.

(1) Suppose the gravity wave is a mesoscale system, $L_H \sim 100$ km, $H \sim 10$ km, then $(k_Z^2 + k_H^2) \approx k_Z^2$, hence from equation (3.2.10) and (3.2.8), it can get

$$C_H^2 = \frac{\sigma^2}{k_H^2} = \frac{k_Z^2 f^2 + k_H^2 N^2}{(k_Z^2 + \lambda k_H^2) k_H^2} \approx \frac{k_Z^2 f^2 + k_H^2 N^2}{k_Z^2 f k_H^2} = \frac{f^2}{k_H^2} + \frac{N^2}{k_Z^2} \qquad (3.2.14)$$

Since $N \sim 10^{-2} - 10^{-3} \cdot s^{-1}$, $f \sim 10^{-4} \cdot s^{-1}$, $f \ll N$, so equation (3.2.14) may be further simplified as

$$C_H^2 \approx \frac{N^2}{k_Z^2} \text{ or } C \approx \frac{N}{k_Z} \sim \frac{10^4}{\pi} N \qquad (3.2.15)$$

This shows that for the mesoscale gravity waves the wave speed does not related with the horizontal wave length, in other words the mesoscale gravity waves are non-dispersion waves. While for the subsynoptic gravity waves, $f \neq 0$, hence it is known from equation (3.2.14) that C_H is related with k_H, in other words it related with horizontal wave length. Hence this kind of gravity waves are the dispersion waves.

(2) For the non static balance ($\lambda = 1$) and the situation omitting the Coriolis force ($f \simeq 0$), equation (3.2.8) may be simplified as

$$\sigma^2 = \frac{k_H^2 N^2}{k_Z^2 + k_H^2} \quad \text{or} \quad \frac{\sigma^2}{N^2} = \frac{k_H^2}{k_H^2 + k_Z^2} \qquad (3.2.16)$$

The above equation shows that, in contrast with the acoustic waves, which is isotropous, the gravity waves have strong non isotropic propagation features. The propagation direction of the gravity waves is depending on their wave frequency. For the long period waves, $\sigma^2 \ll N^2$, then the equation (3.2.16) can be simply written as $\sigma^2/N^2 = k_H^2/k_Z^2$, and get $k_H^2 \ll k_Z^2$. Therefore under this situation, the phase propagation is close to the vertical direction, while the energy propagation direction is close to the horizontal direction, since there is orthogonality between the phase propagation and energy propagation directions. Besides, since the atmosphere is assumed as incompressible, the motions in (x, z) plane can be described by the following equations:

$$\begin{cases} u = u_0 e^{i(k_x X + k_z Z - \alpha)} \\ w = w_0 e^{i(k_x X + k_z Z - \alpha)} \\ \dfrac{\partial u}{\partial x} + \dfrac{\partial w}{\partial z} = 0 \end{cases} \quad (3.2.17)$$

And get
$$k_x u + k_z w = 0 \quad (3.2.18)$$

This showed in the waves the motion of particle is perpendicular to that of the moving direction of the waves.

Fig. 3.4 is the illustration of the phase propagation directions and energy propagation direction of a gravity wave pocket. As shown in the diagram, at the time $t = t_1$, there is a gravity wave pocket originated from the low part of the left side of the diagram. After that, at the time $t = t_2$, $(t_2 > t_1)$, the wave pocket moved upward to the right side departing from the original source place. At the time $t = t_3$, $(t_3 > t_2)$, the wave pocket further moved toward up and right side, therefore the wave energy propagated along the direction either, while at the meantime the waves in the wave pocket propagated toward right side and downward.

In the atmosphere this type of wave propagation can be observed under the situation when the gravity waves occurred near the point sources, such as earthquake, volcanic eruption, nuclear explosion and cumulonimbus etc. and met the boundary inversion or stratification.

Fig. 3.5 is a gravity wave excited by cumulonimbus. The short period gravity waves can be directly observed near the area over the point source. Under this situation, the gravity waves are dispersion waves and the wave pocket seems propagate to the direction toward the place far away from the source region. Larger scale waves propagate quicker, smaller scale waves propagate relative slowly. In the actual atmosphere the situation maybe more complicated, since many factors, such as the vertical wind shear and temperature stratification of the environment etc., can influence comprehensively on the propagation of the phase and energy of the gravity waves.

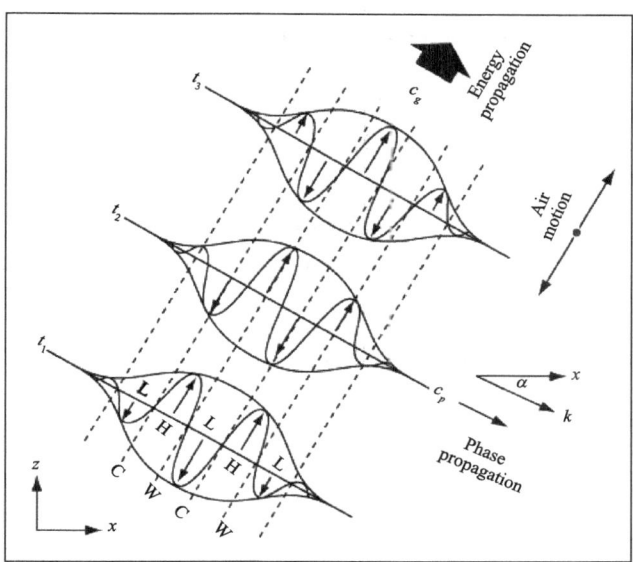

Fig. 3.4 Propagation of the energy and phase of the gravity waves. The energy of the wave group propagates with the group velocity (c_g; thick blunt arrow), while the phase of the wave propagates with the phase speed (c_p). Symbols H and L denote the perturbation high and low pressures, respectively, while W and C denote the warmest and coldest regions, respectively, Symbol α represents the angle of the wave number vector k from the horizontal axis or the wave front (line of constant phase) from the vertical axis. (Adapted from Hooke, 1986)

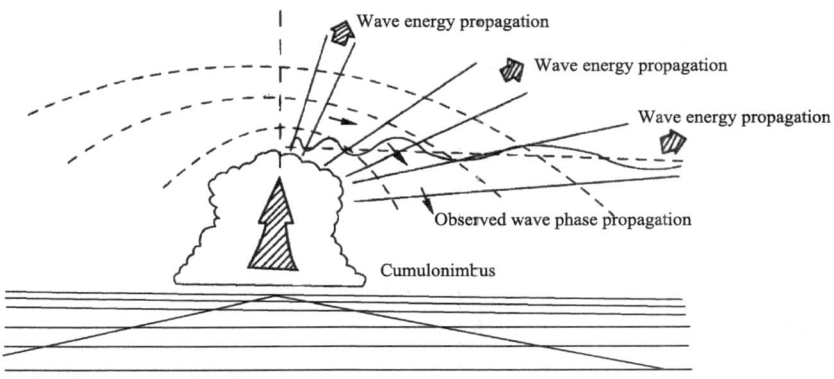

Fig. 3.5 The gravity wave caused by penetrance cumulonimbus.
(Adapted from Hooke, 1986)

§ 3.3 Structure and influences of gravity waves

Under the boundary condition: $w=0$ at $z=0$ and $z=H$, the solution of the vertical velocity of the gravity inertial waves of the equation (3.2.7) can be expressed as follows:

$$w = -w_0 \sin k_z Z \sin(k_x x + k_y y - \sigma t) \qquad (3.3.1)$$

Substituting the solution (3.3.1) into the equation (3.2.3), after integration it can get

$$\begin{cases} \theta = \dfrac{s}{\sigma} w_0 \sin k_z Z \cos(k_x x + k_y y - \sigma t) \\ D = k_z w_0 \cos k_z Z \sin(k_x x + k_y y - \sigma t) \\ \zeta = -\dfrac{w_0 f k_z}{\sigma} \cos k_z Z \cos(k_x x + k_y y - \sigma t) \\ p' = -\sigma \left(1 - \dfrac{f^2}{\sigma^2}\right)(k_x^2 + k_y^2)^{-1} k_z w_0 \cos k_z z \, \cos(k_x x + k_y y - \sigma t) \end{cases} \qquad (3.3.2)$$

By using stream function ψ and velocity potential φ to indicate the wind velocity, then we have

$$\begin{cases} u = -\dfrac{\partial \psi}{\partial y} + \dfrac{\partial \varphi}{\partial x} \\ v = \dfrac{\partial \psi}{\partial x} + \dfrac{\partial \varphi}{\partial y} \end{cases} \qquad (3.3.3)$$

Hence
$$\xi = \nabla^2 \psi, \qquad D = \nabla^2 \varphi \qquad (3.3.4)$$

Based on the first, second and fourth equations in the equation set (3.2.3), and by using the relationship of equation (3.3.4), the following equation set can be derived.

$$\begin{cases} \dfrac{\partial \psi}{\partial t} + f\varphi = 0 \\ \dfrac{\partial \varphi}{\partial t} - f\psi = -p' \\ \nabla^2 \varphi + \dfrac{\partial w}{\partial z} = 0 \end{cases} \qquad (3.3.5)$$

By substituting the equation (3.3.1) into the equation set (3.3.5), and after integrating, the following solutions can be derived.

$$\begin{cases} \varphi = -(k_x^2 + k_y^2)^{-1} k_z w_0 \cos k_z z \sin(k_x x + k_y y - \sigma t) \\ \psi = \dfrac{f}{\sigma}(k_x^2 + k_y^2)^{-1} k_z w_0 \cos k_z z \cos(k_x x + k_y y - \sigma t) \end{cases} \qquad (3.3.6)$$

According to the equations (3.3.1), (3.3.2) and (3.3.6) the chart of the gravity wave structure can be drawn as shown in Fig. 3.6. In the chart, it can be seen that in the troposphere, the surface pressure disturbance field and fluid field are most clear, and then decrease with height; while the potential temperature field has no disturbance, it is getting clear with height. The phase difference between the pressure field center and the divergence center is $\pi/2$. The pressure field center coincided with the vorticity center. i. e. , the high pressure center coincided with anticyclonic vorticity center, the cyclonic vorticity center coincided with low pressure center. At the lower part of the wave, ascending flow and convergence have same phase, while it lag behind a phase of $\pi/2$; the descending flow has a same phase with divergence, and lag behind a phase of $\pi/2$ comparing with the center of high pressure center.

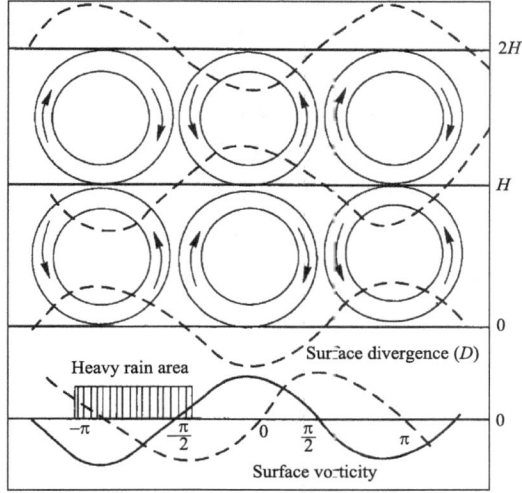

Fig. 3.6 Structure of gravity waves. The solid line is stream line, dashed line is constant pressure surface distribution; The relationship between the phases of surface divergence and vorticity are presented in the bottom part of the diagram; The slashed area represented the heavy rain area. (After Li, 1978)

The above analysis shows that there is a divergence and descending area in the front part of the low pressure disturbance and a convergence and ascending area in the rear part of the low pressure disturbance. For the high pressure disturbance, the situation is just opposite, the divergence and descending area locates at its rear part and the convergence and ascending area locates at its front part. Therefore the high pressure moves toward the convergence area, while the low pressure moves toward divergence area, the pressure disturbance propagates along the air flow. If the atmosphere is convective instability, then after the gravity wave trough passed, i. e. in the ascending area, the convection will develop. The most strong convection will occur in the wave ridge area. When the gravity wave occurs before the convective weather development, it plays a role as a triggering mechanism. When the gravity waves pass a existed convective weather area, it will cause the intensity of convection varied periodically. In the rear part of the wave trough the convection intensified, the strongest convection will occur near the wave ridge. When the next wave trough approaches, the convection will weaken, while it will intensify again when another wave ridge approaches.

On the Meiyu front there are very often many propagating mesoscale rain bands. Based on mesoscale analysis of the surface pressure and wind fields after filtered out the large scale field, it can be discovered that the Mesoscale convergence area associated with the mesoscale rain bands is located between the low and high pressure and more closer to the high pressure as shown in Fig. 3.7. This shows the Mesoscale rain bands are related with the gravity waves. Furthermore, researches in recent years revealed that the activities of the gravity waves can be very often founded in many systems such as frontal cyclones, landed typhoons, and low level jet streams etc. They are very closely related with torrential rain processes. The above features of the structure of gravity waves are also schematic represented in Fig. 3.8.

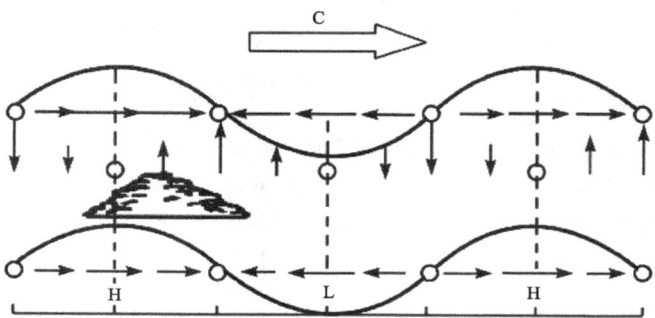

Fig. 3.7 Schematic diagram of the gravity wave structure.
(C, H, L denote moving direction of wave, high pressure and low pressure respectively.)
(After Uccellini and Koch, 1987)

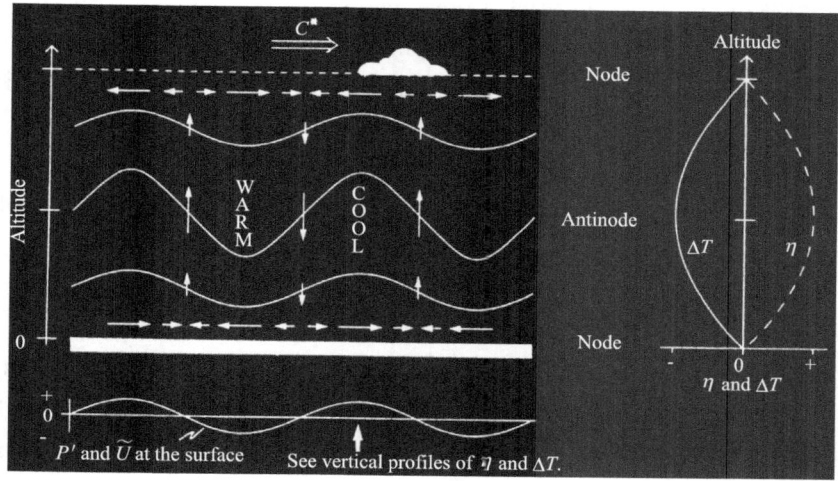

Fig. 3.8 Schematic representation of a Ducted Gravity Wave with 1/2 vertical wavelength trapped between the ground and a Critical Level. (After Ralph et al., 1993)

§ 3.4 Development of gravity waves

The development of gravity waves is closely related with the vertical wind shear and static instability. Furthermore it is also related with temperature advection and non-uniform horizontal distribution of the instability. The following are discussions on the relationship between these factors and the development of the gravity waves.

3.4.1 Vertical wind shear and development of gravity waves

In a stationary atmospheric environment the gravity waves are caused by gravity force and buoyancy force. In other words, the oscillation of the disturbance is realized through the inter conversion between disturbance kinetic energy and gravity potential energy. In the environment that the basic air flow $\bar{u}$ is not changed with height, the mechanism of the disturbance oscillation is same with that in stationary atmosphere. While in the environment that the basic air flow $\bar{u}$ changes with height, i.e. $\frac{\partial \bar{u}}{\partial z} \neq 0$, the disturbance oscillation will suffer the influences significantly induced by the vertical wind shear. When the vertical wind shear is stronger the influences will be more significantly.

Miles (1961) and Howard (1961) investigated the relationship between the vertical wind shear and the gravity wave development. They concluded that when the vertical wind shear of the mean air flow is larger and presents shear instability, the wave will grow up by obtaining the dynamic instability energy from the dynamic unstable air flow. Especially in the area near the jet stream, where the vertical wind shear is strong, it is easier to cause gravity waves. Therefore the area in vicinity of the jet stream is the energy source to generate mesoscale gravity waves. The results of theoretical analysis showed that the necessary condition to generate the gravity waves under this situation is $Ri<1/4$. The following is an explanation of this rule.

Suppose the atmospheric background with wind velocity $\overline{U}$ and potential temperature Θ is disturbed, the air parcel moves upward. Due to the momentum exchange in the atmospheric layer from z to $z+\Delta z$, the wind velocity in the layer is homogenized. Setting the average wind velocity in the layer Δz is $U+\frac{1}{2}\Delta U$. If the disturbance energy is converted from the kinetic energy of the mean motion of the basic flow, then the kinetic energy obtained by the disturbed air parcel is as follows:

$$\frac{1}{2}\left\{\frac{1}{2}[U^2+(U+\Delta U)^2]\right\}-\frac{1}{2}\left(U+\frac{1}{2}\Delta U\right)^2=\frac{1}{8}\Delta U^2=\frac{1}{8}\left(\frac{\partial U}{\partial z}\right)^2\Delta z^2 \quad (3.4.1)$$

During the process that the air parcel is lifted, it has to do work for overcoming gravity force under the stable stratification condition. The acceleration of the air parcel induced by the atmospheric stratification is

$$\frac{dw}{dt}=-\frac{g}{\Theta}\frac{\partial\Theta}{\partial z}dz \quad (3.4.2)$$

Setting the initial balance height is $z=0$, then

$$\frac{dw}{dt}=-\frac{g}{\Theta}\frac{\partial\Theta}{\partial z}z \quad (3.4.3)$$

In the layer Δz, the work acting by gravity force on the air parcel is

$$\int_0^{\Delta z}\frac{dw}{dt}dz=-\frac{g}{\Theta}\frac{\partial\Theta}{\partial z}\int_0^{\Delta z}zdz=-\frac{1}{2}\frac{g}{\Theta}\frac{\partial\Theta}{\partial z}\Delta z^2 \quad (3.4.4)$$

Therefore the work acting by the lifting air parcel for overcoming the gravity force is

$$\frac{1}{2}\frac{g}{\Theta}\frac{\partial\Theta}{\partial z}\Delta z^2 \quad (3.4.5)$$

When the energy converted from the mean motion is greater than the work done for overcoming the gravity force, i. e.

$$\frac{1}{8}\left(\frac{\partial U}{\partial z}\right)^2\Delta z^2>\frac{1}{2}\frac{g}{\Theta}\frac{\partial\Theta}{\partial z}\Delta z^2 \quad (3.4.6)$$

or
$$Ri = \frac{g}{\Theta} \frac{\frac{\partial \Theta}{\partial z}}{\left(\frac{\partial U}{\partial z}\right)^2} < \frac{1}{4} \qquad (3.4.7)$$

The vertical disturbance will continue to develop, that means the amplitude of the gravity waves will be magnified.

The above analysis showed that the gravity waves are influenced by vertical wind shear significantly. When the vertical wind shear is bigger, i. e. the Richardson number is smaller, $Ri<1/4$, the small disturbance will be exponential growth with time.

Furthermore, in the atmospheric layer with vertical wind shear it is possible to form a "critical level" and therefore will influence the gravity waves. So called critical level is a special altitude. On the altitude the horizontal phase velocity C of the wave is equal with the component along the horizontal direction of the basic flow velocity as shown in Fig. 3. 9. The critical level is a zone with strong interaction between wave and environmental flows. When $Ri<1/4$ at the critical level, i. e. there is the shear instability, the kinematic energy of the basic flow may convert into wave energy to cause the gravity waves increased. While if $Ri>1/4$ at the critical level, the situation will be opposite. When the gravity wave approaches the critical level the wave energy will be absorbed and converted into the kinetic energy of the basic flows. Hence the critical level plays a role of a source or sink area.

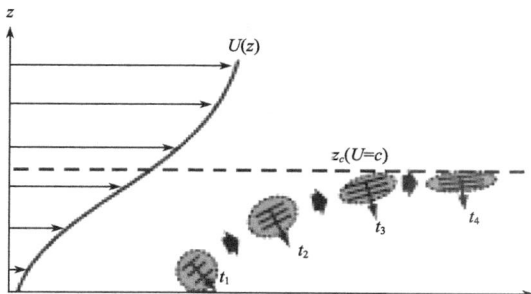

Fig. 3. 9 The propagation of a wave packet upward toward a critical level located at $Z=Z_c$. The particle motions are parallel to the wave crests, which are denoted by straight lines. Note that the vertical wavelength decreases as the wave packet approaches the critical level. The phase lines are horizontally oriented at the critical level in this case. (Adapted after Bretherton, 1966)

3.4.2 Thermal wind adjustment and development of gravity waves

The development of gravity wave is closely related with the thermal wind adjustment. The following is a discussion.

To consider the following equations:

Vorticity equation

$$\frac{\partial \zeta}{\partial t}+\mathbf{V}\cdot\nabla(\zeta+f)+fD=0 \qquad (3.4.8)$$

Divergence equation

$$\frac{\partial D}{\partial t}+\nabla^2\phi-f\zeta=0 \qquad (3.4.9)$$

Thermodynamic equation

$$\frac{\partial}{\partial t}\left(\frac{\partial \phi}{\partial p}\right)+\mathbf{V}\cdot\nabla\left(\frac{\partial \phi}{\partial p}\right)+\frac{C^2}{p^2}\omega+\frac{R}{C_p p}\frac{dQ}{dt}=0 \qquad (3.4.10)$$

where ζ and D are vertical component of vorticity and horizontal divergence respectively. ϕ is geopotential, f is Coriolis parameter, $\omega=\frac{dp}{dt}$ is vertical velocity in p coordinate system, $\frac{dQ}{dt}$ is non adiabatic heating rate. $C^2=\pm\frac{R^2 T}{g}(\gamma'-\gamma)$, γ is the vertical lapse rate of temperature of the stratification, γ' is the vertical lapse rate of temperature of the air parcel, for the dry adiabatic process $\gamma'=\gamma_d$, for the moist adiabatic process $\gamma'=\gamma_m$. For dry adiabatic process, taking the coefficient of C^2 as positive, for the moist adiabatic process, taking the coefficient as negative. All other signs in the equations are the common signs.

By using mathematical decomposition analysis method, the equations (3.4.8)—(3.4.10) can be broken into two equation sets as follows:

$$\begin{cases} \dfrac{\partial \zeta}{\partial t}=-\mathbf{V}\cdot\nabla(\zeta+f) \\ \dfrac{\partial D}{\partial t}=0 \\ \dfrac{\partial}{\partial t}\left(\dfrac{\partial \phi}{\partial p}\right)=-\mathbf{V}\cdot\nabla\left(\dfrac{\partial \phi}{\partial p}\right) \end{cases} \qquad (3.4.11)$$

$$\begin{cases} \dfrac{\partial \zeta}{\partial t}=-fD \\ \dfrac{\partial D}{\partial t}=f\zeta-\nabla^2\phi \\ \dfrac{\partial}{\partial t}\left(\dfrac{\partial \phi}{\partial p}\right)=-\dfrac{C^2}{p^2}\omega, \quad \text{or} \quad \dfrac{\partial \phi}{\partial t}=-\dfrac{C^2}{p}\omega \end{cases} \qquad (3.4.12)$$

The equation set (3.4.11) describes the advection variations, while the equation set (3.4.12) describes the adjustment variations. The adiabatic approximation is used in the third equation in (3.4.12).

Inducing a two layers model as shown in Fig. 3.10, and to write the equation set (3.4.12) on the first and third layers, i. e. the 250 hPa and 750 hPa constant pressure surfaces respectively.

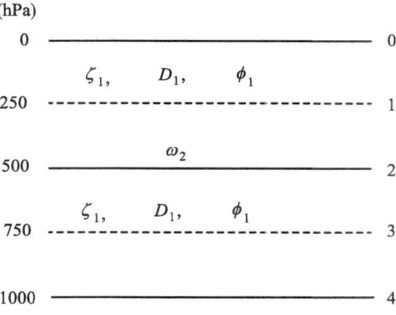

Fig. 3.10 Two leyers model

Let
$$\begin{cases} \hat{\zeta} = \dfrac{1}{2}(\zeta_1 - \zeta_3) \\ \hat{D} = \dfrac{1}{2}(D_1 - D_3) \\ \hat{\phi} = \dfrac{1}{2}(\phi_1 - \phi_3) \end{cases} \quad (3.4.13)$$

Then we can get

$$\begin{cases} \dfrac{\partial \hat{\zeta}}{\partial t} = -f\hat{D} \\ \dfrac{\partial \hat{D}}{\partial t} = f\hat{\zeta} - \nabla^2 \hat{\phi} \\ \dfrac{\partial \hat{\phi}}{\partial t} = \dfrac{C^2}{p} \omega_2 \end{cases} \quad (3.4.14)$$

Where $p = 1000$ hPa, ω_2 is the vertical velocity at 500 hPa constant pressure surface.

Based on the continuity equation $D = -\dfrac{\partial \omega}{\partial p}$ and the boundary conditions:

$$\begin{cases} p = 0 & \omega_0 = 0 \\ p = 1000 \text{ hPa} & \omega_4 = 0 \end{cases} \quad (3.4.15)$$

We can get

$$\hat{D} = -\dfrac{2\omega_2}{p} \quad (3.4.16)$$

Put equation (3.4.16) into the second equation in the equation set (3.4.13), get

$$\frac{\partial \omega_2}{\partial t} = -\frac{p}{2}(f\hat{\zeta} - \nabla^2 \hat{\phi}) \tag{3.4.17}$$

Take the time derivative for the above equation, and using the relationship of (3.4.14), we can get

$$\frac{\partial^2 \omega_2}{\partial t^2} = \frac{C^2}{2}\nabla \omega_2 - f^2 \omega_2 \tag{3.4.18}$$

When the stratification is stable, i. e. $\gamma < \gamma_d$, $C^2 > 0$, the equation (3.4.18) is in hyperbolic type, it is gravity inertial wave equation, which has the solution in the following form:

$$\omega_2 \sim e^{i(mx+ny-\sigma t)} \tag{3.4.19}$$

where m and n are the wave numbers in the x and y direction respectively. Substituting equation (3.4.19) into equation (3.4.18), to get

$$\sigma^2 = \frac{C^2}{2}(m^2+n^2) + f^2 \tag{3.4.20}$$

Based on the equation (3.4.18), the development of gravity waves can be explained as follows. Suppose at the initial state, the atmosphere at upper level (say 250 hPa) is under geostrophic balance, i. e. $\zeta_1 = \zeta_{g_1} = \frac{1}{f}\nabla^2 \phi_1$, the low level (say 750 hPa) is also under geostrophic balance, i. e. $\zeta_3 = \zeta_{g_3} = \frac{1}{f}\nabla^2 \phi_3$, the vorticity difference between two layers is according to the relationship of thermo wind, i. e. $\hat{\zeta} = \frac{1}{f}\nabla^2 \hat{\phi} = \zeta_{g_T}$, or $f\hat{\zeta} = \nabla^2 \hat{\phi}$. Obviously, it can be seen that at this time $\frac{\partial \omega_2}{\partial t} = 0$, hence there is no perturbation and no gravity waves.

Suppose at the time t, there is warm advection between layer 1 and layer 3. Then $\hat{\phi}$ will increase, if let ϕ_1 keep constant, then ϕ_3 will decrease and $\nabla^2 \phi_3$ will increase, $\nabla^2 \hat{\phi}$ will decrease. Suppose the thermo wind vorticity on the fluid field keep constant, then due to the thermo wind vorticity on the thickness field ($\frac{1}{f}\nabla^2 \hat{\phi}$) decreases, the situation $\hat{\zeta} > \frac{1}{f}\nabla^2 \hat{\phi}$ will occur. Therefore due to the thermo wind balance is destroyed, the vertical velocity will increase, i. e. $\frac{\partial \omega_2}{\partial t} < 0$. While due to the vertical velocity increasing, the adiabatic cooling will cause $\hat{\phi}$ decrease, ($\frac{\partial \hat{\phi}}{\partial t} = \frac{C^2}{p}\omega_2$). Under the condition that ϕ_1 keeps constant, ϕ_3 will increase, $\nabla^2 \phi_3$ will decrease, $\nabla^2 \hat{\phi}$ will increase, so that the situation $\hat{\zeta}=\frac{1}{f}\nabla^2\hat{\phi}$, i. e. $\frac{\partial \omega_2}{\partial t}=0$ will be recovered. However, due to the ascending intensity reaches maximum at this time, hence $\hat{\phi}$ will continue to decrease, φ_3 will continue to increase, $\nabla^2\hat{\phi}_3$ will continue to increase, so that the situation $\hat{\zeta}<\frac{1}{f}\nabla^2\hat{\phi}$ and $\frac{\partial \omega_2}{\partial t}>0$ will occur, after that the opposite variation process will begin. This circulating process will cause the atmosphere perturbation and propagation outward and therefore to form the gravity waves. In conclusion, we can say that the gravity waves are formed in the thermal wind adjustment process, vice versa, we can also say that the atmosphere achieved the thermal wind adjustment by stimulating the gravity waves.

§ 3.5 Development of the gravity waves near the upper level jet streak

Uccellini and Koch (1987) analyzed 13 cases of the gravity wave processes related with severe precipitation and summarized the characteristics of the synoptic background situation when the gravity waves occurred and provided a synoptic conceptual model for the development of the mesoscale gravity waves as shown in Fig. 3.11. From the figure, it can be seen that in the upstream of the wave activity region (indicated by the shading area) there is a low pressure system (marked by L) or an inverted trough on the surface map. In the low pressure there is a distinguished front stretched to the south or southeast of the wave activity region. The gravity waves are commonly found in the region with inversion layer in the low level of troposphere. While in the high level of troposphere, the waves are found in the exit region of the upper level jet streak and majorly to the right side of the axis of the jet streak, in other words to the side with anticyclonic shear. The gravity wave activity region is located in the area between the inflection axis between trough and ridge lines on 300 hPa constant pressure surface and the ridge line at the downstream.

Thus it can be seen that the development of the gravity waves is closely related with the upper level jet streak. In which, if the atmospheric stratification is stable, when the vertical wind shear is strong enough to let $Ri<1/4$, the gravity waves are able to develop through absorbing the energy from environmental wind field. Except the shear instability, the geostrophic adjustment is also one of the mechanisms for the development of the gravity waves. When the mass and momentum are out of balance and the motion is under non geostrophic balance state, the gravity wave or the inertial

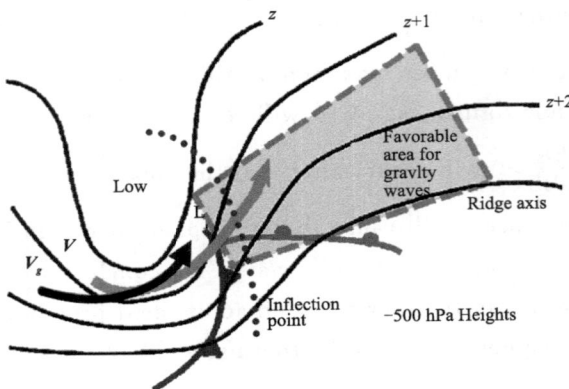

Fig. 3.11 Schematic of the environment conducive for mesoscale inertia-gravity wave generation by geostrophic adjustment. Depicted are the 300 hPa jet, 500 hPa height field (solid contours), and surface synoptic features. Positions of all atmospheric features shown on the schematic are approximate means during first half of wave episode, when wave generation mechanisms are assumed to operate most efficiently. The shaded region represents the area favorable for gravity wave activity during entire wave episode. Jet streak positions representing core of maximum wind speeds within the 300 hPa jet stream before (V_g) and after (V) the wave generation are denoted by thick arrows. (Adapted after Uccellini and Koch, 1987)(see the color illustrations)

gravity waves will be formed in the geostrophic adjustment process. Especially, under the situations when there is strong wind core propagating in the upper or low level jet stream, or in the frontogenesis and intense cyclone development processes, it is very often to form the large amplitude mesoscale gravity waves.

In the exit region of the upper level jet streak in front of the upper air trough, when the strong real wind core (V) breaks away from the strong wind core (V_g) located on the bottom of the trough and moves toward the inflection axis in the front of the trough, due to the geostrophic adjustment the mesoscale gravity waves will begin to form in the region near the inflectional axis and move forward and finally disappear near the ridge line. The south boundary of the activity region of gravity waves is the surface warm front or quasi-stationary front, while the north boundary is the upper level jet stream axis.

In the exit region of the upper level jet stream, the non-geostrophic balance feature of the motion can be denoted by the Lagrange Rossby number R_α:

$$R_{\alpha} = \frac{\left|\frac{d\mathbf{V}}{dt}\right|}{f|\mathbf{V}|} \tag{3.5.1}$$

It indicates the proportion between the acceleration of the air parcel and the earth rotational deflection acceleration. The small value of R_α denotes the motion closed to the quasi-geostrophic motion, while the big value of R_α denotes the non-geostrophic motion features, the bigger value of R_α denotes the stronger imbalance of the air flow. According to research, $R_\alpha > 0.5$ is the dynamic condition for generating the mesoscale gravity waves by geostrophic adjustment processes.

Based on the acceleration formula:

$$\mathbf{k} \times \frac{d\overline{\mathbf{V}}}{dt} = f\mathbf{V}_{ag}$$

We can change the above formula into the following:

$$R_{\alpha} \cong \frac{|\overline{V}_{ag}|}{|\mathbf{V}|} \tag{3.5.2}$$

In the right side of the above formula, $\overline{V}_{ag}$ is the lateral component of the non-geostrophic wind across the constant geostrophic potential height, which shows directly the degree of unbalance of the air flow and can be calculated by the difference between real wind and the geostrophic wind. Uccellini and Johnson (1979) pointed out that in the strong wind core of the straight line jet stream which is under geostrophic balance, the decelerated motion of the air parcel in the exit region occurred the lateral non-geostrophic motion pointed to the high pressure side. Under the situation when the strong real wind core left the geostrophic jet stream core moving to the downstream, the air parcel accelerated in the exit region, occurring the non-geostrophic air flow pointed from the anti-cyclonic side to the low pressure side due to the atmospheric mass and momentum have lost balance. In such a exit region with strong divergence, if the Rossby number, which denotes the non-geostrophic motion feature, is greater than 0.5 ($R_\alpha > 0.5$), then there is possibility to form large amplitude Mesoscale gravity waves.

Ferretti et al. (1988) used the nonlinear balance equation (NBE) for quantitative diagnosing the unbalance of the motion in the region where the gravity waves occurred. The non-balance equation can be expressed as follows:

$$-\nabla^2 \Phi + 2J(u,v) + f\zeta - \beta u = 0 \tag{3.5.3}$$

where Φ is geopotential height, ∇^2 is two dimensional Laplace operator; u, v are the components of wind velocity, J is Jacobi operator; f is Coriolis parameter, ζ is relative vorticity; $\beta = \dfrac{\partial f}{\partial y}$. NBE included four terms, which may all be calculated based on the data of upper air wind field and geopotential height field. Zack and Kaplan (1987) pointed out according to their calculations that in the exit region of the upper level jet stream, the sum of total four terms in the nonlinear balance equation has significant non zero value ($\sim 10^{-8}$ s^{-2}). Therefore, both the calculation results of NBE and R_α can be used as the quantitative indicators for analyzing the unbalance of the airflows and diagnosing the possibility to form large amplitude Mesoscale gravity waves.

References

BOOKER J R, BRETHERTON F P, 1967. The critical layer for internal gravity waves in a shear flow[J]. J Fluid Mech, 27:513-539.

GOSSARD E E, HOOKE W H, 1975. Waves in the Atmosphere:Developments in Atmospheric Sciences[M]. Vol II. Amsterdam:Elsevier Scientific Publishing Co:456.

HOOKE W H, 1986. Gravity wave[M]//Ray P S. Mesoscale Meteorology and Forecasting. Boston: Am Meteor Soc: 272-288.

HOWARD L N, 1961. Note on a paper of John W Miles[J]. J Fluid Mech, 10:509-512.

KOCH S E, DORIAN P B, 1988. A mesoscale gravity waves event observed during CCOPE. Part III:Wave environment and probable source mechanisms[J]. Mon Wea Rev, 116:2570-2592.

KOCH S E, EINAUDI F, DORIAN P B, et al, 1993. A mesoscale gravity waves event observed during CCOPE. Part IV:Stability analysis and Doppler derived vertical structure[J]. Mon Wea Rev, 121:2483-2510.

KOCH S E, et al, 1985. Observed interactions between strong convection and internal gravity waves [R]. Preprints, 14th Conf on Severe Local Storms, Indianapolis, Am Meteor Soc:198-201.

KOCH S, ZHANG F, KAPLAN M L, et al, 2001. Numerical simulation of a gravity wave event observed during CCOPE. Part 3:Mountain-plain solenoids in the generation of the second wave episode[J]. Mon Wea Rev, 129:909-932.

LINDZEN R S, TUNG K K, 1976. Banded convective activity and ducted gravity waves[J]. Mon Wea Rev, 104:1602-1617.

LUDLAM F H, 1967. Characteristics of billow cloud and their relation to clear-air turbulence[J]. Q J Roy Meteor Soc, 93:419-435.

MILES J W, 1961. On the stability of heterogeneous shear flows[J]. J Fluid Mech, 10:496-508.

QIN W J, SHOU S W, GAO S T, et al, 2010. Observational and numerical studies of the inertial gravity waves in a hailstorm process[J]. Journal of Geophysics, 53(5):1039-1049.

RAMAMURTHY M K, RAUBER R M, COLLINS B P, et al, 1993. A comparative study of large amplitude gravity waves events[J]. Mon Wea Rev, 121:2951-2974.

REED R J, HARDY K R, 1972. A case study of persistent, intense clear air turbulence in upper level frontal zone[J]. J Appl Meteor, 11:541-549.

STOBIE J G, EINAUDI F, UCCELLINI L W, 1983. A case study of gravity waves convective storms interaction:9 May 1979[J]. J Atmos Sci, 40:2804-2830.

SUN Y H, LI Z C, SHOU S W, 2015. The features and impacting of mesoscale gravity waves in a heavy snow storm precess[J]. Acta Meteorologica Sinica, 73(4):697-710.

UCCELLINI L M, KOCH S E, 1987. The synoptic setting and possible energy source for mesoscale wave disturbance[J]. Mon Wea Rev, 115:721-729.

ZACK J W, KAPLAN M L, 1987. Numerical simulation of the subsynoptic features associated with the AVE-SESAME I case. Part I: The preconvective environment[J]. Mon Wea Rev, 115: 2367-2393.

ZHANG F, 2004. Generation of mesoscale gravity waves in the upper-tropospheric jet-front systems [J]. J Atmos Sci, 61:440-457.

ZHANG F, KOCH S E, DAVIS C A, et al, 2001. Wavelet analysis and the governing dynamics of a large-amplitude gravity wave event along the East Coast of the United States[J]. Q J Roy Meteor Soc, 127: 2209-2245.

ZHANG F, KOCH S E, KAPLAN M L, 2003. Numerical simulations of a large-amplitude gravity wave event[J]. Meteorology and Atmospheric Physics, 84:199-216.

ZHANG Y, SHOU S W, WANG Y Q, et al, 2008. The mesoscale feature of a heavy snow process over Shandong peninsula[J]. Journal of Nanjing University of Information Science and Technology, 31(1):51-60.

Chapter 4
Front and Jet Stream

Although the front and jet stream are normally regarded as the large scale systems, while they are also regarded as the mesoscale systems since they are characterized by mesoscale structure and mesoscale dynamic features, and their related secondary circulations are closely related with mesoscale weather such as torrential rain and severe local storms etc. Therefore in recent years the problem on front and jet stream systems is an active mesoscale meteorological research area. In this chapter the structure of front, the frontogenesis and jet stream as well as the secondary circulations in the vicinity of jet stream systems etc. will be discussed.

§ 4.1 Structure of the front

4.1.1 The scale of front

Normally a long and narrow area with strong horizontal temperature gradient and bigger static stability as well as distinct cyclonic vorticity is called "front". Its length is normally one order bigger than width. The frontal length posses the order of Rossby deformation radius (L_R):

$$L_R = NH/f \qquad (4.1.1)$$

where N is buoyancy frequency, the characteristic value in troposphere is 10^{-2} s^{-1}; H is the depth scale of troposphere (10 km); f is Coriolis parameter, in mid-latitude $f \sim 10^{-4}$ s^{-1}. According to these characteristic values, we have $L_R \sim 1000$ km, hence the length of front is about 1000 km, the width is about 100 km. According to the definition of scale given by Orlanski (1975), the length of front belongs meso-α scale and the width belongs to meso-β scale.

4.1.2 The model of front

The earlier polar frontal theory given by Bjerknes (1919) regarded the atmospheric front as a sharply interface between two air masses with different density. In two sides of the front the density is discontinuous, but the pressure is continuous, hence the temperature is discontinuous according to the state equation. The frontal model regarded the temperature as discontinuous is called as "wedge" model, based on which the Margules frontal slope formula for the dry, non-viscosity, fluid static atmosphere can be get as follows:

$$\mathrm{tg}\psi = \partial h/\partial y = (f\overline{T}/g)[(U_{gw} - U_{gc})/(T_w - T_c)] \tag{4.1.2}$$

where $\overline{T}$ is the mean temperature in the frontal zone across the front; U_g is the component of geostrophic wind along the front; the subscripts w and c present the warm and cold side respectively; other signals are the common signs.

After the upper air sounding data have been used in operational analysis since 1930s, it is found that the front is a tilt transition zone rather than a sharply discontinuous surface. Therefore the wedge model is replaced by the frontal zone model, according to which the front is a finitude area. Inner the area and in two sides of the boundary the temperature is continuous, while the horizontal temperature gradient and the vertical shear of the geostrophic wind are discontinuous. Moreover, the geostrophic wind is continuous, but the horizontal geostrophic wind shear is discontinuous, hence the slops of the two sides of the frontal zone are different.

In recent years more people made detail observational research for the structure of the front (Keyser et al., 1986). They found that the feature of the front can be developed at different level of the troposphere. Generally the fronts characterized by the low troposphere features are called "low level front" or "surface front", while the fronts with distinct upper level frontal features are called as "upper level front" or "upper troposphere front".

Shapiro (1983) investigated the detail structure of the surface front based on the data recorded by a meteorological tower and mesoscale network. He selected a front without convective precipitation companying as an object for observational research, for the purpose to avoid confusing with squall lines. His research pointed that the strong temperature gradient concentrated in a very narrow area as shown in Fig. 4.1, which is a time-height cross-section of the potential temperature variation at a meteorological tower during the passage of the front. According to the moving speed of the front (~ 15 m·s^{-1}), the estimated width of the front is about 900 m, and the height of frontal "head" is about 600 m. The structure is quite similar to the model of the gravity flow or density current. Shapiro calculated the moving speed of the front according to the speed of gravity flow and got the result of 14 m·s^{-1}, which is quite close to the real situation. The above analysis showed that the early wedge model is still consist with fact in a sense.

Reed et al. (1955) discovered by means of sounding data analysis that there was

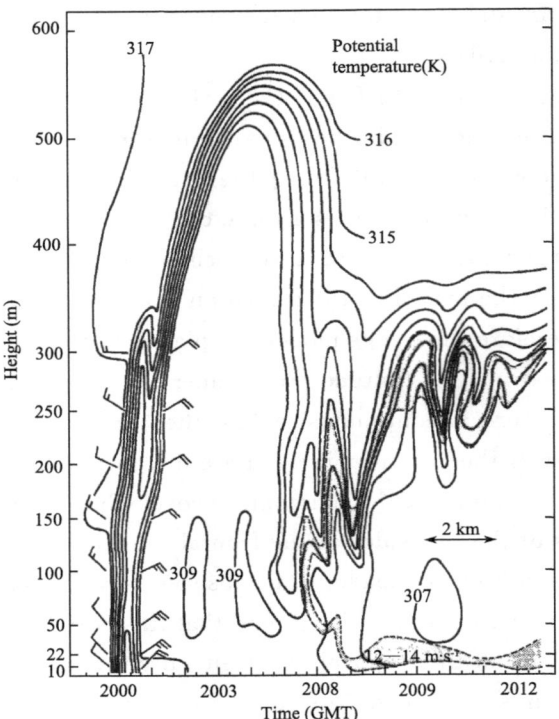

Fig. 4.1 The time-height cross-section of the potential temperature variation during the front passing the meteorological tower at BAO in the time period 1959—2013 GMT, Sept. 19, 1983. (the shading area is a 12—14 m·s^{-1} wind surge area) (After Shapiro, 1985)

a big area with great thermodynamic contrast and cyclonic wind shear in the mid-level of troposphere and presented the concepts on upper level frontogenesis and tropopause folding based on the analysis. So called the tropopause folding means the process that the air in stratosphere is crowded into the middle layer of the troposphere, sometimes it may reach down to 700—800 hPa. For demonstrating the air is descending from the stratosphere, Reed analyzed the potential vorticity p_θ.

$$p_\theta = -(\zeta_\theta + f)\frac{\partial \theta}{\partial p} \qquad (4.1.3)$$

where ζ_θ is the relative vorticity on the isentropic surface. By using p_θ the air flows from stratosphere and troposphere may be distinguished since in the adiabatic and non-frictional processes the potential vorticity is conservation along the trajectory of the air motion. Due to the static stability is different the magnitude of the potential vorticity in stratosphere is about one order bigger than that in the troposphere. Since it is discovered that the typical value of the potential vorticity in stratosphere is involved in the frontal zone in troposphere, hence it may be regarded as that the frontal zone is composed by the air from stratosphere.

The concept of tropopause folding has been proved by various observational factors, including that the nucleus radioactive fallouts from stratosphere is found near the surface; the ozone distribution presented an ozone tongue stretching downward from stratosphere, in the lifecycle of the front, ozone may be regarded as a conserved value. Moreover, Shapiro (1978) also proved the objective existing of the tropopause folding based on the observations by aircraft flying in upper air frontal zone. Meanwhile he also pointed that the cyclonic shear in the stratosphere in only concentrated actually in a meso-β scale area rather than in a meso-α scale area like Reed pointed out.

The formation of tropopause folding can be regarded as the result of the joint action effect combining the direct circulation in warm side and the indirect circulation in cold side of the front as shown in Fig. 4.2. The fact of troposphere folding has changed the traditional concepts that the polar front is a substance surface separated tropical air and polar air as well as the concept that tropopause is the substance surface separated the stratosphere and troposphere.

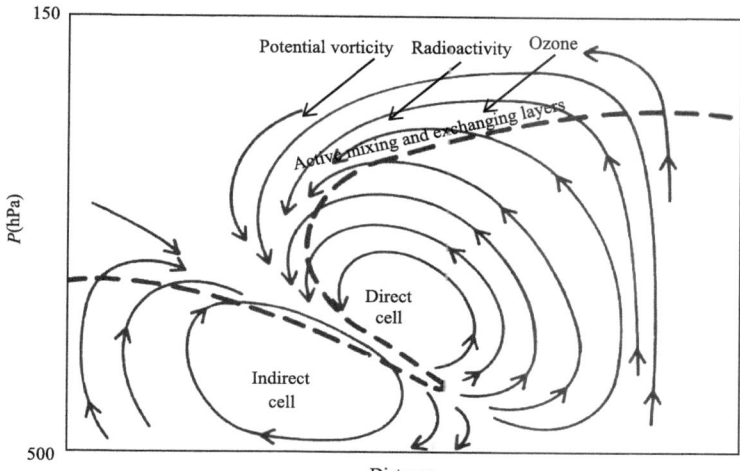

Fig. 4.2 The schematic diagram of the transverse circulation related with the tropopause folding process. (After Danielsen, 1968)

§ 4.2 Kinematic and thermodynamic frontogenesis

The process of front formed or intensified is called frontogenesis, oppositely, the process of front disappeared or weakened is called frontolysis. Petterssen (1936) used "frontogenesis function" F for quantitatively describing the processes. The two dimensional frontogenesis function is defined as follows:

$$F_p = \frac{D}{Dt}|\nabla_p \theta| \tag{4.2.1}$$

where $\frac{D}{Dt} = \frac{\partial}{\partial t} + \mathbf{V} \cdot \nabla_p + \omega \frac{\partial}{\partial p}$, the subscript p denotes on the constant pressure surface. If the long and narrow frontal zone is along the x axis, the constant potential temperature surface is parallel to the front, the wind along x axis is constant and suppose the temperature on the constant pressure surface decreases northward, then we have

$$F_p = \underbrace{\left(\frac{\partial v}{\partial y}\right)_p \left(\frac{\partial \theta}{\partial y}\right)_p}_{①} + \underbrace{\left(\frac{\partial w}{\partial y}\right)_p \frac{\partial \theta}{\partial p}}_{②} - \underbrace{\frac{1}{C_p}\left(\frac{P_0}{P}\right)^\kappa \frac{\partial}{\partial y}\left(\frac{dQ}{dt}\right)_p}_{③} - \underbrace{\frac{\partial}{\partial y}\left(K\frac{\partial^2 \theta}{\partial y^2}\right)}_{④} \tag{4.2.2}$$

where K is turbulence diffusion coefficient; Q is heating; and

$$C_p \frac{d\ln\theta}{dt} = \frac{1}{T}\frac{dQ}{dt} \tag{4.2.3}$$

In the equation (4.2.2), the terms ①, ②, ③ and ④ in the right side are convergence term, slant term, the horizontal gradient of non-adiabatic heating term and the turbulence diffusion coefficient term.

In the z coordinate system, the three dimensional frontogenesis function may be defined as follows:

$$F = \frac{D}{Dt}|\nabla \theta| = \frac{D}{Dt}\left[\left(\frac{\partial \theta}{\partial x}\right)^2 + \left(\frac{\partial \theta}{\partial y}\right)^2 + \left(\frac{\partial \theta}{\partial z}\right)^2\right]^{\frac{1}{2}}$$

$$= \frac{1}{|\nabla \theta|}\left\{\frac{\partial \theta}{\partial x}\left[\underbrace{\frac{1}{C_p}\left(\frac{P_0}{P}\right)^\kappa \frac{\partial}{\partial x}\left(\frac{dQ}{dt}\right)}_{①} + \underbrace{\frac{\partial}{\partial x}\left(K_h \nabla_h^2\theta + K_z \frac{\partial^2 \theta}{\partial z^2}\right)}_{②} - \underbrace{\frac{\partial u}{\partial x}\frac{\partial \theta}{\partial x}}_{③} - \underbrace{\frac{\partial v}{\partial x}\frac{\partial \theta}{\partial y}}_{④} - \underbrace{\frac{\partial w}{\partial x}\frac{\partial \theta}{\partial z}}_{⑤}\right]$$

$$+ \frac{\partial \theta}{\partial y}\left[\underbrace{\frac{1}{C_p}\left(\frac{P_0}{P}\right)^\kappa \frac{\partial}{\partial y}\left(\frac{dQ}{dt}\right)}_{⑥} + \underbrace{\frac{\partial}{\partial y}\left(K_h \nabla_h^2\theta + K_z \frac{\partial^2 \theta}{\partial z^2}\right)}_{⑦} - \underbrace{\frac{\partial u}{\partial y}\frac{\partial \theta}{\partial x}}_{⑧} - \underbrace{\frac{\partial v}{\partial y}\frac{\partial \theta}{\partial y}}_{⑨} - \underbrace{\frac{\partial w}{\partial y}\frac{\partial \theta}{\partial z}}_{⑩}\right]$$

$$+\frac{\partial\theta}{\partial z}\left[\frac{P_0^\kappa}{C_p}\frac{\partial}{\partial z}\left(P^{-\kappa}\frac{dQ}{dt}\right)+\frac{\partial}{\partial z}\left(K_h\,\mathbf{V}_h^2\theta+K_z\frac{\partial^2\theta}{\partial z^2}\right)-\frac{\partial u}{\partial z}\frac{\partial\theta}{\partial x}-\frac{\partial v}{\partial z}\frac{\partial\theta}{\partial y}-\frac{\partial w}{\partial z}\frac{\partial\theta}{\partial z}\right]$$

⑪ ⑫ ⑬ ⑭ ⑮ (4.2.4)

In the equation (4.2.4), the terms ①, ⑥ and ⑪ are non-adiabatic terms; the terms ②, ⑦ and ⑫ are the diffusion terms; the terms ③, ④, ⑧ and ⑨ are horizontal deformation terms including stretching deformation and shearing deformation; the terms ⑬ and ⑭ are the vertical deformation terms; the terms ⑤ and ⑩ are the slant terms; and the term ⑮ is the vertical divergence term. The two dimensional frontogenesis function is only a special case of the equation (4.2.4).

Hoskins (1978) introduced a vector Q, which is called as quasi-geostrophic Q vector, and it can be expressed as

$$Q=-\frac{R}{\sigma P}\left(\frac{P}{P_0}\right)^\kappa\begin{bmatrix}\frac{\partial\mathbf{V}_g}{\partial x}\cdot\mathbf{V}_p\theta\\ \frac{\partial\mathbf{V}_g}{\partial y}\cdot\mathbf{V}_p\theta\end{bmatrix} \qquad (4.2.5)$$

where $\sigma=\frac{RT\partial\ln\bar{\theta}(P)}{P\,\partial P}$, $\kappa=\frac{AR}{C_p}$; $\mathbf{V}_p\theta$ is the ascendant of the potential temperature on the constant pressure surface; $\mathbf{V}_g$ is the geostrophic wind vector. The meaning of Q vector will be further discussed in the Chapter 8.

By using Q we can denote the two dimensional frontogenesis function conveniently. Suppose the non-adiabatic heating and the diffusion effects can be neglected, then it can be seen from (4.2.4) that for the surface level, where $w=0$, the two dimensional frontogenesis function

$$\begin{aligned}F_p&=\frac{D}{Dt}|\mathbf{V}_p\theta|=\frac{1}{|\mathbf{V}_p\theta|}\left[\frac{\partial\theta}{\partial x}\frac{d}{dt}\left(\frac{\partial\theta}{\partial x}\right)+\frac{\partial\theta}{\partial y}\frac{d}{dt}\left(\frac{\partial\theta}{\partial y}\right)\right]\\ &=\frac{1}{|\mathbf{V}_p\theta|}\left[\frac{\partial\theta}{\partial x}\left(-\frac{\partial u}{\partial x}\frac{\partial\theta}{\partial x}-\frac{\partial v}{\partial x}\frac{\partial\theta}{\partial y}\right)+\frac{\partial\theta}{\partial y}\left(-\frac{\partial u}{\partial y}\frac{\partial\theta}{\partial x}-\frac{\partial v}{\partial y}\frac{\partial\theta}{\partial y}\right)\right]\\ &=\frac{1}{|\mathbf{V}_p\theta|}\left[\frac{\partial\theta}{\partial x}\left(\frac{\partial\mathbf{V}}{\partial x}\cdot\mathbf{V}_p\theta\right)+\frac{\partial\theta}{\partial y}\left(\frac{\partial\mathbf{V}}{\partial y}\cdot\mathbf{V}_p\theta\right)\right]\\ &\cong\frac{1}{|\mathbf{V}_p\theta|}\mathbf{V}_p\theta\begin{bmatrix}\frac{\partial\mathbf{V}_g}{\partial x}\cdot\mathbf{V}_p\theta\\ \frac{\partial\mathbf{V}_g}{\partial y}\cdot\mathbf{V}_p\theta\end{bmatrix}\end{aligned} \qquad (4.2.6)$$

Considering the equation (4.2.5), then the equation (4.2.6) can be written as follows:

$$F_p=\frac{\sigma P}{R}\left(\frac{P_0}{P}\right)^\kappa\frac{1}{|\mathbf{V}_p\theta|}\mathbf{V}_p\theta\cdot\mathbf{Q} \qquad (4.2.7)$$

Hence from equation (4.2.7) we can see that if Q and $\nabla_p \theta$ are in the same direction, i. e., the angle between two vectors is less than 90°, then $F_p > 0$, it means frontogenesis. The above frontogenesis defined by equation (4.2.1) and (4.2.4) is only related with the distribution of temperature and wind, this is the kinematic and thermodynamic frontogenesis.

§ 4.3 Dynamic frontogenesis

4.3.1 Concept of dynamic frontogenesis and the quasi-geostrophic frontogenesis model

It can be seen from the equations (4.2.2) or (4.2.4) and (4.2.6) that the horizontal deformation is able to cause the horizontal temperature gradient increasing, in other words to cause the isotherms getting denser. While the horizontal temperature gradient increasing will lead to the original thermo wind balance or geostrophic balance destroyed and meanwhile generating the geostrophic departure and as a result to give rise the secondary circulations, which will in turn influence the frontogenesis or frontolysis. This effect is called dynamic frontogenesis, which is caused by the acceleration. Therefore the complete frontogenesis function should include not only the frontogenesis on temperature field, but also that on momentum field.

When we discuss the dynamic frontogenesis problems, we have to consider the motion equation and the thermodynamic equation. Under the condition that the frictionless, the atmospheric motion equation can be written as

$$\frac{du}{dt} = -\frac{1}{\rho}\frac{\partial p}{\partial x} + fv = f(v - v_g) = f v_a \qquad (4.3.1)$$

$$\frac{dv}{dt} = -\frac{1}{\rho}\frac{\partial p}{\partial y} - fu = -f(u - u_g) = -f u_a$$

or

$$\frac{d\mathbf{V}}{dt} = -f(\mathbf{k} \times \mathbf{V}_a) \qquad (4.3.2)$$

where

$$\frac{d}{dt} = \frac{\partial}{\partial t} + u\frac{\partial}{\partial x} + v\frac{\partial}{\partial y} + \omega\frac{\partial}{\partial p} \qquad (4.3.3)$$

$$\mathbf{V} = \mathbf{V}_g + \mathbf{V}_a \qquad (4.3.4)$$

$\mathbf{V}_g$ is the geostrophic wind vector, $\mathbf{V}_a$ is the ageostrophic wind vector or the geostrophic departure, $\mathbf{k}$ is the unit vector in vertical direction.

The thermodynamic equation is

$$\frac{d\theta}{dt} = H \qquad (4.3.5)$$

where H is non-adiabatic heating. For processing the motion equation and the thermodynamic equation, one of the common used model is the quasi-geostrophic model.

Under the geostrophic approximation, neglected friction and non-adiabatic heating as well as the β effect, then

$$\frac{D_g}{Dt}\mathbf{V}_g = 0 \tag{4.3.6}$$

$$\left(\frac{D_g\theta}{Dt}\right) = 0 \tag{4.3.7}$$

where

$$\frac{D_g}{Dt} = \frac{\partial}{\partial t} + \mathbf{V}_g \cdot \nabla \tag{4.3.8}$$

and existing the thermo wind balance

$$\frac{\partial \mathbf{V}_g}{\partial(-p)} = -\frac{R}{f_0 p}\left(\frac{p}{p_0}\right)^\kappa (\mathbf{k} \times \nabla_p \theta) \tag{4.3.9}$$

From equation (4.3.9) we can also get

$$\frac{D_g}{Dt}\left[\frac{\partial \mathbf{V}_g}{\partial(-p)}\right] = -\frac{R}{f_0 p}\left(\frac{p}{p_0}\right)^\kappa \frac{D_g}{Dt}(\mathbf{k} \times \nabla_p \theta) \tag{4.3.10}$$

Under the conditions of quasi-geostrophic approximation, neglected friction, non-adiabatic heating and β effect, the motion equation and thermodynamic equation can be written as

$$\frac{D_g}{Dt}\mathbf{V}_g = -f_0(\mathbf{k} \times \mathbf{V}_a) \tag{4.3.11}$$

$$\left(\frac{D_g\theta}{Dt}\right) + \omega\frac{\partial\theta}{\partial p} = 0 \tag{4.3.12}$$

Comparing the equations (4.3.11) and (4.3.12) with the equations (4.3.6) and (4.3.7) respectively, their differences can be found that in the quasi-geostrophic model the effect of $\mathbf{V}_a$ is kept, and in the thermodynamic equation the vertical advection term $\omega(\partial\theta/\partial p)$ is kept. From equations (4.2.5) and (4.3.8) we can get

$$Q = \frac{R}{\sigma p}\left(\frac{p}{p_0}\right)^\kappa \frac{D_g}{Dt}(\nabla_p \theta) \tag{4.3.13}$$

Neglecting the horizontal variation of $\partial\theta/\partial p$, then from equation (4.3.12) we can get

$$\frac{\partial}{\partial y}\left(\frac{D_g\theta}{Dt}\right) = \frac{\partial\theta}{\partial p}\frac{\partial w}{\partial y} \tag{4.3.14}$$

Or

$$\frac{D_g}{Dt}\left(-\frac{\partial\theta}{\partial y}\right) = -\frac{\partial\theta}{\partial p}\frac{\partial w}{\partial y} \tag{4.3.15}$$

Hence, it can be seen from equations (4.2.7) or (4.3.13) and (4.3.15) that if Q and $(\nabla_p\theta)$ are in the same direction, then frontogenesis and a thermal direct circulation formed (Fig. 4.3), vise versa (Hoskins and Pedder, 1980).

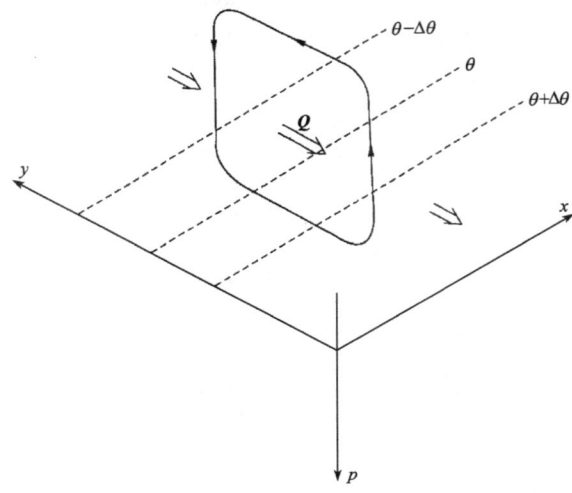

Fig. 4.3 The schematic diagram of the thermal direct circulation formed when the vectors Q and $(\nabla_p \theta)$ are in the same direction. (Hoskins and Pedder, 1980)

4.3.2 The quasi-geostrophic two dimensional frontogenesis equations expressed by potential temperature and absolute momentum M

The intensity of front and frontogenesis are normally expressed by the potential temperature gradient $|\nabla \theta|$ and absolute momentum gradient $|\nabla M|$ (where $M = u_g - fy$), and their rate of changing with time $\frac{d}{dt}\nabla \theta$ and $\frac{d}{dt}\nabla M$. Taking the (x, y, p) coordinate system and to set the axis x along the direction parallel to the front and the axis y perpendicular to the axis x and pointed to the cold side. The motion equation in the x direction is taken as quasi-geostrophic approximation, while in the y direction the geostrophic approximation is taken. Suppose the fluid is the Boussinesq fluid, f is constant, under these conditions, the dynamic equations can be written as follows:

$$\frac{\partial u_g}{\partial t} + \frac{u_g \partial u_g}{\partial x} + \frac{v_g \partial u_g}{\partial y} = f v_a \text{ (the momentum equation)} \quad (4.3.16)$$

$$\frac{\partial \theta}{\partial t} + \frac{u_g \partial \theta}{\partial x} + \frac{v_g \partial \theta}{\partial y} + \frac{\omega \partial \theta}{\partial p} = H \text{ (the thermodynamic equation)} \quad (4.3.17)$$

$$u_g = -\frac{1}{f}\frac{\partial \phi}{\partial y} \quad (u \text{ component of geostrophic wind}) \quad (4.3.18)$$

$$v_g = \frac{1}{f}\frac{\partial \phi}{\partial x} \quad \text{(v component of geostrophic wind)} \tag{4.3.19}$$

$$\frac{\partial u_g}{\partial x} + \frac{\partial v_g}{\partial y} = 0 \quad \text{(non divergence of geostrophic wind)} \tag{4.3.20}$$

$$\frac{\partial v_a}{\partial y} + \frac{\partial \omega}{\partial p} = 0 \quad \text{(the continuity equation)} \tag{4.3.21}$$

$$\frac{\partial u_g}{\partial p} = \frac{\partial M}{\partial p} = r\frac{\partial \theta}{\partial y} \quad \text{(the relationship of thermo wind)} \tag{4.3.22}$$

$$\frac{\partial v_g}{\partial p} = -r\frac{\partial \theta}{\partial x} \quad \text{(the relationship of thermo wind)} \tag{4.3.23}$$

$$\frac{\partial \phi}{\partial p} = -fr\theta \quad \text{(the static equation)} \tag{4.3.24}$$

where
$$r = \frac{R}{fp_0}\left(\frac{p_0}{p}\right)^{\frac{C_v}{C_p}} \tag{4.3.25}$$

H is the non adiabatic heating rate. From the equation (4.3.21) to introducing the stream function ψ

$$v_a = -\frac{\partial \psi}{\partial p} \tag{4.3.26}$$

$$\omega = \frac{\partial \psi}{\partial y} \tag{4.3.27}$$

for the equation (4.3.16) to take $\frac{\partial}{\partial y}$ and $\frac{\partial}{\partial p}$ respectively; for the equation (4.3.17) to take $r\frac{\partial}{\partial y}$ and $\frac{\partial}{\partial p}$ respectively. Using the equations (4.3.18)—(4.3.23) without regard to the vertical change of γ (i.e. $\partial r/\partial p \cong 0$), and taking Boussinesq approximation, then we can get the quasi-geostrophic frontogenesis forecasting equation corresponding to the three dimensional air parcel trajectory as follows:

$$\frac{D_g}{Dt}\left(\frac{\partial M}{\partial y}\right) = f\frac{\partial v_a}{\partial p} \tag{4.3.28}$$

$$\frac{D_g}{Dt}\left(\frac{\partial M}{\partial p}\right) J_{y,p}(u_g, v_g) + f\frac{\partial v_a}{\partial p} \tag{4.3.29}$$

$$\frac{D_g}{Dt}\left(r\frac{\partial \theta}{\partial y}\right) = J_{y,p}(u_g, v_g) - r\frac{\partial \theta}{\partial p}\frac{\partial \omega}{\partial y} + r\frac{\partial H}{\partial y} \tag{4.3.30}$$

$$\frac{D_g}{Dt}\left(\frac{\partial \theta}{\partial p}\right) = -\frac{\partial \theta}{\partial p}\frac{\partial \omega}{\partial p} - \frac{\partial H}{\partial p} \tag{4.3.31}$$

where
$$\frac{D_g}{Dt} = \frac{\partial}{\partial t} + u_g\frac{\partial}{\partial x} + v_g\frac{\partial}{\partial y} \tag{4.3.32}$$

$$J_{y,p}(\alpha,\beta)=\frac{\partial\alpha}{\partial y}\frac{\partial\beta}{\partial p}-\frac{\partial\alpha}{\partial p}\frac{\partial\beta}{\partial y} \qquad (4.3.33)$$

$J_{y,p}$ is the lateral Jacobi operator of the (y,p) coordinate system.

4.3.3 Semi-geostrophic frontogenesis model

Hoskins (1975) pointed out that the non-viscidity motion equation may be expressed as,

$$V_a=\frac{1}{f}k\times\frac{DV_g}{Dt}-\frac{1}{f^2}\frac{D^2V}{Dt^2} \qquad (4.3.34)$$

For the situation that the time scale is longer than $1/f$, $\left|\frac{1}{f}\frac{\partial^2 V}{\partial t^2}\right|\ll\left|\frac{DV_g}{Dt}\right|$, hence the term $\frac{1}{f^2}\frac{D^2V}{Dt^2}$ may be neglected (Hoskins, 1982), then the motion equation can be written as

$$\frac{DV_g}{Dt}=-f(k\times V_a) \qquad (4.3.35a)$$

or

$$V_a=\frac{1}{f}k\times\frac{DV_g}{Dt} \qquad (4.3.35b)$$

where

$$\frac{D}{Dt}=\frac{\partial}{\partial t}+(V_g+V_a)\cdot V_p+\omega\frac{\partial}{\partial p} \qquad (4.3.36)$$

The equation (4.3.35a) is similar to the quasi-geostrophic motion equation (4.3.11), the only difference is the non-geostrophic advection and the vertical advection of the geostrophic momentum terms are kept. Meanwhile, the thermodynamic equation can be written as

$$\frac{D_p\theta}{Dt}+\omega\frac{\partial\theta}{\partial p}=0 \qquad (4.3.37)$$

where

$$\frac{D_p}{Dt}=\frac{\partial}{\partial t}+(V_g+V_a)\cdot V_p \qquad (4.3.38)$$

The only difference of the equation (4.3.37) different from the quasi-geostrophic thermodynamic equation (4.3.13) is that in the equation (4.3.37) the temperature advection caused by non-geostrophic wind and the vertical component of wind is kept. Comparing them with the equations (4.3.11) and (4.3.12), the distinction between semi-geostrophic and quasi-geostrophic models can be clearly seen.

Then, let us discuss the semi-geostrophic frontogenesis equations. To set the front is along x direction and to take the geostrophic momentum approximation, i. e. $|u_a|\ll|u_g|$, $\left|\frac{du_a}{dt}\right|\ll\left|\frac{du_g}{dt}\right|$, $\left|\frac{\partial u_a}{\partial x}\right|\ll\left|\frac{\partial v_a}{\partial y}\right|=\left|\frac{\partial\omega}{\partial p}\right|$; neglecting the non-geostrophic advection along the front, and taking Boussinesq approximation, setting $f=$constant, and suppose the front is a straight line ($\partial u_g/\partial x=0$). Based on the above approximations the dynamic equations can be written as follows:

$$\frac{\partial u_g}{\partial t}+u_g\frac{\partial u_g}{\partial x}+(v_g+V_a)\frac{\partial u_g}{\partial y}+\omega\frac{\partial u_g}{\partial p}=-\frac{\partial \phi}{\partial x}+fv \quad (4.3.39)$$

$$\frac{\partial v_g}{\partial t}+(v_g+V_a)\frac{\partial v_g}{\partial y}+\omega\frac{\partial v_g}{\partial p}=-fu_a\cong 0 \quad (4.3.40)$$

$$\frac{\partial \theta}{\partial t}+u_g\frac{\partial \theta}{\partial x}+(v_g+V_a)\frac{\partial \theta}{\partial y}+\omega\frac{\partial \theta}{\partial p}=H \quad (4.3.41)$$

Additional equations are (4.3.18)—(4.3.27), (4.3.33). Taking the partial differential of the equations (4.3.39) and (4.3.41) with respective to y and p respectively, and then we can get two dimensional frontogenesis forecasting equation corresponding to the three dimensional air parcel moving trajectory under the geostrophic momentum approximation can be written as follows:

$$\frac{D}{Dt}\left(\frac{\partial M}{\partial y}\right)=J_{y,p}(M,\omega) \quad (4.3.42)$$

$$\frac{D}{Dt}\left(\frac{\partial M}{\partial p}\right)=-J_{y,p}(u_g,v_g)-J_{y,p}(M,v_a) \quad (4.3.43)$$

$$\frac{D}{Dt}\left(r\frac{\partial \theta}{\partial y}\right)=J_{y,p}(u_g,v_g)+rJ_{y,p}(\theta,\omega)+r\frac{\partial H}{\partial y} \quad (4.3.44)$$

$$\frac{D}{Dt}\left(\frac{\partial \theta}{\partial p}\right)=-J_{y,p}(\theta,v_a)+\frac{\partial H}{\partial p} \quad (4.3.45)$$

Spreading the Jacobi operator included non-geostrophic wind component in the equations (4.3.43) and (4.3.44), and by using the equations (4.3.21)—(4.3.23), then we can get another form of the equations (4.3.43) and (4.3.44) as follows:

$$\frac{D}{Dt}\left(\frac{\partial M}{\partial p}\right)=-\left[J_{y,p}(u_g,v_g)-\frac{\partial M\partial v_g}{\partial p\partial y}\right]-\frac{\partial M\partial v_a}{\partial y\partial p} \quad (4.3.46)$$

$$\frac{D}{Dt}\left(r\frac{\partial \theta}{\partial y}\right)=\left[J_{y,p}(u_g,v_g)-\frac{\partial M\partial v_a}{\partial p\partial p}\right]-r\frac{\partial \theta\partial \omega}{\partial p\partial y}+r\frac{\partial H}{\partial y} \quad (4.3.47)$$

Where

$$\frac{D}{Dt}=\frac{\partial}{\partial t}+u_g\frac{\partial}{\partial x}+(v_g+V_a)\frac{\partial}{\partial y}+\omega\frac{\partial}{\partial p} \quad (4.3.48)$$

4.3.4 The two dimensional frontogenesis equation in original equation form

Decomposing the wind field as follows:

$$\begin{cases}u=u_g+u_a\\v=v_g+v_a\\\omega=\omega\end{cases} \quad (4.3.49)$$

Suppose the frontal non-geostrophic vertical circulation is restricted in the cross section traversed the front, and the non-geostrophic wind component along the front, u_a is non-divergent, i. e.

$$\frac{\partial u_a}{\partial x} = 0 \qquad (4.3.50)$$

The original equations for describing the dynamic and thermodynamic frontogenesis processes can be written as

$$\frac{\partial u}{\partial t} + u\frac{\partial u}{\partial x} + v\frac{\partial u}{\partial y} + \omega\frac{\partial u}{\partial p} = -\frac{\partial \phi}{\partial x} + fv \qquad (4.3.51)$$

$$\frac{\partial v}{\partial t} + u\frac{\partial v}{\partial x} + v\frac{\partial v}{\partial y} + \omega\frac{\partial v}{\partial p} = -\frac{\partial \phi}{\partial y} - fu \qquad (4.3.52)$$

$$\frac{\partial \theta}{\partial t} + u\frac{\partial \theta}{\partial x} + v\frac{\partial \theta}{\partial y} + \omega\frac{\partial \theta}{\partial p} = H \qquad (4.3.53)$$

And the equations (4.3.18)—(4.3.27), (4.3.49), (4.3.50) as well as (4.3.33).

Use equation (4.3.49) to develop (4.3.51) and (4.3.53), then taking the partial derivative of equation (4.3.51) with respect to y and p, i. e. $\frac{\partial}{\partial y}$ and $\frac{\partial}{\partial p}$, respectively; taking the partial derivative of equation (4.3.53) with respect to y and p, i. e. $\frac{\partial}{\partial y}$ and $\frac{\partial}{\partial p}$, respectively also; by using the equations (4.3.18)—(4.3.27), $M = u_g - fy$ and equation (4.3.33). Suppose the equation (4.3.50) is set up, and the terms $(\partial v_a/\partial p) \cdot (\partial u_a/\partial p)$ and $(\partial \omega/\partial p) \cdot (\partial u_a/\partial p)$, as well as the terms $(\partial v_a/\partial y) \cdot (\partial u_a/\partial y)$ and $(\partial \omega/\partial y) \cdot (\partial u_a/\partial p)$ are neglected, finally we can get a two dimensional frontogenesis forecasting equation set under the original equation model as follows:

$$\frac{d}{dt}\left(\frac{\partial M}{\partial y}\right) = J_{y,p}(M,\omega) - \frac{d}{dt}\frac{\partial u_a}{\partial y} \qquad (4.3.54)$$

$$\frac{d}{dt}\left(\frac{\partial M}{\partial p}\right) = -J_{y,p}(u_g, v_g) - J_{y,p}(M, v_a) - \frac{\partial u_a}{\partial p}\frac{\partial u_g}{\partial x} - \frac{\partial v_g}{\partial p}\frac{\partial u_a}{\partial y} - \frac{d}{dt}\frac{\partial u_a}{\partial p} \qquad (4.3.55)$$

$$\frac{d}{dt}\left(r\frac{\partial \theta}{\partial y}\right) = J_{y,p}(u_g, v_g) + rJ_{y,p}(\theta,\omega) - r\frac{\partial \theta}{\partial x}\frac{\partial u_a}{\partial y} + r\frac{\partial H}{\partial y} \qquad (4.3.56)$$

$$\frac{d}{dt}\left(\frac{\partial \theta}{\partial p}\right) = -J_{y,p}(\theta, v_a) - \frac{\partial \theta}{\partial x}\frac{\partial u_a}{\partial p} + \frac{\partial H}{\partial y} \qquad (4.3.57)$$

where

$$\frac{d}{dt} = \frac{\partial}{\partial t} + (u_g + u_a)\frac{\partial}{\partial x} + (v_g + V_a)\frac{\partial}{\partial y} + \omega\frac{\partial}{\partial p} \qquad (4.3.58)$$

It can be seen from the equations (4.3.54)—(4.3.57) that the frontogenesis is mainly determined by the factors including horizontal deformation of the geostrophic flow field, the vertical deformation of the lateral non-geostrophic floe field, the non-geostrophic deformation along the front and the individual change of the vertical and horizontal gradient of momentum as well as the non-adiabatic heating etc.

§ 4.4 The factors influencing frontogenesis

Based on the equations (4.3.54)—(4.3.58), under the adiabatic situation, the factors for influencing the horizontal and vertical momentum and potential temperature frontogenesis will be discussed respectively as following.

4.4.1 Factors influencing horizontal momentum frontogenesis

To expand the term $J(M, \omega)$ in equation (4.3.54) and get

$$\frac{d}{dt}\left(\frac{\partial M}{\partial y}\right) = -\frac{\partial M \partial v_a}{\partial y \partial y} - \frac{\partial M \partial \omega}{\partial p \partial y} - \frac{d}{dt}\frac{\partial u_a}{\partial y} \qquad (4.4.1)$$

The three terms in the right side of equation (4.4.1) show the factors influencing the variation of the horizontal momentum gradient. The first term indicates the effect of horizontal convergence of the non-geostrophic motions. For instance, when $\frac{\partial M}{\partial y}<0$, $\frac{\partial v_a}{\partial y}<0$, then $\frac{d}{dt}\left(\frac{\partial M}{\partial y}\right)_1<0$, i.e., the horizontal momentum frontogenesis. The second term indicates the tilt effect of the non-geostrophic motion. For instance, when $\frac{\partial M}{\partial p}<0$, $\frac{\partial \omega}{\partial y}<0$, then $\frac{d}{dt}\left(\frac{\partial M}{\partial y}\right)_2<0$, frontogenesis. The third term indicates the effect of the individual change of the non-geostrophic horizontal momentum gradient, $\frac{d}{dt}\left(\frac{\partial M}{\partial y}\right)_3 = -\frac{d}{dt}\frac{\partial u_a}{\partial y}$, when there is no horizontal convergence and vertical tilt effect, $\frac{d}{dt}\left(\frac{\partial M}{\partial y}+\frac{\partial u_a}{\partial y}\right)=0$.

From the equation (4.3.55) we can get

$$\frac{d}{dt}\left(\frac{\partial M}{\partial p}\right) = -\frac{\partial u_g \partial v_g}{\partial y \partial p} - \frac{\partial u_g \partial u_g}{\partial p \partial x} - \frac{\partial M \partial v_a}{\partial y \partial p} - \frac{\partial M \partial \omega}{\partial p \partial p} - \frac{\partial M \partial u_a}{\partial x \partial p} - \frac{\partial v_g \partial u_a}{\partial p \partial y} - \frac{d}{dt}\frac{\partial u_a}{\partial p}$$

(4.4.2)

The above equation shows that the factors influencing the vertical momentum frontogenesis include the follows.

(1) the effect of the vertical shear of the geostrophic wind in x direction, $\frac{d}{dt}\left(\frac{\partial M}{\partial p}\right)_1 = -\frac{\partial u_g \partial v_g}{\partial y \partial p}$. For instance, when $\frac{\partial u_g}{\partial y}<0$, $\frac{\partial v_g}{\partial p}<0$, then $\frac{d}{dt}\left(\frac{\partial M}{\partial p}\right)_1<0$, i.e. the vertical momentum frontogenesis occurs;

(2) the effect of the vertical shear of the component in x direction of the geostrophic flow, i. e., $\frac{d}{dt}\left(\frac{\partial M}{\partial p}\right)_2 = -\frac{\partial u_g}{\partial p}\frac{\partial u_g}{\partial x}$. For instance, when $-\frac{\partial u_g}{\partial p}<0$, $\frac{\partial u_g}{\partial x}<0$, then frontogenesis occurs;

(3) the effect of tilting horizontal momentum gradient due to the vertical shear of the non-geostrophic horizontal motion, i. e., $\frac{d}{dt}\left(\frac{\partial M}{\partial p}\right)_3 = -\frac{\partial M}{\partial y}\frac{\partial v_a}{\partial p} - \frac{\partial M}{\partial x}\frac{\partial u_a}{\partial p}$. For instance, when $\frac{\partial M}{\partial y}<0$, $\frac{\partial v_a}{\partial p}<0$ and $\frac{\partial M}{\partial x}<0$, $\frac{\partial u_a}{\partial p}<0$, then frontogenesis occurs;

(4) the convergence effect of the non-geostrophic flows, i. e., $\frac{d}{dt}\left(\frac{\partial M}{\partial p}\right)_4 = -\frac{\partial M}{\partial p}\frac{\partial \omega}{\partial p}$, for instance, when $\frac{\partial M}{\partial p}<0$, $\frac{\partial \omega}{\partial p}<0$, then frontogenesis occurs;

(5) the effect of the horizontal shear of the non-geostrophic flow along the front, i. e., $\frac{d}{dt}\left(\frac{\partial M}{\partial p}\right)_5 = -\frac{\partial v_g}{\partial p}\frac{\partial u_a}{\partial y}$. For instance, when $\frac{\partial v_g}{\partial p}<0$, $\frac{\partial u_a}{\partial y}<0$, then frontogenesis occurs;

(6) the effect of the individual change of the vertical momentum gradient along the front, i. e., $\frac{d}{dt}\left(\frac{\partial M}{\partial p}\right)_6 = -\frac{d}{dt}\frac{\partial u_a}{\partial p}$.

4.4.2 The factors influencing the horizontal temperature frontogenesis

To expand the equation (4.3.56) taking no account of heating term, we have

$$\frac{d}{dt}\left(r\frac{\partial \theta}{\partial y}\right) = -\frac{\partial u_g}{\partial y}\left(r\frac{\partial \theta}{\partial x}\right) - r\frac{\partial \theta}{\partial y}\frac{\partial v_g}{\partial y} - r\frac{\partial \theta}{\partial y}\frac{\partial v_a}{\partial y} - r\frac{\partial \theta}{\partial p}\frac{\partial \omega}{\partial y} - r\frac{\partial \theta}{\partial x}\frac{\partial u_a}{\partial y} \quad (4.4.3)$$

It can be seen from the above equation, there are five factors influencing the horizontal temperature frontogenesis.

(1) The effect of horizontal shear of the geostrophic flows, $\frac{d}{dt}\left(r\frac{\partial \theta}{\partial y}\right)_1 = -\frac{\partial u_g}{\partial y}\left(r\frac{\partial \theta}{\partial x}\right)$, for instance, when $\frac{\partial \theta}{\partial x}>0$, $\frac{\partial u_g}{\partial y}>0$, $\frac{d}{dt}\left(r\frac{\partial \theta}{\partial y}\right)_1<0$, i. e., horizontal temperature frontogenesis;

(2) The effect of the horizontal convergence of geostrophic flows, $\frac{d}{dt}\left(r\frac{\partial \theta}{\partial y}\right)_2 = -r\frac{\partial \theta}{\partial y}\frac{\partial v_g}{\partial y}$, for instance, when $\frac{\partial \theta}{\partial y}<0$, $\frac{\partial v_g}{\partial y}<0$, frontogenesis;

(3) The effect of the convergence of non-geostrophic flows, $\frac{d}{dt}\left(r\frac{\partial \theta}{\partial y}\right)_3 = -r\frac{\partial \theta}{\partial y}$

$\frac{\partial v_a}{\partial y}$, for instance, when $\frac{\partial \theta}{\partial y}<0$, $\frac{\partial v_a}{\partial y}<0$, then frontogenesis;

(4) The tilt effect of the non-geostrophic flows, $\frac{d}{dt}\left(r\frac{\partial \theta}{\partial y}\right)_4 = -r\frac{\partial \theta}{\partial p}\frac{\partial \omega}{\partial y}$, for instance, when $\frac{\partial \theta}{\partial p}<0$, $\frac{\partial \omega}{\partial y}<0$, then frontogenesis;

(5) The effect of the horizontal shear of the non-geostrophic flow along the front, $\frac{d}{dt}\left(r\frac{\partial \theta}{\partial y}\right)_5 = -r\frac{\partial \theta}{\partial x}\frac{\partial u_a}{\partial y}$, for instance, when $\frac{\partial \theta}{\partial x}>0$, $\frac{\partial u_a}{\partial y}>0$, then frontogenesis.

4.4.3 The factors influencing the vertical temperature frontogenesis

To expand the equation (4.3.57) taking no account of the heating term, then we have

$$\frac{d}{dt}\left(\frac{\partial \theta}{\partial p}\right) = -\frac{\partial \theta}{\partial p}\frac{\partial \omega}{\partial p} - \frac{\partial \theta}{\partial y}\frac{\partial v_a}{\partial p} - \frac{\partial \theta}{\partial x}\frac{\partial u_a}{\partial p} \qquad (4.4.4)$$

According to the above equation, it can be seen that there are following factors influencing the vertical temperature frontogenesis:

(1) The effect of non-geostrophic convergence, $\frac{d}{dt}\left(\frac{\partial \theta}{\partial p}\right)_1 = -\frac{\partial \theta}{\partial p}\frac{\partial \omega}{\partial p}$. For instance, when $\frac{\partial \theta}{\partial p}<0$, $\frac{\partial \omega}{\partial p}<0$, then $\frac{d}{dt}\left(\frac{\partial \theta}{\partial p}\right)_1<0$, i.e., the vertical temperature frontogenesis;

(2) The tilting effect of the non-geostrophic, $\frac{d}{dt}\left(\frac{\partial \theta}{\partial p}\right)_2 = -\frac{\partial \theta}{\partial y}\frac{\partial v_a}{\partial p} - \frac{\partial \theta}{\partial x}\frac{\partial u_a}{\partial p}$, for instance, when $\frac{\partial \theta}{\partial y}<0$, $\frac{\partial \theta}{\partial x}>0$, $\frac{\partial v_a}{\partial p}>0$, $\frac{\partial u_a}{\partial p}<0$, then $\frac{d}{dt}\left(\frac{\partial \theta}{\partial p}\right)_2<0$, i.e., vertical temperature frontogenesis.

Above discussing about the effects of divergence, wind shear and tilting on the frontogenesis are based on kinematics view. In actual frontogenesis process, the flow field interacts with the temperature field each other. Therefore the frontogenesis is a dynamical interaction process.

§ 4.5 Frontal lateral secondary circulation

4.5.1 Frontal lateral secondary circulation equation

Sawyer (1956) and Eliassen (1962) derived the equations for describing the frontal lateral non-geostrophic vertical circulation. Shapiro (1981) developed the theory describing frontogenesis circulation near the upper level jet and front system based on the geostrophic momentum approximation. Eliminating the time derivative terms of $\left(\frac{\partial M}{\partial p}\right)$ and $\left(r\frac{\partial \theta}{\partial y}\right)$ in the frontogenesis forecasting equations by using the thermo wind equation (4.3.22), in addition to use the equations (4.3.26) and (4.3.27), then we can get the Sawyer-Eliassen frontal lateral secondary circulation equations under the quasi-geostrophic approximation, the geostrophic momentum approximation and the original equation model respectively.

Subtracting equation (4.3.29) from (4.3.30) and after neatening, we can get the diagnosis equation in quasi-geostrophic form for describing the secondary circulations as follows:

$$-r\frac{\partial \theta}{\partial p}\frac{\partial^2 \psi}{\partial y^2}+f\frac{\partial^2 \psi}{\partial p^2}=Q_g+Q_H \qquad (4.5.1)$$

Based on the equations (4.3.43) and (4.3.44) or (4.3.46) and (4.3.47), we can get the diagnosis equation under the geostrophic momentum approximation as follows:

$$-r\frac{\partial \theta}{\partial p}\frac{\partial^2 \psi}{\partial y^2}+2\left(\frac{\partial M}{\partial p}\right)\frac{\partial^2 \psi}{\partial y \partial p}-\frac{\partial M}{\partial y}\frac{\partial^2 \psi}{\partial p^2}=Q_g+Q_H \qquad (4.5.2)$$

By eliminating the time derivative terms in equation (4.3.55) and (4.3.56), we can get the diagnosis equation for the frontal lateral secondary circulation under the original equation model as follows:

$$-r\frac{\partial \theta}{\partial p}\frac{\partial^2 \psi}{\partial y^2}+2\left(\frac{\partial M}{\partial p}\right)\frac{\partial^2 \psi}{\partial y \partial p}-\frac{\partial M}{\partial y}\frac{\partial^2 \psi}{\partial p^2}=Q_g+Q_{ag}+Q_H \qquad (4.5.3)$$

where

$$Q_g=-2J_{y,p}(u_g,v_g)=-2J_{x,y}(u_g,\theta) \qquad (4.5.4)$$

$$Q_{ag}=-\frac{\partial}{\partial y}\frac{\partial u_a}{\partial p}+2r\frac{\partial \theta}{\partial x}\frac{\partial u_a}{\partial y}-\frac{d}{dt}\frac{\partial u_a}{\partial p} \qquad (4.5.5)$$

$$Q_H=-r\frac{\partial H}{\partial y} \qquad (4.5.6)$$

Q_g, Q_{ag} and Q_H are the geostrophic forcing term, the non-geostrophic forcing term and the non-adiabatic heating forcing term respectively.

Equation (4.5.1) is a stable second order elliptic equation. Equations (4.5.2) and (4.5.3) are the linear second order partial differential equations contained mixture partial differential term. The type of equation can be determined by using the coefficient criterion $\Delta=B^2-AC$. Where $B=\left(\frac{\partial M}{\partial p}\right)$ is a parameter representing baroclinicity; $A=-r\dfrac{\partial \theta}{\partial p}$ is the parameter equivalent to static stability; $C=-\dfrac{\partial M}{\partial y}$ is the parameter of inertial instability indicated by horizontal wind shear. Substituting the above parameters into the criterion, we can get

$$\Delta = \left(\dfrac{\partial M}{\partial p}\right)^2 - r\dfrac{\partial \theta}{\partial p}\dfrac{\partial M}{\partial y} = \left(r\dfrac{\partial \theta}{\partial y}\right)\left(\dfrac{\partial M}{\partial p}\right) - r\dfrac{\partial \theta}{\partial p}\dfrac{\partial M}{\partial y}$$

$$=rJ_{y,p}(\theta,M)=rP \qquad (4.5.7)$$

where P is potential vorticity,

$$P=J_{y,p}(\theta,M)=-J_{y,p}(M,\theta)=-\dfrac{\partial M}{\partial y}\bigg|_\theta \dfrac{\partial \theta}{\partial p} \qquad (4.5.8)$$

Hence, the equation type can be distinguished by using potential vorticity P

$$P\begin{cases}<0 & \text{elliptic equation}\\ =0 & \text{parabolic equation}\\ >0 & \text{hyperbolic equation}\end{cases}$$

Actually, the atmosphere is inertial stable ($\dfrac{\partial M}{\partial y}\big|_\theta<0$) and static stable ($\dfrac{\partial \theta}{\partial p}<0$), therefore, $P<0$, the equation (4.5.3) is an elliptic equation.

Since the equation we discussed is an elliptic equation, so that when the forcing term Q in the right side of the equation (4.5.3) is positive ($Q>0$), ψ has minimum, when Q is negative ($Q<0$), ψ has maximum. According to the definition of the stream function based on the equation (4.3.26) and (4.3.27) we know that the minimum and maximum of ψ are corresponding to the positive (or thermal direct) circulation and negative (or thermal indirect) circulation respectively as shown in Fig. 4.4.

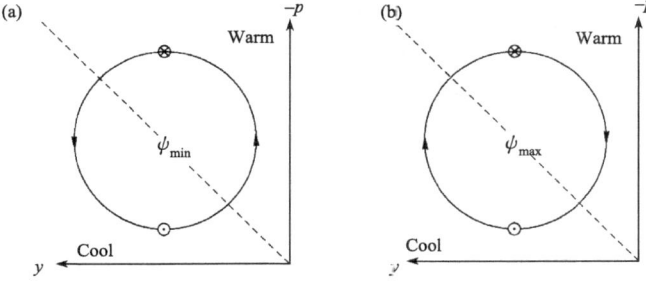

Fig. 4.4 Illustration of the relationship between the lateral non-geostrophic circulation and the sign of the forcing term. (a) $Q>0$, ψ_{min}, positive circulation; (b) $Q<0$, ψ_{max}, negative circulation. The subscripts max and min represent maximum and minimum respectively; the broken lines represent the isotropic line separating the cold and warm air; the signs ⊙ and ⊗ represent the components along the front of the flow flowing out and into the vertical section.

4.5.2 The forcing factors of the frontal secondary circulation

According to the equation (4.5.3), it can be seen that the forcing terms of the frontal secondary circulation include (1) geostrophic flow; (2) the non-geostrophic motion along the front; and (3) non adiabatic heating, which are discussed respectively as follows.

(1) The geostrophic forcing term $-2J_{y,p}(u_g, v_g)$

$$Q_g = -2J_{y,p}(u_g, v_g) = -2J_{x,y}(u_g, \theta)$$

$$= -2r\frac{\partial u_g}{\partial x}\frac{\partial \theta}{\partial y} + 2r\frac{\partial u_g}{\partial y}\frac{\partial \theta}{\partial x} = Q_{g_1} + Q_{g_2} \quad (4.5.9)$$

where $Q_{g_1} = -2r\frac{\partial u_g}{\partial x}\frac{\partial \theta}{\partial y} = 2r\frac{\partial v_g}{\partial y}\frac{\partial \theta}{\partial y}$ is the geostrophic stretching deformation forcing; $Q_{g_2} = 2r\frac{\partial u_g}{\partial y}\frac{\partial \theta}{\partial x}$ is the geostrophic shear deformation forcing; the Fig. 4.5a shows that the effect of the geostrophic stretching deformation forcing term when $\frac{\partial u_g}{\partial x} > 0$ and $\frac{\partial \theta}{\partial y} < 0$. The Fig. 4.5b shows the effect of the geostrophic shear deformation forcing term when $\frac{\partial \theta}{\partial x} > 0$ and $\frac{\partial u_g}{\partial y} > 0$. The Fig. 4.6 shows the temperature and pressure field situations to cause the frontolysis of warm front and frontogenesis of the cold front caused by the shear deformation.

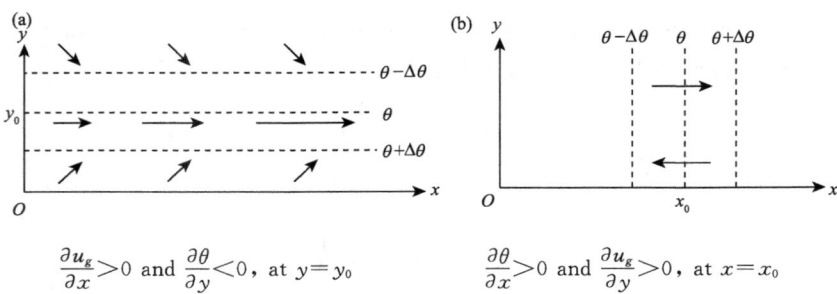

Fig. 4.5 (a) Frontogenesis caused by geostrophic stretching deformation; (b) frontogenesis caused by geostrophic shear deformation. (Adapted from Bluestein, 1984)

(2) Ageostrophic forcing term Q_{ag}

$$Q_{ag} = -\frac{\partial u_a}{\partial p}\frac{\partial u_g}{\partial x} + 2r\frac{\partial \theta}{\partial x}\frac{\partial u_a}{\partial y} - \frac{d}{dt}\left(\frac{\partial u_a}{\partial p}\right) = \frac{\partial u_a}{\partial p}\frac{\partial v_g}{\partial y} + 2r\frac{\partial \theta}{\partial x}\frac{\partial u_a}{\partial y} - \frac{d}{dt}\left(\frac{\partial u_a}{\partial p}\right)$$

$$(4.5.10)$$

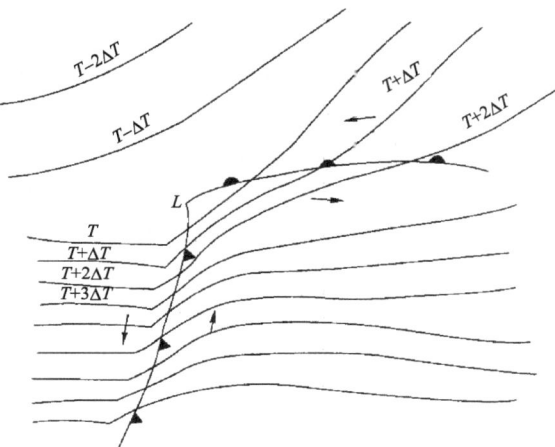

Fig. 4.6 A favorable situation for frontolysis of warm front and frontogenesis of cold front caused by shear. (Adapted from Gidel, 1978)

The ageostrophic forcing term includes three terms: the first term, $Q_{ag_1} = \dfrac{\partial u_a}{\partial p} \dfrac{\partial v_g}{\partial y}$, is the forcing related with the vertical distribution of the ageostrophic wind component along the front and the gradient of the ageostrophic wind component perpendicular to the front; the second term, $Q_{ag_2} = 2r \dfrac{\partial \theta}{\partial x} \dfrac{\partial u_a}{\partial y}$, is the shear deformation forcing term of the ageostrophic components along the front; the third term, $Q_{ag_3} = -\dfrac{d}{dt}\left(\dfrac{\partial u_a}{\partial p}\right)$, is the forcing induced by the individual change of the vertical gradient of the ageostrophic momentum along the front.

(3) the forcing term related with the diabatic heating effect, Q_H

$$Q_H = -r \dfrac{\partial H}{\partial y} \qquad (4.5.11)$$

This term is mainly the secondary circulation which is caused by the frontogenesis forcing induced by the non uniform distribution of diabatic heating in horizontal direction. When there is convective precipitation on the front, the latent heating caused by condensation should be considered. When there is condensation latent heating, the ascending branch of the frontogenesis secondary circulation occurs along the maximum heating axis. In the south side of the maximum heating axis $\dfrac{\partial H}{\partial y} > 0$, hence $Q_H < 0$, the indirect thermal circulation is induced; while in the north side of the axis of the maximum heating center $\frac{\partial H}{\partial y}<0$, a thermal direct circulation is induced. The distributions of the heating rate of the condensation latent heating and the secondary circulations caused by frontogenesis forcing are as shown in the Fig. 4. 7. In Fig. 4. 7a the solid lines are the isolines of the heating rate H, which can be calculated by using the scheme of Kuo or others; in the Fig. 4. 7b, the solid lines are the isolines of the stream function ψ. The arrows show the directions of the air flows.

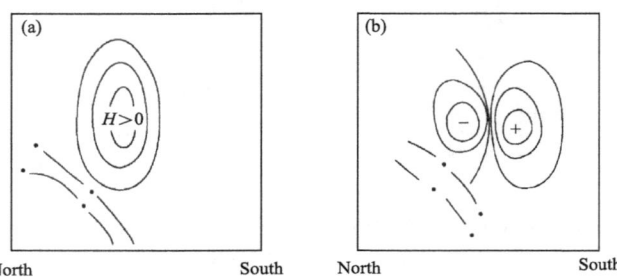

Fig. 4. 7 (a) The distribution of the rate of heating caused by convective condensation latent heating; and (b) the distribution of the secondary circulation forced by latent heating forcing.

§ 4. 6 Jet stream

4. 6. 1 Formation of the jet stream

Along the polar front jet stream and subtropical jet stream axis normally there are some strong wind velocity centers with the lengths about 1000 km and width about several hundred kilometers. These kind of meso-systems are called jet stream core. The wind field strength of the core can be expressed by $\mathbf{V}^2 U$. In the jet core $\mathbf{V}^2 U$ is negative. The generation function of the jet stream can be defined as the strength change rate of the jet stream as shown in the followsing:

$$J_s = \frac{D}{Dt}(-\mathbf{V}^2 U) \qquad (4.6.1)$$

The above formula can also be written as

$$J_s = -\mathbf{V}^2 \left[\frac{DU}{Dt} + \mathbf{V}^2 (\mathbf{V} \cdot \mathbf{V} U) - \mathbf{V} \cdot \mathbf{V}(\mathbf{V}^2 U) \right] \qquad (4.6.2)$$

Substituting the x component of the motion equation into the equation (4. 6. 2), and assuming $\mathbf{V}U$ is uniform near the air parcel (in the center of the jet stream $\mathbf{V}U=0$), then we have

$$J_s = -f\nabla^2 v_a - 2\beta\frac{\partial v_a}{\partial y} \tag{4.6.3}$$

where v_a is the ageostrophic component in y direction of the wind. For the typical situation in mid-latitude, the first term in the right side of the equation is the primary.

4.6.2 The secondary circulation near the upper level jet stream core

At the exit region of the upper level jet stream there is a latent secondary circulation with ascending in left side and descending in right side. At the entrance region there is a latent secondary circulation with ascending in right side and descending in the left side. The formation of the circulations can be expressed by using the kinetic energy equation under the condition of frictionless as shown in the followsing:

$$\frac{dK}{dt} = -\mathbf{V}\cdot\nabla\phi \tag{4.6.4}$$

where $K = \frac{1}{2}(u^2 + v^2)$ is the kinetic energy of the air with unit mass, ϕ is the geopotential. If $\frac{dK}{dt} = -\mathbf{V}\cdot\nabla\phi > 0$, it denotes $\mathbf{V}$ is deflected to the low pressure, in other words there is ageostrophic wind component pointed to the low pressure side. Due to the air parcel is accelerated along the stream line in the entrance region of the straight jet stream, hence there is southerly ageostrophic wind. Meanwhile due to the air parcel obtains the largest incrementation of the kinetic energy in the area near the jet stream axis, the southerly ageostrophic wind should be the maximum. Therefore in the entrance region of the upper level jet stream, there are divergence at the right side and convergence at the left side of the jet stream. So that the positive circulation is formed in the y-z vertical section. It is just opposite in the exit region of the jet stream. Where the air parcel is decelerated along the stream line direction, therefore the northerly ageostrophic wind component occurs, so that there are divergence at the left side and convergence at the right side of the jet stream (see Fig. 4.8) and form an negative circulation with the ascending in the north side and subsidence in the south side in the y-z vertical section. Since it is cold in the left side and warm in the right side of the westerly jet stream, so that it is a lateral direct thermal circulation in the entrance region causing the atmospheric available potential energy changes into kinetic energy, while it is a lateral indirect thermal circulation in the exit region causing the kinetic energy changes into available potential energy and decelerates the air velocity.

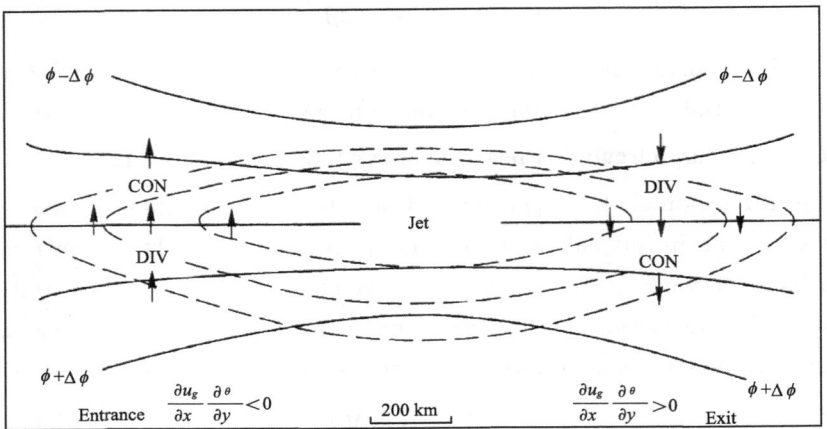

Fig. 4.8 Distributions of the divergence of ageostrophic winds in entrance and exit regions of the jet stream. (After Shapiro and Kennedy, 1981)

The upper level jet stream is normally closely related with the upper level frontal zone. In the entrance region of the jet stream, $\frac{\partial u_g}{\partial x}>0$, $\frac{\partial \theta}{\partial y}<0$, $\frac{\partial u_g}{\partial x}\frac{\partial \theta}{\partial y}<0$, so that the geostrophic stretching deformation forcing to produce direct thermal circulation with the subsidence in cold side and ascending in warm side, while oppositely in the exit region, $\frac{\partial u_g}{\partial x}\frac{\partial \theta}{\partial y}>0$, hence to produce an indirect thermal circulation with ascending in cold side and subsidence in warm side according to the equation (4.5.9). Therefore we can also explain the secondary circulation near the jet stream by using the theory of frontogenesis secondary circulation.

4.6.3 Coupling of the upper level and low level jet streams

The low level jet stream can be divided into three categories: the large scale low level jet stream, the jet stream related with disturbance and the boundary layer low level jet stream. The first category of the low level jet stream is the basic flow related with the planetary scale systems in low level of troposphere. For instance the prevailing southwest jet flow over East Asia continent is such kind of low level jet stream. It is related with monsoon and moves with season. The boundary layer jet stream occurs in the atmospheric boundary. It is characterized by strong vertical wind shear but weak horizontal wind shear and distinct diurnal variation. The low level jet stream related with disturbance is the common low level jet stream. The height of its core is located near 850—700 hPa constant pressure surface. It can maintain longer time with lesser diurnal variation and its formation is closely related with the development of the synoptic systems. The warm conveyor belt defined by Browning is kind of the low level jet stream.

Recent years, people found that the low level jet stream is very often coupled with upper level jet stream.

Uccellini and Johnson (1979) explained the causes of the coupling of upper level jet and low level jet. As said above, in the entrance region of the upper level jet stream at the high pressure side of the upper level jet stream the air flow is divergent, which leads low level pressure decreasing, i. e. $\frac{\partial p}{\partial t} < 0$; while at the low pressure side of the upper level jet stream the air flow is convergent, which leads low level pressure increasing, i. e. $\frac{\partial p}{\partial t} > 0$, the induced variation of the low level geostrophic wind, $\frac{\partial V_g}{\partial t}$, has opposite direction with the geostrophic wind V_g. Based on the quasi geostrophic approximation the ageostrophic wind equation can be written as follows:

$$V_a = \frac{1}{f} k \times \left(\frac{\partial V_g}{\partial t} + V_g \cdot \nabla V_g \right) \tag{4.6.5}$$

At low level in the entrance region of the upper level jet stream, due to the $\frac{\partial V_g}{\partial t}$ the transverse wind component V_a pointed to the anticyclone (high pressure) side of the upper level jet stream will be caused. For the same reason, in the exit region the direction of the variation of the low level geostrophic wind, $\frac{\partial V_g}{\partial t}$, is same with the geostrophic wind, so that the $\frac{\partial V_g}{\partial t}$ leads to the transverse wind component V_a pointed to the low pressue side of the upper level jet stream. Since the low level V_a is mainly caused by the allobaric gradient, therefore it is actually the allobaric wind. In the exit region of the core of the upper level jet stream, due to the action of allobaric wind the actual wind will deflect to the low pressure side. This is the situation $\frac{dK}{dt} = -V \cdot \nabla \phi > 0$. The increasing of the low level kinetic energy will cause the formation of low level jet.

The coupling of high and low level jet can be formed in both the entrance and exit regions of the upper level jet stream, but their types are different. In the exit region, as shown in Fig. 4.9a, the low level jet axis intersects with the upper level jet axis, while in the entrance region the low and upper level jet axises are parallel each other. The relationship between the secondary circulations and the low and high level jet in the entrance and exit regions are as shown in Fig. 4. 9b and Fig. 4. 9c respectively. The low level jet in exit region is a constituent part of the indirect thermal circulation near the upper level jet core, while the low level jet in entrance region and the upper level jet are respectively in two dependent secondary circulations, which have the coincided ascending branches. From Fig. 4. 9 it can be seen that the secondary circulations related with the low level jet are all located in the left side of the low level jet stream, where is the location favorable to form heavy rain and severe convective storms.

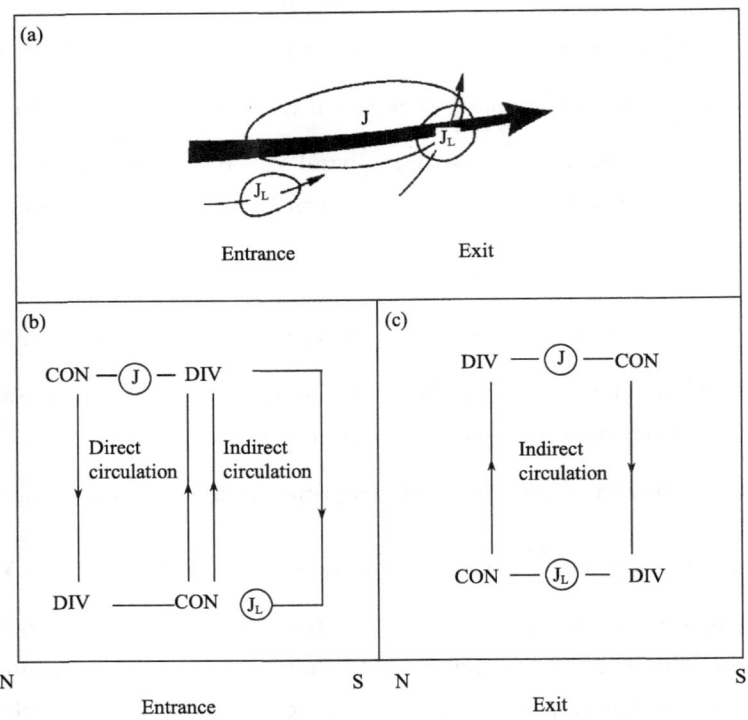

Fig. 4. 9 The coupling types of upper level and low level jet streams and the secondary circulation. (J denotes the core of upper level jet stream, J_L denotes the core of the low level jet stream)

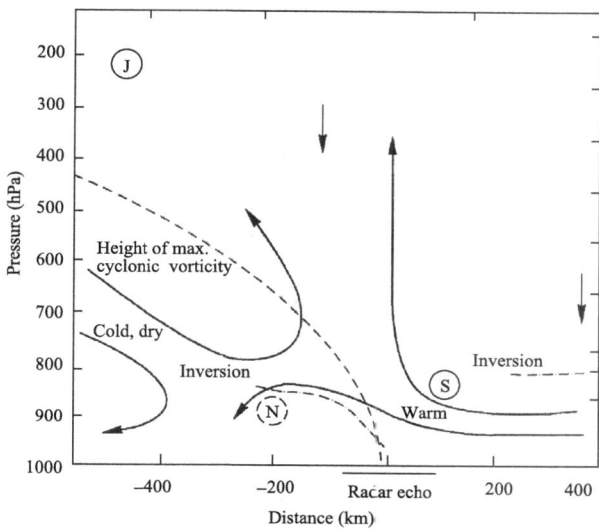

Fig. 4.10 A schematic diagram of the frontogenesis circulation. J: jet stream; N: northeast wind behind front; S: southwest wind ahead of front; dot-dashed line: the location of maximum surface temperature gradient behind front. (After Ogura and Portis, 1982)

4.6.4 The effects of the frontogenesis circulations and the coupling of low and high level jet streams on the development of severe convective storms

Ogura and Portis (1982) summarized a two dimensional frontal feature model as shown in Fig. 4.10, in which the features such as the maximum temperature axis, the frontal zone and low level inversion layer, the low level southwest jet flow and the maximum northerly wind behind the cold front are described. From the picture it can be seen that there is a indirect circulation ahead of the front and a direct circulation behind of the front and there is a strong ascending motion above the ground surface, These are the major factors to cause the development of frontal thunderstorms.

Under the situation with the coupling of low and high level jet streams especially the coupling in the exit region of the upper level jet stream it is normally favorable for the development of the severe convective storms. Under such situations, the low level jet stream can cause the warm and moist air transportation, while the upper level jet will cause the dry and cold air advection and to increase the atmospheric potential instability. Moreover the ascending branch of the secondary circulation caused by the coupling of the low and high level jet streams will trigger the releasing of the potential instability energy.

§ 4.7 Mesoscale fronts in boundary layer

In the actual atmosphere beside the large scale front there are boundary layer mesoscale fronts with smaller size and shorter life cycle. They are related with ageostrophic deformation background field and also related with the nonuniform dynamic and thermodynamic features of local underneath surface. They have important influences on local weather variations and thus they are the important research objects for short and very short range forecasts.

4.7.1 Dry line

Dry lines are also called dry line fronts or dew point fronts. Their vertical scope are normally only about 1—3 km high above the surface. They can be observed in many places in the world. In the great plains area of the United States in spring and early summer seasons the dry lines occur very often. The number of dry line days in April to June is about 41% of the total number of days; In India, dry lines are also the important feature in the months before the monsoon break out; In middle to west Africa the tropical convergence zones often play the role similar to dry lines; In China the dry line activity is also quite frequently. The dry lines may give rise to severe convective storms. In forecasting of convective weather the dry lines and the dry warm lid are very often regarded as the earlier stage conditions for diagnosing the convective activities. Thus it is valuable to pay high attention on the dry line analyzing.

(1) General characters of dry line

Dry line is a discontinuity line of humidity in horizontal direction. It is dry and warm air in one side of the dry line, while in another side it is the cold and moist air. Across the dry line the surface horizontal dew point temperature gradient can be 5 ℃·km^{-1} or more. In the afternoon, the dew point temperature difference in 2 km distance between two sides of the dry line can be as great as 15 ℃. Above the cold and moist air it is ususlly covered by a temperature inversion layer, which is normally called as the capping inversion or the dry and warm lid. It plays important role for storing potential instability energy. Fig. 4.11 shows the features of the dry line occurred at 1200 GMT Apr. 3, 1981 in the east slope of the Rocky Mountain in the United States. In the Fig. 4.11a the surface dew point temperature decreases and stream lines are depicted, Fig. 4.11b is the vertical section along the line BB' in Fig. 4.11a. From the figures the features of dry line can be seen clearly.

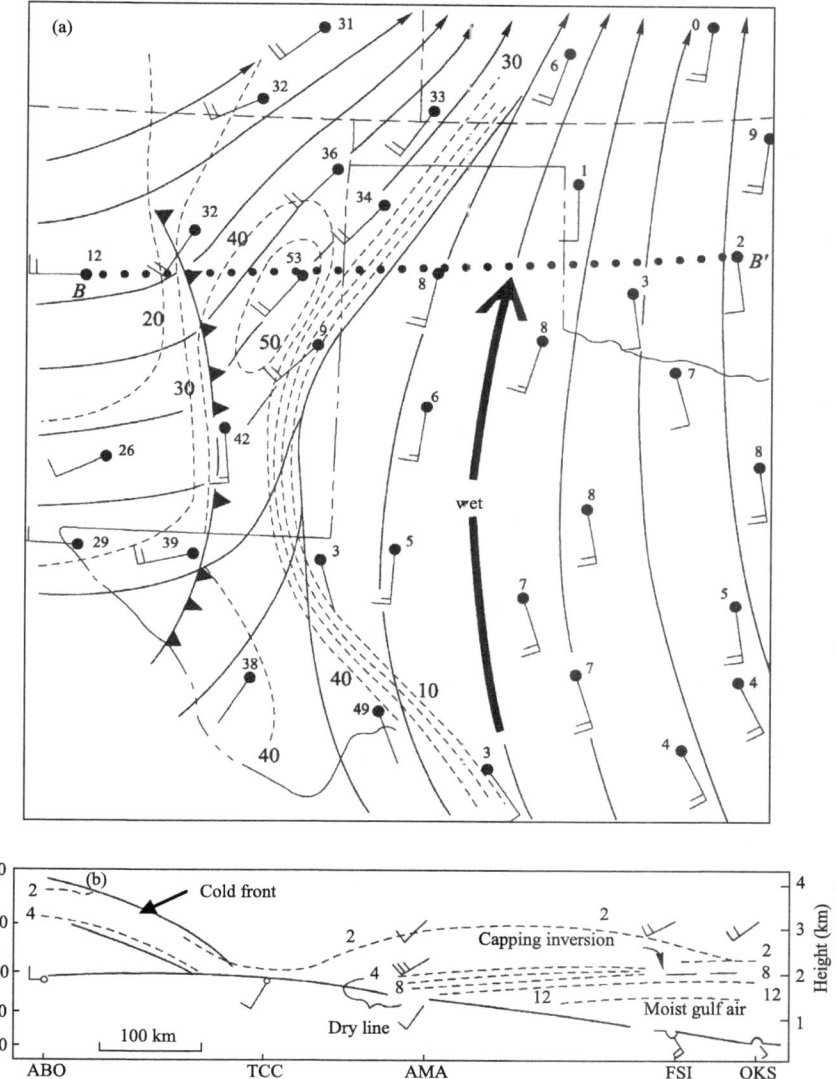

Fig. 4.11 The features of the dry line occurred at 1200 GMT Apr. 3, 1981 in east slope of the Rocky Mountain in the United States. (a) surface dew point temperature decreasing (°F, the break lines) and stream lines (solid arrow lines); (b) the vertical section along the line BB′ in Fig. 4.11a. (After Shapiro, 1983)

Dry line has important influences on the convective activities. Cumulus occur in the area near dry lines very often and can develop into severe storms, which can propagate leaving the dry line and then develop new cumulus belts near the dry line. The cumulus belts can be regarded as the direct result from dry line, which therefore plays a role as a source of disturbing source for convective activities (Rhea, 1966). Researches show that in the great plain of the United States in the spring season there are about 60% of new radar echoes occur in the 200 km area apart from dry line. They then can evolve into a squall lines. The squall line occurred in Jianghui area of China on Jun. 17, 1974 was also influenced by the dry line which played a trigger mechanism of the convection. The dry and warm lid is favorable to store and accumulate instability energy. Once the lid is opened the energy will release vigorously and the severe convective weather will develop. Severe convective weather will take place frequently in the narrow area along the boundary of dry and warm lid (Carlson, 1980). Some people made a comparison between the dry and warm lids in east China area and in the great plains area of the United States, and found that they all played important roles on the development of severe convective weather but there were some differences on their structure and formation mechanism. In east China the dry and warm lids are mainly caused by the subsidence from midlevel of troposphere and the lid inclined from southeast to northwest direction. While in the great plains area of the United States the dry and warm lids are mainly caused by the warm air advection and the height of the lid presented high in east and low in west direction. As to the intensity, the dry and warm lid in the great plains is normally stronger, hence the severe weather is more violent (Yang and Shu, 1985).

(2) Formation of the dry lines

The formation of the dry line may be related with many factors such as the synoptic scale situation, the features of the underlying surface, the turbulent mixing and sky status etc. (Fujita, 1958; Schaefer, 1974). In the rear of the synoptic scale low pressure trough, it normally prevails subsidence flows and may form subsidence inversion and constitute the dry and warm lid. The intersecting line that the lid intersects with the ground surface is the dry line at surface. In the different areas with different features of the underlying surface, for instance in the east slope of the Rocky Mountain in the United States, where it is moist plain area in east side but it is partial desiccation or desert highland in west side, when the southerly wind blows in east side under the controlling of the moist oceanic air from Gulf of Mexico, and the southerly wind blows in west side bringing up the dry continent air, then the dry line will be formed when the two air masses are connected each other as shown in Fig. 4.11a.

In the dry line area the direct circulation occurs in daytime. If considering the moist land as a sea surface, then this kind of circulation is similar to the sea-land breeze circulation, hence it is also called as inland sea breeze circulation. If it prevails "offshore wind" in the dry side, then the strong ascending motion can occur near the dry line. In the morning at the precipitation or cloudy district, the temperature at noon time will be lower 5—8 ℃ than the clear air district. The zone connected the two districts posses the feature similar to the dry line. In the developing process, the role played by turbulence mixing could not be ignored. The equations (4.7.1) and (4.7.2) show the factors influencing line frontogenesis of the dry line.

$$\frac{d}{dt}\left(\frac{\partial q}{\partial y}\right)=\left(\frac{\partial u}{\partial x}\frac{\partial q}{\partial y}-\frac{\partial u}{\partial y}\frac{\partial q}{\partial x}\right)+\left(\frac{\partial q}{\partial y}\frac{\partial \omega}{\partial p}-\frac{\partial \omega}{\partial y}\frac{\partial q}{\partial p}\right)-\frac{\partial}{\partial y}\left(\overline{\frac{\partial \omega' q'}{\partial p}}\right) \quad (4.7.1)$$
$$\qquad\qquad\qquad ①\qquad\qquad\qquad ②\qquad\qquad\qquad ③$$

$$\frac{d}{dt}\left(\frac{\partial q}{\partial p}\right)=-\left(\frac{\partial u}{\partial p}\frac{\partial q}{\partial x}+\frac{\partial v}{\partial p}\frac{\partial q}{\partial y}\right)+\left(\frac{\partial q}{\partial p}\frac{\partial v_a}{\partial y}-\frac{\partial v_a}{\partial p}\frac{\partial q}{\partial y}\right)-\frac{\partial}{\partial p}\left(\overline{\frac{\partial \omega' q'}{\partial p}}\right) \quad (4.7.2)$$
$$\qquad\qquad\qquad ①\qquad\qquad\qquad ②\qquad\qquad\qquad ③$$

where q is the water vapor mixing ratio, the equations (4.7.1) and (4.7.2) are the horizontal and vertical frontogenesis of the dry line respectively. It can be seen from equation (4.7.1), the horizontal gradient of the moisture $(\partial q/\partial y)$ is forced by the following factors: ① the geostrophic stretching and shearing deformations; ② the horizontal divergence and vertical tilting caused by the secondary ageostrophic circulation; ③ horizontal gradient of the vertical flux divergence of the moisture turbulence. The forcing factors to cause the variation of moisture vertical gradient $(\partial q/\partial p)$ as shown in equation (4.7.2) are as follows: ① the vertical shear of the geostrophic motions; ② the horizontal divergence and vertical shear of the horizontal ageostrophic motion; ③ the vertical gradient of the vertical flux divergence of the moisture turbulence. It can be seen obviously that all the factors including the synoptic scale situation, the ageostrophic secondary circulations related with the dry lines, as well as the turbulence mixing processes related with the underlying surfaces with different features etc. will contribute to the formation and development of the dry lines.

Sun and Ogura (1981) made the simulation of the development of dry line, of which the integrating started from 0630 CST in the early morning, by using the non-elasticity approximation and considering the hydrostatic relation as well as the turbulence diffusion effects, under the initial condition with the horizontal uniform distribution of the potential temperature. The results are as shown in Fig. 4.12. In the figure the thick solid line indicates the temperature inversion layer, J indicates the center of southerly wind low level jet across the vertical section, U and D denote the ascending and descending centers respectively, the break lines are the constant value lines of the vertical velocity of 2 cm · s^{-1}. After integrating 4 h, since the difference of the surface warming, when the temperature in the mixing layer located at the left side of the cross connect area of the warm and cold air is higher that at the right side, the horizontal pressure gradient pointed to the negative direction of X axis will increase and therefore the wind will blow from the cold side to the warm side just like the onshore wind. For satisfying the continuity of mass, in the warm air side will occur ascending motion, while in the cold side the descending motion will occur. In the earlier stage of the development, the vertical circulation in the cross connect area of the cold and warm air is nearly symmetric and the strength of the ascending and descending flow are nearly equally each other. As time goes by, the mixing layer in warm air side is getting thicker and thicker comparing with that in cold side, and the vertical circulation is also getting stronger to the extent that stretching into the stable cold air layer above temperature inversion layer. Since the constant pressure surface elevating leads to cause the horizontal pressure gradient above the temperature inversion layer point to the positive direction of the X axis (offshore acceleration), obtaining the local maximum of the x component of the velocity in this layer and leading to the convergence in cold air side. The subsidence flowing from the top of the ascending branch of the circulation divergent along the inversion layer. So that the center of the subsidence motion moves to the location above the inversion layer. In the mixing layer, in the cold air side there is only wider but weaker subsidence, meanwhile the ascending motion center is still maintaining in the mixing layer as shown in Fig. 4.12. From the Fig. 4.12 it can be seen that the ascending motion is almost supplied from the mixing layer in the cold air side (actually it is the moist air). Obviously, this type of developing dry line circulation is favorable to the formation and maintaining of the squall lines.

4.7.2 Coast front

(1) The features of coast front

Coast front is also a kind of the mesoscale front in boundary layer. It usually occurs near the coast and nearly parallels to the coast line. The coast front is usually formed in later autumn and earlier winter seasons, when the temperature difference between sea and land is biggest. In the coast district of New England of the United

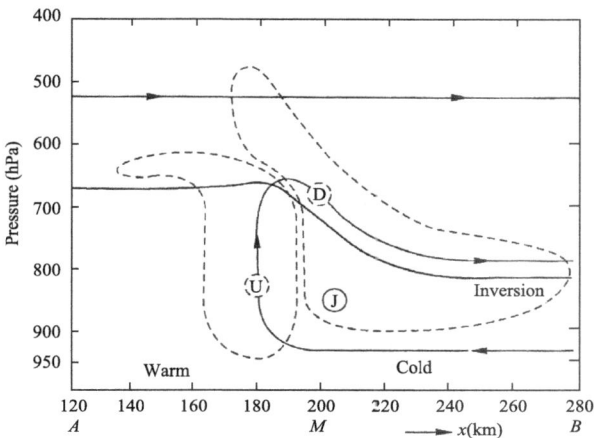

Fig. 4.12 The schematic diagram of the vertical section across the dry line J, U, D denote the low level jet center, the centers of ascending and descending motion respectively, the break lines are the isolines of the vertical velocity in unit 2 cm · s^{-1}. (After Sun and Ogura, 1981)

States, the coast fronts can occur 5—10 per year. Typically, the length of the coast front may range from 200 km to 600 km, and the width ranges from 50 km to 100 km, the time scale is about 12 hours. It is characterized by the cyclonic wind shear and fairly strong horizontal temperature gradient. In the land side of the coast front it is normally the weaker northerly or northwest winds, while in the sea side it blows usually the stronger easterly wind. In the vertical section the temperature difference in the distance of 5—10 km may be as large as 5—10 ℃, the strongest temperature gradient can be 1 ℃ · km^{-1}. The distinct baroclinicity mainly occurs in the lowest 1 km of the boundary layer.

A case of the vertical structure of the coast front is shown in Fig. 4.13. From the figure it can be seen that the front surface is confined in a shallow layer with thickness about 300 m above surface ground. In the cold air area behind and below the front it blows northerly wind, while in the area ahead of and above the front it blows southeasterly wind. Near the front the constant lines of potential temperature are concentrated. The front slope is steep. It can be seen from Fig. 4.13 that the coast front is shallow and strong similar to a strong cold front or a density current. During the existing duration of the coast front, there is clouds in the ascending flows on the front and behind of the front and there is fog in the cold air behind of the front. It is normally the border line of the frozen precipitation and non-frozen precipitation. Although the precipitation in the existing duration of the coast front is mainly caused by the large scale circulations, while the mesoscale circulations induced by the coast front may cause local enhancement of the precipitation. Marks and Austin (1979) believed that the precipitation enhancement was probably caused by the low clouds induced from the coast front circulation. Such mesoscale weather features related with the coast front is one of the factors that the forecasters have to consider when they make a local forecasting.

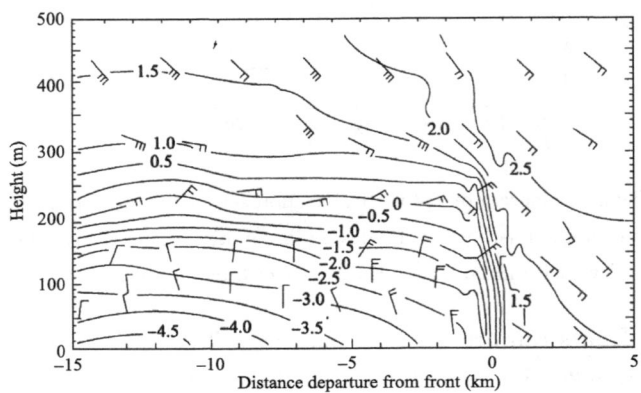

Fig. 4.13   The vertical section across the coast front. Solid lines are constant lines of potential temperature, with the interval of 0.5 ℃; the every scribing line of the wind vanes representing 5 kn. (After Emanuel, 1984)

(2) Frontogenesis of the coast front

With the same common meaning of the frontogenesis, the frontogenesis of the coast front implies the whole process of the formation and maintaining as well as the frontolysis. There are many factors may affect the formation of coast front. It is related with the large scale situation. It is also related with the diabatic heating that the ocean exerts to the eruptive cold air and the frictional effects between sea and land. Meanwhile, the topography of the coast region and the shape of the coast line can also affect the formation of the coast front. The coast fronts are normally formed in the favorable synoptic environment. Taking the coast front in New England for example, the typical situation is that when the cold anticyclone moves eastward passing the north part of New England and meanwhile the wind direction near the coast area changes from the northerly wind (offshore wind) to the easterly wind (onshore wind), and as the result the coast front will be formed. Analysis shows that the ageostrophic effect is the dominant factor in the frontogenesis process of the coast front.

The ageostrophic forces for the frontogenesis of the coast front are the combination of the frictional difference between sea and land, the resisting action the mountain on the cold air, and the different non-adiabatic heating in the boundary. The effects of these factors can be expressed by the physical model as shown in Fig. 4.14. For idealization, the coast line is presented as from south to north toward; the horizontal temperature gradient is perpendicular to the coast and points to the land reflecting the non-adiabatic heating effect; the uniform geostrophic flow is perpendicular to the coast and flows from the sea to the land; the frictional forces are different in the sea and land regions while they are set as uniform in the same region and the force in the land region is stronger than that in the sea region. The models of the types of the surface constant pressure lines and the constant temperature lines are approximately agreed with the actual situation of the observational coast front.

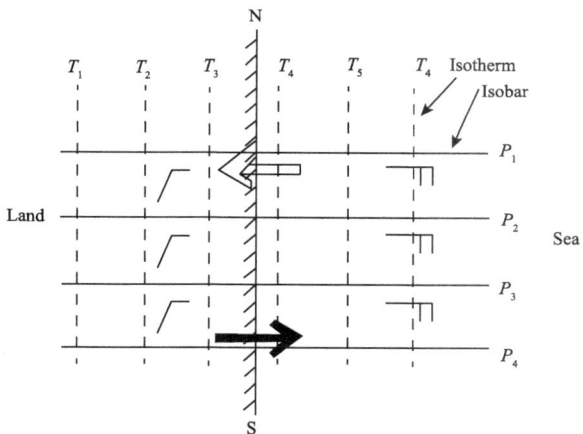

Fig. 4.14 The physical model of the frontogenesis of the coast front. The black and blank arrows indicate the ageostrophic flows pointed to the sea and land respectively. (After Bosart, 1975)

The difference of the frictional effect between sea and land will cause the mesoscale convergence in the area near the coast and company with a real wind deformation field causing the previous existed temperature gradient increase and leads to the frontogenesis of the coast front. In this process, the non-geostrophic flows toward the sea and land will occur in the low and high level of the boundary layer. The low level offshore acceleration will be help for maintaining of the northerly flow in the cold air and thus increase the frontogenesis of the coast front. Obviously, the previously existed horizontal temperature gradient and different frictional effects are necessary for maintaining the physical model.

Analysis shows that the process of the coast front frontolysis will be caused by the diurnal variation of land heating, which will decrease the temperature gradient. The frontolysis can also be caused under the situations that the onshore wind is intensified or the wind direction is changed into the offshore wind. The onshore wind intensifying will cause the front move to inland mountain region and thus lead to frontolysis and the wind direction changes into the offshore wind will lead the front move to the sea surface and thus to cause frontolysis too. Moreover, the background situation variation will probably provide the conditions favorable to frontogenesis.

4.7.3 Sea breeze front

Sea breeze front is another kind of mesoscale front in boundary layer. Similar to the seashore front, the sea breeze front has features of density current. But there are important difference in their intensity and formation mechanism. The seashore front is a phenomenon which normally occurs in late autumn and earlier winter season and in the area between warm sea and cold land, while the sea breeze front is normally formed in the area between cold sea and warm land and in the time periods of spring and summer seasons when the sea breeze is strong. The sea breeze front is closely related with the sea breeze phenomenon. It has mesoscale weather features. The low level convergence related with the sea breeze front is normally the important mechanism for triggering convection.

(1) The general features of the sea breeze front

The sea breeze front occurs under special environmental wind condition. The environmental wind has distinct influences on the formation and movement of the sea breeze front. In general speaking, the occurring of the sea breeze front is related with the steady or weaker offshore wind, say less than $4 \text{ m} \cdot \text{s}^{-1}$, on the surface ground. Occasionally, the sea breeze circulation can also be self-organized into a clear frontal structure under the situation with weaker onshore wind.

The seasons that sea breeze fronts occur frequently are the spring and summer when the sea breeze effects are most strong. According to the statistics, for instance, in the south area of England the sea breeze front occurred mainly concentrated in the duration of May to August, especially in June. In these days, the sea breeze fronts easily occur in afternoon and disappear in the night showing the important effects of the non-adiabatic heating on the formation of the sea breeze fronts.

During the passage of the sea breeze front, the meteorological elements will present various mesoscale variations. Typically the temperature will decrease and dew point temperature will increase. The amplitude can be as 10 ℃ or more in a short time duration. Before and after the passage of the sea breeze front the wind direction shifting can be 180°. Normally there is cloud systems near the sea breeze front. When the front moves slower than the sea breeze, the oceanic air will be lifted in the front edge of the sea breeze. If the air can reach the condensation level then the distinct cloud line can be formed. When the front invades toward inland area, the air over the land will be lifted, then a denser cumulus line can be formed along the sea breeze front. There are many factors, such as the direction and intensity of the low level prevailing wind, the shape of the coast line, the temperature difference between sea and land, the atmospheric instability and the frictional effect etc. may impact the cumulus convection near the sea breeze front. Sometimes the shape of coast line is probably the major factor to determine the distribution of the cumulus clouds along the front. This is because the different curves of the coast line may cause divergence or convergence of the local sea breeze and hence to lead the local cumulus activity intensified or weakened. In the projecting part of the coast line, when the sea breeze front moves toward inland the convergence will be intensified, while in the dented part of the coast line the situation will be opposite, the gulf divergence will be intensified. When the convergence caused by the local heating and the convergence caused by the shape of the coast line are superimposed each other the vertical motion will be most strong and the convection will be most vigorous.

(2) Structure of the sea breeze front

The sea breeze front is very similar to the density current, there is also a "head" with the height about 1 km. It is a common feature of the sea breeze front.

Fig. 4.15 shows a two dimensional flow field of the sea breeze front in a vertical section made by Simpson based on 5 sounding observational data. In the figure, the type of stream line shows the flow characteristic relative to the front. In vicinity of the front, the low level airs comeing from sea and land are converged toward the front, lead to the strong ascending motion. In the side facing the sea, the oceanic air is rolled up nearby the front. The roll extends from the head to the back forming a distinct cutoff vortex. According to the extent of the zero value stream line, the horizontal scale of the cutoff vortex can be estimated as about 7 km. Many other case analyses also show that most of the sea breeze front intruded inland deeply may also form the cutoff vortexes. Obviously, this is a close sea breeze circulation. The sea breeze front forms in the development process.

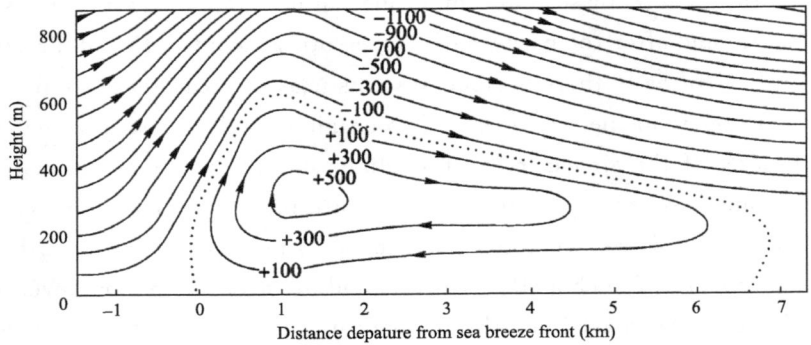

Fig. 4.15　The features of the stream lines related with the sea breeze front Interval of the isolines 200 $m^3 \cdot s^{-1}$, the dot line is the zero value stream line indicated the boundary of the cutoff circulation. (After Simpson, 1977)

4.7.4　Squall front

(1) Structure of the squall front

The front edge of the strong outflow in the bottom of the thunderstorm is normally as "squall front". The outflow related with the squall front is mainly caused by the mid troposphere dry and cold air, which is entrained into the storm and sinks downward due to the negative buoyancy. In this process the precipitation dragging and the raindrop evaporation cooling encourage the subsidence intensity. The subsidence will cause a cold pool in the low level and surface ground and divergent all around. This cold pool corresponding to a Mesoscale high pressure, which is normally called as thunderstorm high. The squall front is located at the front edge of the thunderstorm high. By the impact of the squall front passage, the pressure will rise, wind direction will shift, wind velocity will suddenly increase and temperature will drop rapidly, and sometimes it will company heavy rainfall or other severe weather phenomena.

Like other weather systems the squall front lifecycle has four stages: the formation stage, the developing stage, the mature stage and the dissipating stage. The features of different stages are shown in Fig. 4.16 and Table 4.1. The formation stage of the squall front occurs in the mature stage of the thunderstorm. When the thunderstorm begins to enter dissipating stage, as the front edge of the thunderstorm outflow, the squall front begins its initial mature stage and the precipitation roll will be formed in low level. When the squall front propagates in the cold air under the impacting of the greater hydrostatic pressure, it can change into a density current. When the thunderstorm is close to the total dissipating stage, the squall front reaches stage IV, when the squall front spreads in the front of rain column, no longer obtains the cold air supplying and the total thickness of the squall front will become smaller.

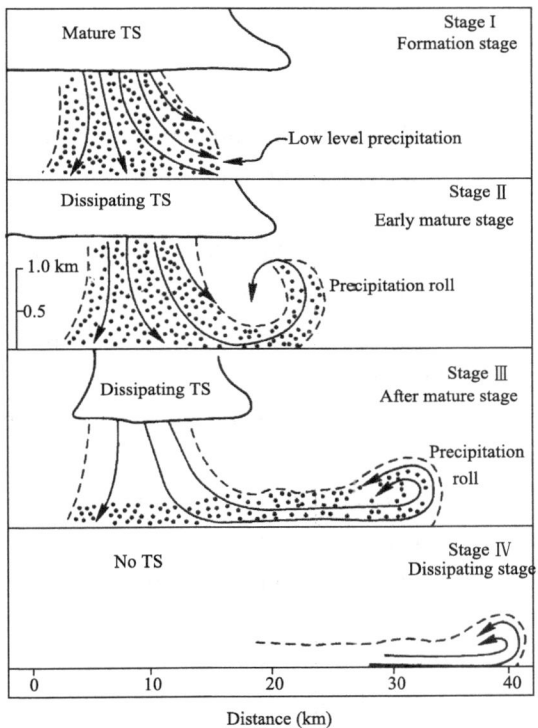

Fig. 4.16 The four stages of the thunderstorm (TS) squall front. (After Wakimoto, 1982)

Table 4.1 Features of the squall front in different stages

Stage	Thickness	Velocity	Radar scattering particles
I formation	$\geqslant 1$ km	10—20 m·s^{-1}	Rain drop comes from squall front
II early mature	1—2 km	10—30 m·s^{-1}	Rain drop in the precipitation roll
III later mature	0.5—1 km	10—25 m·s^{-1}	Rain drop in precipitation roll, dust and insects
IV dissipating	<0.5 km	5—15 m·s^{-1}	Dust and insects

The life cycle of squall front can be divided into stages according to different point of view. For example, Goff (1976) divided the squall front lifecycle into 4 stages based on the intensity of the storm outflows. In the first stage the squall front is related with the increasing storm or the accelerated outflow; in the second stage the squall front is related with the mature severe storm or steady outflows; in the third stage the squall front is related with dissipating storm or the moderating outflow; and in the fourth stage is the last stage or the dissipating stage of the squall front.

Fig. 4. 17 shows the structure model of the squall front based on the data observed by the meteorological tower. According to the figure, the squall front consists of five parts. The most forward part is the cold air nose, which is located at the most leading edge of the cold outflow, extruding forward to the warm air like a nose. The depth of the nose may be different for different cases. Some of them may be located on the height of 750 m and 1. 3 km ahead of the cold air edge, while some of them may be located on the height of 100 m and stretching into the warm air for about 400 m. The second part is the cold air head. in the front of the cold air pool, the shape of the air is like a head. Its height is about 1. 7 km. In the front part of the head there is strong ascending motion. In the rear part of the head, the airflow become weaker subsidence motion furtherly go backward it will be the wake area, i. e. the cold air area behind the head. The third part is the underflow region. This is the high speed air flow running forward in the rear part of the squall front. It is located below the head departing away from surface ground about 100 m. The high speed underflow runs upward and then turns to the rear in the upper boundary and finally goes down to the rear part of the head. The fourth is the cold air backflow, which is a ground-hugging flow leaving the squall front induced by the surface resistance. The fifth part is the squall front, which is the interface between the cold air outflow and the lifted warm air. The interface is coincided with the zero line of the relative horizontal wind component (u). But it has to be noted that the squall front is not agreed with the density current boundary. The squall front is confirmed by analyzing the isolines of wind velocity, while the density current boundary is defined by analyzing the isentropes. Since the squall front boundary is normally clearer than the thermal boundary, therefore the squall front boundary is more often to be used for characterizing the kinematic features of the leading edge of thunderstorm outflow.

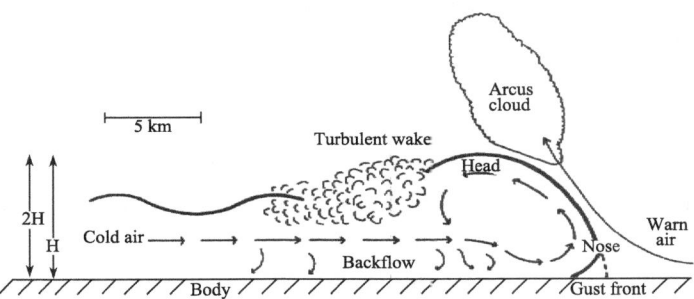

Fig. 4.17 The structure of the squall front. (After Goff, 1976)
(The occurrence of the clouds is depending on the height of lifted condensation level, the flows are relative to the squall front.)

One of the important structure features of the squall line is the fluctuation of the strong wind or the gust surge phenomenon. Observation shows that the density current of the thunderstorm propagates in surge form, hence the distribution of the gust in the cold air flow is nonuniform usually with multi centers of the maximum wind velocity. In the cold air flow, there is serious threaten for the taking off and landing of the airplanes due to the existing of maximum wind velocity center and the strong vertical wind shear in the low level lower than 100 m. The most strong vertical wind shear occurs in the low level lower than 30 m at the location in front part of the head and the rear part of the head. Especially the wind shear in the rear part of the head has strong intensity and lasts longer duration. Moreover, in the rear part of the head, the negative vertical wind shear is also the strongest. Its intensity can be as strong as -0.11 s^{-1}. Therefore at the height of 100—120 m in the rear part of the head is another sudden change area of the positive and negative vertical wind shear. Since the strong turbulence and the sharp variation of the wind velocity, it is extremely danger for flying activities.

Another important feature related with the strong wind fluctuation is the alternate distribution of the ascending and descending motions. The strongest ascending motion occurs in the front of the cold air head, while the strongest descending motion occurs in the rear part of the head.

(2) The movement of squall front

Since the squall front is similar to the density current, so that its movement can be described by the theory of the density current. Setting a two-layers inviscid and steady flowing fluid, in which the fluid with greater density is carrying fortward under the fluid with smaller density. The moving velocity is C. Taking the moving coordinate system moving with the density current is shown in Fig. 4.18. In the coordinate system the downstream larger density air current is regarded as stationary, while the moving velocity of the upstream smaller density current related with the density current is C. The dots 1, 2 and 3 are the control points. Where the point 1 and 2 are located on the surface in upstream and downstream respectively. The point 3 is located on the downstream surface. Along the stream line connected the points 1 and 3, by using the Bernoulli equation, we have

$$\frac{1}{2}U_3^2+gh+\frac{p_3}{\rho_1}=\frac{1}{2}C^2+\frac{p_1}{\rho_1} \qquad (4.7.3)$$

where U_3 is the velocity at the point 3, p_1 and p_3 are the pressures at the points 1 and 3. For solving the moving speed of the squall front, we have to seek the relationship between p_3 and p_1 at first. Therefore the horizontal momentum equation is integrated along the low boundary and get the following equation:

$$\frac{p_2}{\rho_1}=\frac{1}{2}C^2+\frac{p_1}{\rho_1} \qquad (4.7.4)$$

where an approximation is employed that the pressure at the nose of the squall front is equal to p_2, and the pressures at points 2 and 3 are related by the hygrostatic equation. By using the equation (4.7.4), we can get the following equation:

$$p_3=p_2-\rho_2 gh=p_1+\frac{1}{2}\rho_1 C^2-\rho_2 gh \qquad (4.7.5)$$

By using equation (4.7.5), the equation (4.7.3) can be changed as

$$\frac{1}{2}U_3^2+gh+\frac{1}{2}C^2-\frac{\rho_2}{\rho_1}gh=\frac{1}{2}C^2$$

or

$$\frac{1}{2}U_3^2=\frac{\rho_2-\rho_1}{\rho_1}gh \qquad (4.7.6)$$

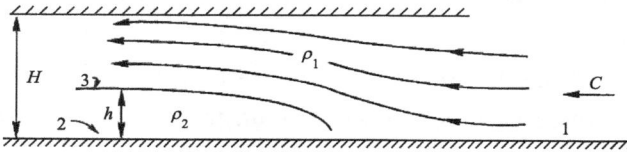

Fig. 4.18　Model of the steady density current. (Afetr Benjamin, 1968)
(The fluid flows between two panels apartted H each other, The fluid with lesser density (ρ_1) flows in from right, and the fluid with larger density (ρ_2) presents stagnating state in the cordinete system. Points 1, 2, 3 are the controlling points)

Because the upstream flow is non-rotational and has no baroclinic effect, hence the downstream flow should also be non-rotational, and $U=U_3$ at the downstream area.

According to the principle of mass continuity, we can get

$$U_3[H-h]=CH$$

Consequently

$$C=\left(1-\frac{h}{H}\right)\sqrt{2gh\left(\frac{\rho_2-\rho_1}{\rho_1}\right)} \qquad (4.7.7)$$

The solution of the moving speed C of the squall front is derived based on the density current theory. It can be seen from equation (4.7.7) that the moving speed is related with ρ_1, ρ_2 and the moving speed is also related with cold air thickness as well as the total thickness H of the two layers of the fluid. If the total thickness H is large. Then h/H is small. Consequently

$$C \sim 1.4\sqrt{gh\left(\frac{\rho_2-\rho_1}{\rho_1}\right)} \qquad (4.7.8)$$

Obviously, the bigger the difference between ρ_2 and ρ_1, the faster the moving of the squall front. If the difference between ρ_2 and ρ_1 is very small or even there is no difference, then $C \sim 0$, or in other words, there is no squall front.

It should be noted that the solution of the moving speed C of the squall front is only suitable for the special situation that $h=(1/2)H$. The following is a demonstrating.

For the steady momentum equation to take integral along the flow tube from 0 to H.

$$\int_0^H \left[\frac{\partial}{\partial x}(u^2)+\frac{\partial}{\partial z}(uw)\right]dz = -\int_0^H \frac{1}{\rho}\frac{\partial p}{\partial x}dz \qquad (4.7.9)$$

Since at $z=0$ and H, $w=0$, the second term in the left side in the above equation can be omitted, Then we can get

$$\frac{\partial}{\partial x}\int_0^H \left[u^2+\frac{p}{\rho}\right]dz = 0 \qquad (4.7.10)$$

The above equation shows the energy conservation along the flow tube, i. e.

$$\frac{\partial}{\partial x}\int_0^H \left[u^2+\frac{p}{\rho}\right]dz = \text{constant} \qquad (4.7.11)$$

This is the Bernoulli energy conservation equation integrated along the flow tube.

Using the equation (4.7.11) at the middle point between upstream and downstream and considering the continuity condition. Then we get

$$h = \frac{3H \pm H}{4} \qquad (4.7.12)$$

Here h has two roots, $h_1 = H$ and $h_2 = (1/2)H$. Obviously the solution is meaningful only the situation when $h = (1/2)H$. That is to say, when making the integral along the stream line at the place $h = (1/2)H$, the equation energy conservation condition is set up, so that the squall front moving velocity equation could be applied.

References

AUSTIN P M, HOUZE R A Jr, 1972. Analysis of the structure of precipitation patterns in New England[J]. J Appl Meteor, 11:926-935.

BLUESTEIN H B, 1986. Fronts and Jet streams: A Theoretical Perspective. Mesoscale Meteorology and Forecasting[M]. Am Meteor Soc.

BOSART L F, 1975. New England coastal frontogenesis[J]. Q J Roy Meteor Soc, 101:957-978.

BOSART L F, VAUDO C J, HELSDON J H Jr, 1972. Coastal frontogenesis[J]. J Appl Meteor, 11:1236-1258.

CARLSON T N, 1980. The role of the lid in severe storm formation: some synoptic examples from SESAME[C]. 12th Conf, On Severe Local Storms, 221-223.

CHARBA J, 1974. Application of gravity current model to analysis of squall line front[J]. Mon Wea Rev, 102:140-156.

ELLROD G P, MARWITZ J D, 1976. Structure and interaction in the subcloud region of thunderstorms[J]. J Appl Meteor, 15:1084-1091.

EMANUEL K A, FANTINI M, THORPE A J, 1987. Baroclinic instability in an environment of small stability to slantwise moist convection, Part I: Two-dimensional models[J]. J Atoms Sci, 44: 1559-1573.

FUJITA T, 1958. Structure and movement of a dry front[J]. Bull Am Meteor Soc, 39:574-582.

GOFF R C, 1976. Vertical structure of thunderstorm outflows[J]. Mon Wea Rev, 104:1429-1440.

HOSKINS B J, 1975. The geostrophic momentum approximation and the semi-geostrophic equations [J]. J Atoms Sci, 32:233-242.

KEYSER D, 1986. Fronts-Observations. Mesoscale Meteorology and Forecasting [M]. Am Meteor Soc.

KEYSER D, SHAPIRO M A, 1986. A review of the dynamics of upper-level frontal zones[J]. Mon Wea Rev, 114:452-498.

MARKS F D Jr, AUSTIN P M, 1979. Effects of the New England coastal front on the distribution of precipitation[J]. Mon Wea Rev, 107:53-67.

REED R J, 1955. A study of a characteristic type of upper level frontogenesis[J]. J Meteor, 12:226-237.

RHEA J O, 1966. A study of thunderstorm formation along dry lines[J]. J Appl Meteor, 5:58-63.

SCHAEFER J T, 1974. A simulative model of dryline motion[J]. J Atmos Sci, 31:956-964.

SCHAEFER J T, 1975. Nonlinear biconstituent diffusion: A possible trigger of convection[J]. J Atmos Sci, 32:2278-2284.

SHAPIRO M A, 1981. Frontogenesis and geostrophically forced secondary circulations in the vicinity of jet stream frontal zone systems[J]. J Atmos Sci, 33:954-973.

SHAPIRO M A, HAMPEL T, ROTZOLL D, et al, 1985. The front hydraulic head: A micro-α scale ($\sim$1 km) triggering mechanism for mesoconvective weather systems[J]. Mon Wea Rev, 113:1166-1182.

SIMPSON J E, 1964. Sea breeze fronts in Hampshire[J]. Weather, 19:208-220.

SIMPSON J E, MANSFIELD D A, MILFORD J R, 1977. Inland penetration of sea-breeze fronts[J]. Q J Roy Meteor Soc, 103:47-76.

UCCELLINI L W, JOHNSON D R, 1979. The coupling of upper and lower tropospheric jet streams and implications for the development of severe convective storms[J]. Mon Wea Rev, 107:682-703.

WAKIMOTO R M, 1982. The life cycle of thunderstorm gust fronts as viewed with Doppler radar and rawinsonde data[J]. Mon Wea Rev, 110:1060-1082.

WALLINGTON C E, 1959. The structure of the sea breeze fronts as revealed by gliding flights[J]. Weather, 14:263-270.

WALLINGTON C E, 1965. Gliding through a sea breeze front[J]. Weather, 20:140-144.

YANG G X, SHU C X, 1985. Large scale environmental conditions for thunderstorm development [J]. Adv Atmos Sci, 2(4):508-521.

Chapter 5
Mesoscale Convective Systems

The mesoscale convective systems (MCSs) refer to the mesoscale weather systems with vigorous convective motions. In mid latitude the MCSs may be roughly divided into three categories according to their organization types. They are the isolated convective systems, the belt-shaped convective systems and the mesoscale convective complex (MCC). In this chapter the different categories of MCS will be discussed respectively.

§ 5.1 Isolated convective systems

The isolated convective systems imply the convective systems existed in the forms such as the individual single cells of common thunderstorms, the multi-cell storms, the supercell storms and the small squall lines etc. According to the classification by Chisholm and Renick (1972), the isolated convective systems can be further divided into three basic types: the common thunderstorms, the multicell storms and the supercell storms. Their basic characteristics will be discussed in this section.

5.1.1 The common thunderstorms

In generally speaking, a thunderstorm is a deep moist convective system (DMC), of which the inner vertical velocity is nearly 10 m $\cdot$ s^{-1} or stronger; the horizontal extent is about ten kilometers; the vertical stretching is almost whole troposphere and the typical companied weather phenomena are lightning, thunders, gust wind, shower and rapid changes of the meteorological elements such as pressure, temperature, humidity, wind velocity and wind direction etc.

The thunderstorms can be divided into two types according to their intensities. The storms companied with one of the severe damage weather phenomena such as strong damage wind, big hailstones, and tornadoes etc. are called the severe thunderstorms. While the thunderstorms with relative less intensity are called the common thunderstorms or ordinary thunderstorms.

During the summer of 1946 and 1947, Byers and Braham et al. organized a field observational research of thunderstorm in the United States. Based on the radar data and the successive data from the dense observational network they studied carefully the structure and development process of the thunderstorms and established a lifecycle model for the ordinary thunderstorms, which is slightly modified by Doswell in 1981 as shown in Fig. 5.1.

As shown in Fig. 5.1, the development of thunderstorm cell goes through three stages including the towering cumulus stage, the mature stage and the dissipating stage. In the towering stage, there is only the ascending motion in the cloud; in the mature stage the ascending becomes strong, an overshooting occurs at the top of cloud and meanwhile the precipitation begins and to cause the subsidence occurred in the rear part of the convective cell due to the dragging effect of the precipitation particles on the air. When the descending flow arrived the surface ground, a cold air dome and horizontal outflow may be formed and a gust front may be formed in the leading edge of the outflow. In the dissipating stage, the descending flow in the cloud is gradually getting dominant and at the end the descending flow totally replaced the ascending flow. The development process normally takes about 30—50 min. during this period, the thunderstorm system moved along the mean air flow of environment at lowest height of 5—8 km above surface. The companied severe weather includes the gust wind, shower rain, small hail etc. in a very short duration.

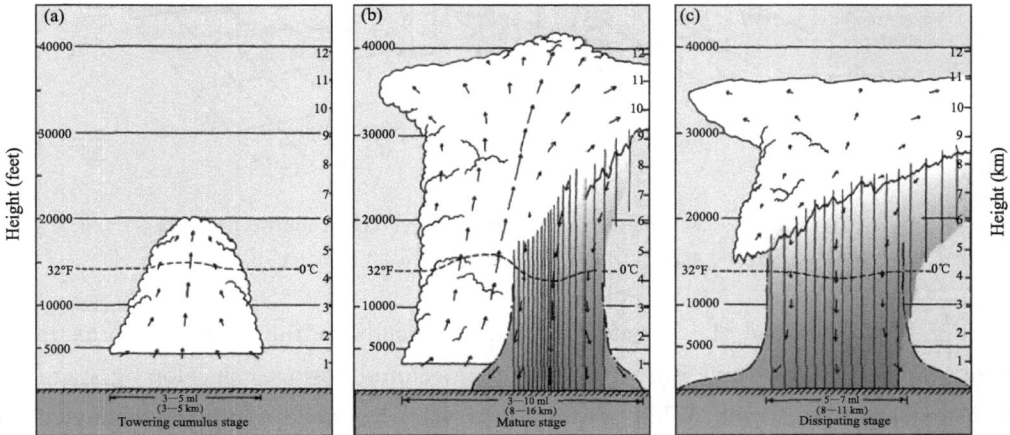

Fig. 5.1 The three stages of an ordinary cell: (a) towering cumulus stage; (b) mature stage; and (c) dissipating stage. (Adapted from Byers and Braham, 1949; Doswell, 1985)

5.1.2 Multicell storms

Multicell storm is a thunderstorm system composed by several convective cells in different developing stages and possess an unified mesoscale circulation. Although all the individual convective cell in the multicell storm possess a cold air outflow, while they can be organized into a larger gust front. Along the leading edge there is air flow convergence, which is the strongest in the moving direction of the storm and favor to promote the development of new convection along the gust front. Every convective cell then will go through self-development process. Thus by the successive replacement process of the single cells the holistic storm will last long time, although the individual cell has only short lifetime. Fig. 5.2 showed the structure of the multicell storm, which consists of four cells in different developing stages. Among them the youngest one is located most southward. The arrows show the trajectory of an air parcel in the developing cell. In the multicell storm there is a couple of organized ascending and descending flows constituting an unified circulation(Fig. 5.3).

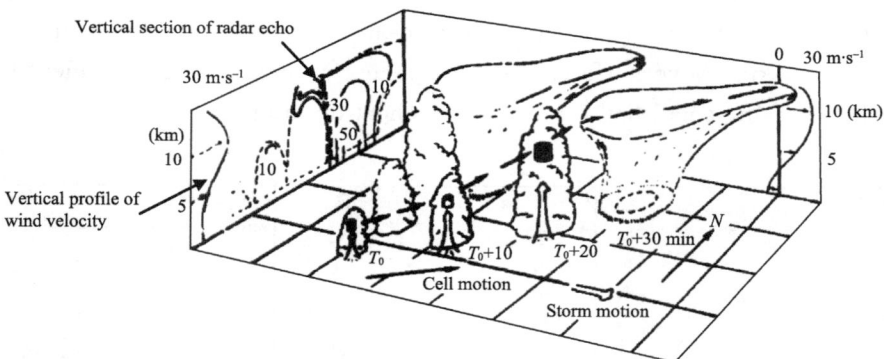

Fig. 5.2 The schematic diagram of the structure of multicell storm
(Adapted from Chisholm and Renick, 1972)

Fig. 5.4 described the variation of the four cells in the storm. At the initial stage, the cell 2, 3, 4 and 5 are at initial stage, cumulus stage, developing stage and mature stage respectively. While after 20 minutes, they are successively changed into developing stage, mature stage and dissipating stage respectively and a new cell generated.

The cells in a multicell storm present organized state, due to the new cells are normally formed in the certain direction. If the new cells can occur in different direc-

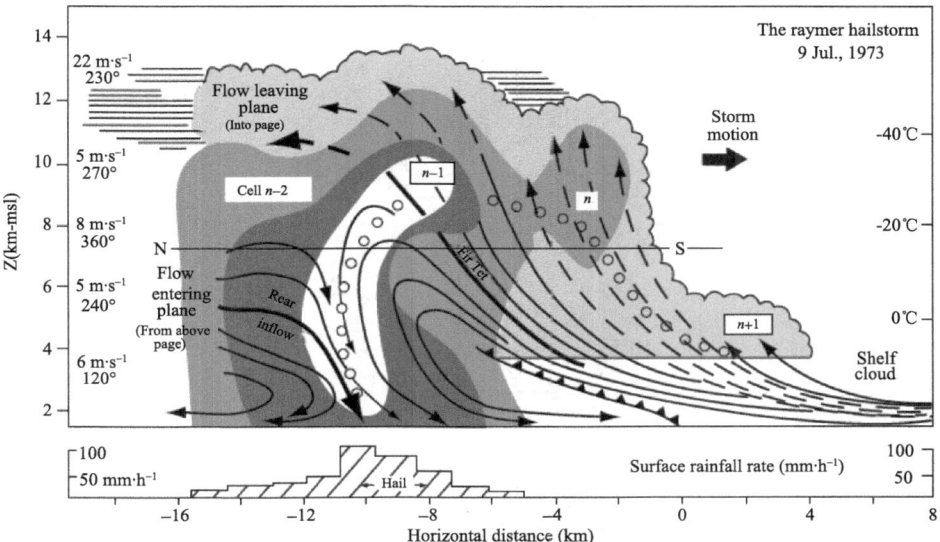

Fig. 5.3 A schematic diagram for a multicell storm in the vertical plane along the direction of the storm's motion. This storm has been referred to as the Raymer hailstorm. A series of convective cells, denoted as $n-2$, $n-1$, n, and $n+1$, were generated at the gust front and moved to the left as they developed. The solid line represented storm-relative streamlines on the vertical plane; the broken lines on the left and right sides of the figure represent flow into and out of the plane and flow remaining within a plane a few kilometers closer to the readers, respectively. FTRJET and RTFJET stand for front-to-rear jet and rear-to-front jet (rear inflow) respectively and are denoted by thick streamlines. The open circles represent the trajectory of a hailstone during its growth from a small droplet at cloud base. (Adapted from Browning et al., 1976)

tions then the cells will be in an unorganized form.

In the organized multicell storm every individual cell moves roughly along the mean wind direction, while the integral storm may deviate from the mean wind direction since every cells in the storm have their own developing processes. The storm's motion and propagation characteristics are shown in the Fig. 5.5. As shown in the figure, there are three different types of propagation of the individual cell in the multicell storm: ① individual cell propagates to the left side of the mean wind direction; ② individual cell propagates to the right side of the mean wind direction; ③ individual cell propagates along the mean wind direction.

The tornadoes with short lifecycle may be found in the vicinity of the gust front

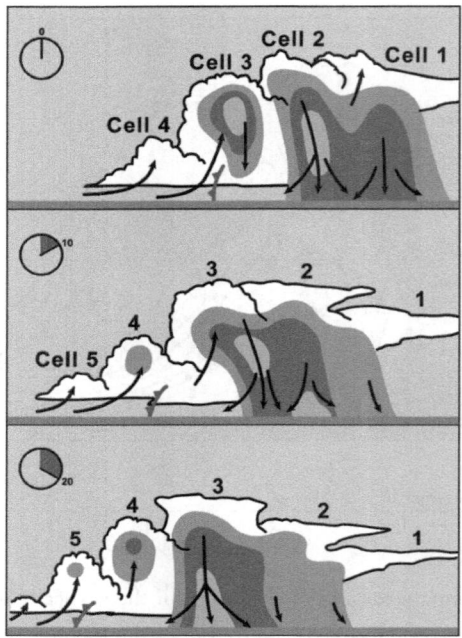

Fig. 5.4 Instant variation of the four cells in the multicell storm. Top: initial state; middle: the state after 10 minutes; bottom: the state after 20 minutes. (Adapted from Marwitz, 1972)

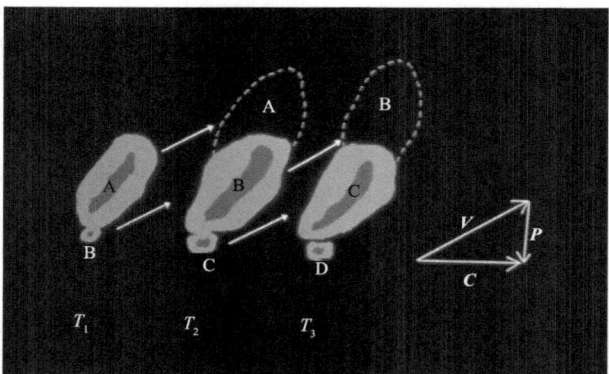

Fig. 5.5 A conceptual model of the movements of the single cell and the bulk of the multicell storm based on the analyses of the multicell storm cases. (Adapted from Marwitz, 1972) A, B, C, D are the convective cells, T_1, T_2, T_3 are the successive time, $\bar{V}, \bar{P}, \bar{C}$ indicate the environmental wind vector, the propagate vector and the storm motion vector respectively.

of the multicell storm. Hails can occur in the central area of the strong updraft. When the storm moves slowly the local heavy rain and flood may be formed.

5.1.3 Supercell storms

The supercell storms imply the huge single cell severe thunderstorm systems with diameter of 20—40 km, lifetime of several hours and strong weather intensity. They are much giant and permanent and violent than the common thunderstorm cells. They rotate distinctly around the vertical axis and posses a nearly steady and highly organized inner circulation. They propagate forward and can move a distance as long as several hundred kilometers. According to the radar observation, the supercell storms possess following distinct characteristics: ① the vault or the Bounded weak echo region (BWER), the overhanging echo and the wall echo on the range-height indicator (RHI); ② the hock echo on the plane position indicator (PPI) as shown in Fig. 5.6, Fig. 5.7 and Fig. 5.8.

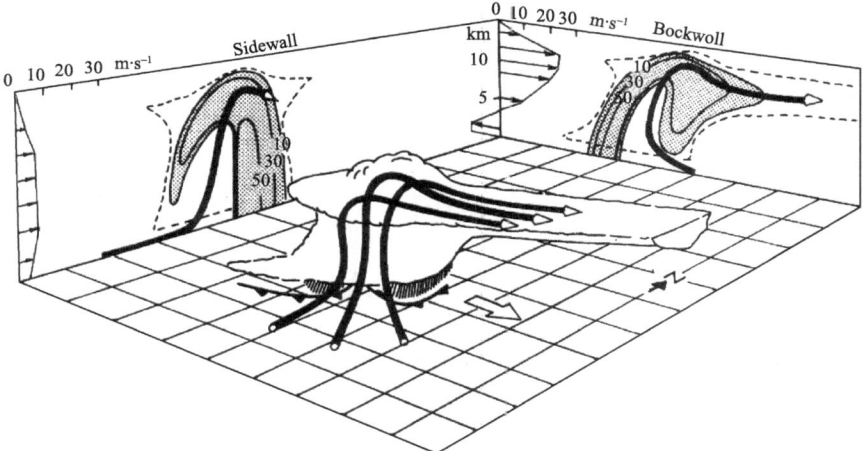

Fig. 5.6 A three dimensional structure diagram of a supercell storm.
(Adapted from Chisholm and Renick, 1972)

The vault represents the area of strong updraft. Since the updraft here is very strong, the water droplets will be quickly carried away by the updraft before they grow so that the radar echo here is weak. A cone-shaped weak echo region (WER) is normally called as a bounded weak echo region (BWER), where the updraft rotates strongly around the vertical axis. Nearby the weak echo region there is a strong echo column which represents a strong precipitation area. Between the WER and the

• 134 •   An Introduction to Mesoscale Meteorology

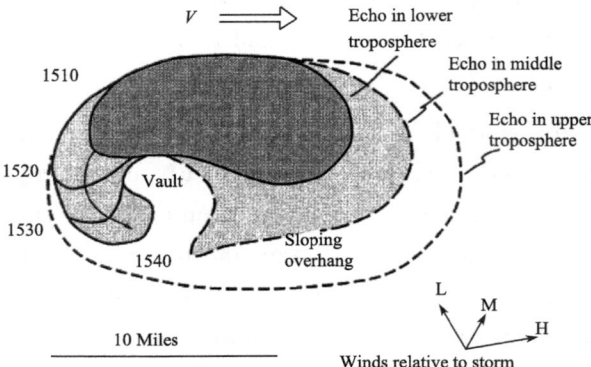

Fig. 5.7 Evolution of the hook echo in an Oklahoma supercell on 26 May, 1963 studied by Browning. (After Browning, 1965b)

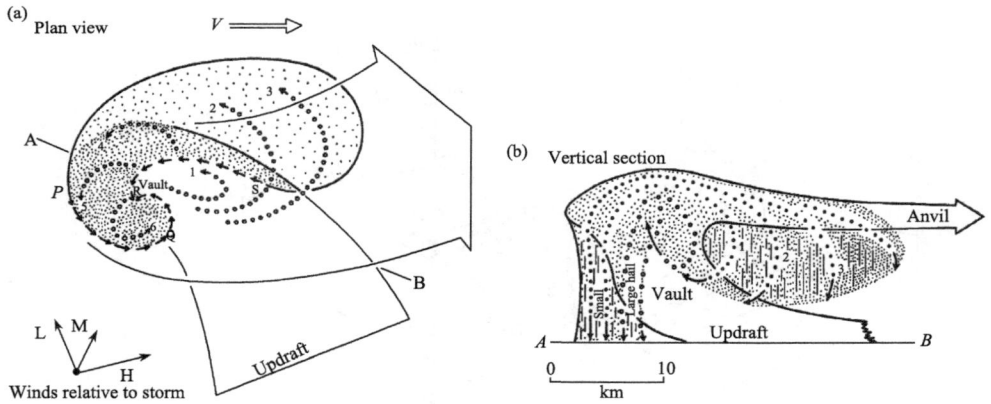

Fig. 5.8 A plan surface view of a severe storm at steady stage with moving velocity V and the vertical section along line AB. (The small hollow circles indicate the trajectories of the precipitation particles) (After Browning, 1976)

strong echo column there is a strong gradient of the radar reflectivity. The strong echo region spreading forward above the WER is called overhanging echo, in which there are abundant hail embryo. The supercell storms can stretch in vertical direction taller than 12—15 km above surface. Their horizontal scales can be larger than 20—40 km. The top of the cloud presents a large smooth round shape dome indicated the updraft in the cloud is not changed with time distinctly. The above typical features of the supercell storm are all clearly expressed in Fig. 5.9.

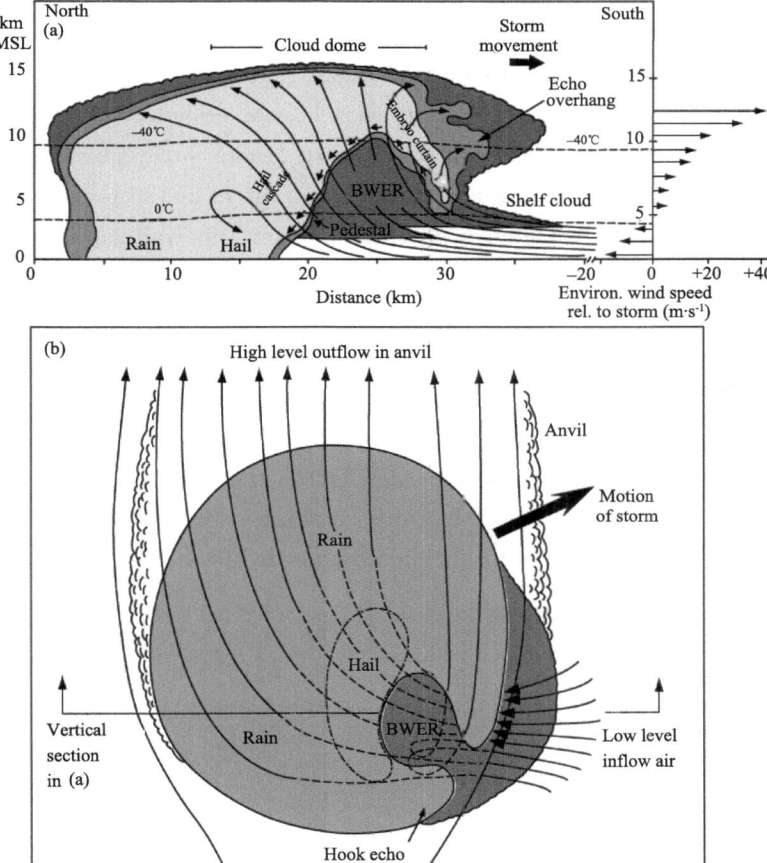

Fig. 5.9 Vertical and horizontal structure of a mature supercell storm. (a) The vertical cross section is taken along the direction of storm movement through the center of the main updraft as depicted in (b). The shaded regions represented two levels of radar reflectivity. Areas of weak echo region or bounded weak echo region (BWER) are shaded. Arrows to the right of the figure indicate the environmental wind relative to the storm movement. Arrows within the figure denote projections of streamlines of airflow relative to the storm movement. (b) The horizontal view at the height of 5.2 km of (a) seen from above. The major region of radar reflectivity in (b) is light shaded. (Adapted after Browning and Foote, 1976)

5.1.4 Tornadic storm

Supercell storm is a kind of the most violent convective storm. It can often give rise severe weather such as surface damage wind, massive hailstones, and tornadoes etc. Following is a discussion about the tornadic supercell storm, the supercell can produce tornadoes.

The tornadic supercell has distinct rotation. Fig. 5.10 depicts appearance of the storms. It can be seen from Fig. 5.10 that there is an overshooting cloud top at the area where strongest updraft located in the centre of the supercell storm. The anvil cloud stretches forward. There is a rotational wall cloud at the bottom of the cloud. The tornado funnel cloud stretches downward from the wall cloud to the surface ground. The Fig. 5.11 denotes the surface features of the tornadic supercell storm. It can be seen from Fig. 5.11 that at the location that the hock echo located there is a mesoscale wave similarly with a synoptic scale occlusion wave, which is a strong circulation related with the surface mesoscale cyclone. The locations where marked with FFD and RFD are the downdraft areas in the front side and rear side of the storm respectively. The location where marked with UD is the position of the updraft. The tornadoes are usually formed near the occlusion point in the edge of the hock echo and in the transition zone between updraft and downdraft but in the updraft.

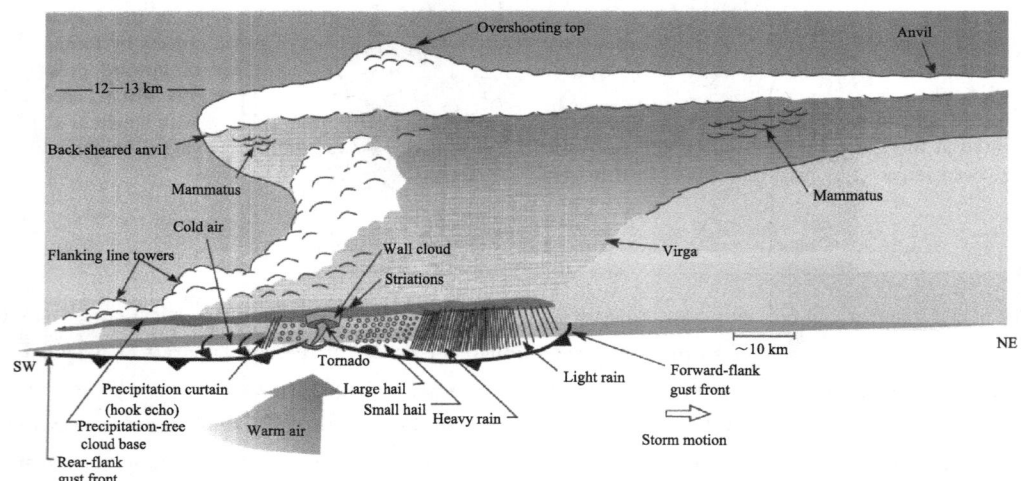

Fig. 5.10 The appearance of a supercell storm accompanied with tornado. Motion of the warm air is relative to the ground. (After Houze, 1977, 2014; Wallace, 2006) (see the color illustrations)

As discussed above, the mesoscale cyclone near the hock echo of the supercell storm is the area easy to form tornadoes, therefore the mesoscale cyclone is called tornadic cyclone or the tornado nest. The tornado refers to the high speed rotational fun-

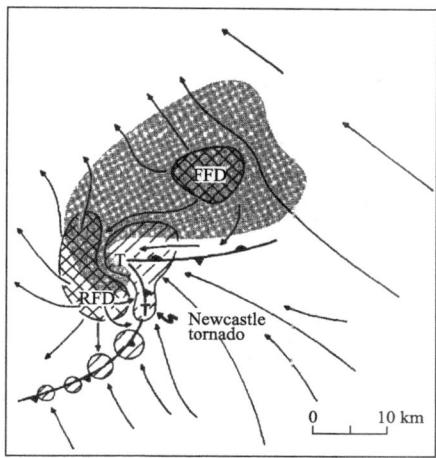

Fig. 5.11 The plan sketch of the surface structure of the tornadic storm. The gray shade encompasses the radar echo. The gust front structure is depicted using a solid line and frontal symbols, Surface positions of the updraft are hatched. Forward-flank downdraft (FFD) and the rear-flank downdraft (RFD) are crosshatched. Streamlines relative to the ground are also indicated. Likely tornado locations are shown by the encircled T's. The location of the Newcastle tornado is indicated. (Adapted after Lemon and Doswell, 1979; Wakimoto et al., 1996)

nel-shaped cloud column stretched downward from the bottom of the thunderstorm cloud. It is a violent small vortex. Its inner structure is similar to a typhoon. It is a small depression. There are downdraft in the centre and updraft in the walls. The intensity of the vertical velocity can be as great as $45-100$ m·s^{-1} or more. The diagram in Fig. 5.12 shows the pressure distribution and the vertical circulation in a tornado.

As said above that the tornado is a fast-rotating micro-cyclone accompanied by a damaging wind on or near the ground and nearly always observable as a funnel cloud pendant from the parent cloud. The most destructive winds in the tornado occurs inside the swatch of one to six suction vortices which orbit around the tornado center. The suction vortex is a mososcale vortex embedded inside a tornado, sometimes observable as a swirling column of dust or debris with or without a funnel cloud inside. A traveling suction vortex produces a narrow swath of extreme winds, often leaving behind a low pile of small debris along the locus of the vortex center on the ground. Some transient suction vortices remain stationary on the ground while the parent tor-

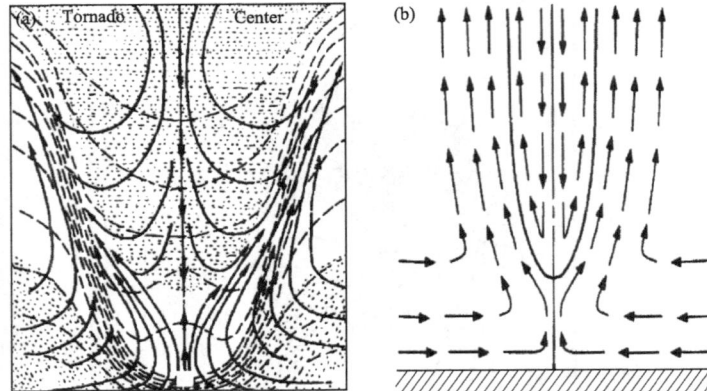

Fig. 5.12 The pressure distribution (a) and the vertical circulation (b) in a tornado funnel. (The dashed lines indicate the isobars and the arrow lines are the stream lines) (After Davics-Jones, 1986)

nado aloft keeps traveling. The suction vortexes rotate around the vertical axis of the tornado (Fig. 5.13 and Fig. 5.14). When the vortex suctions pass through their moving trajectories will be recorded on the surface by leaving traces (Fig. 5.15). A schematic diagram of the mesoscale cyclone with two tornado vortexes and the damage path caused by the tornadoes is presented in Fig. 5.16.

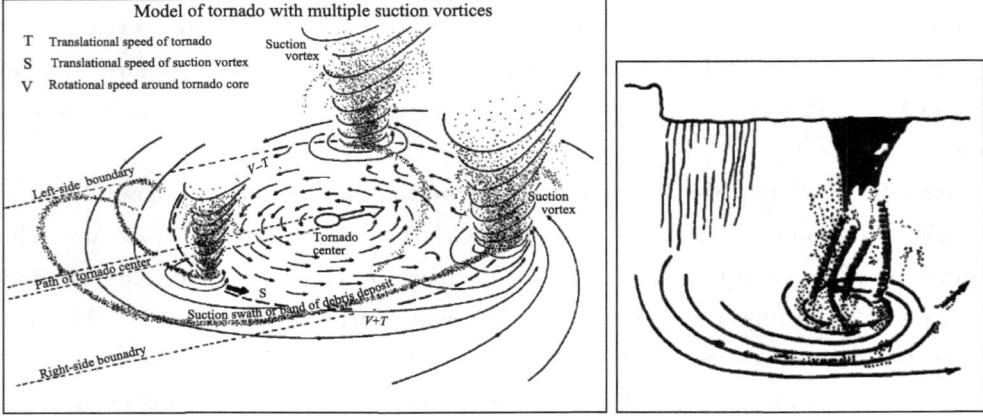

Fig. 5.13 Fujita's conceptual model of multiple suction vortices within a tornado. (After Fujita, 1971)

Chapter 5　Mesoscale Convective Systems　　• 139 •

Fig. 5.14　Six suction vortices in various stages of development Courtesy of Mr. Floyd Styles, Wichita Falls, Texas. (After Fujita, 1981)

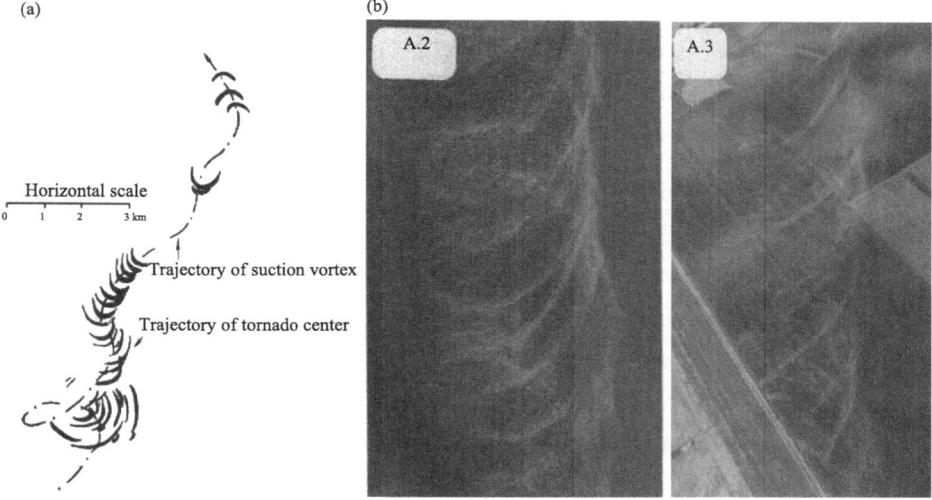

Fig. 5.15　(a) The model of trajectory of the suction vortex of tornado; (b) The examples of ground marks left behind by suction vortices embedded inside tornadoes. Location and dates of occurrences are Magnet, Nebraska tornado, 6 May, 1975 (A. 2) and Homer Lake, Indiana tornado, 3 Apr., 1974 (A. 3) respectively. (Adapted from Fujita, 1981)

To sum up, a tornado storm may include several vortexes in different scales. The situation of the vortexes coexisting is delineated in Fig. 5.17. From the diagram it can be seen that a larger cyclone may include several mesoscale cyclones, which may include several tornadoes rotated around the axis of cyclone, in a tornado there are possibly

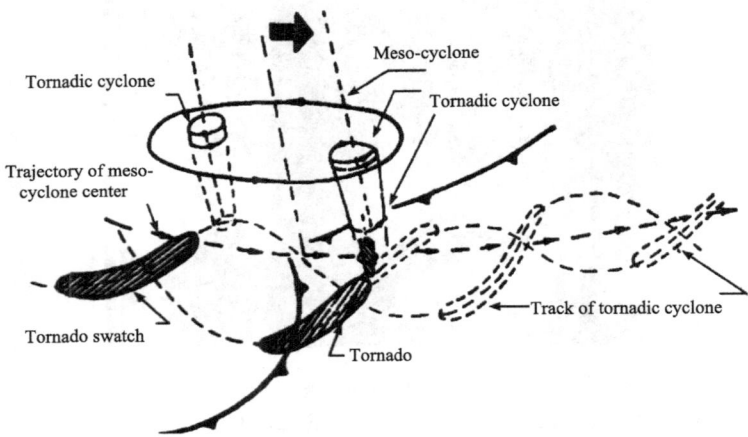

Fig. 5.16 A schematic diagram of the mesoscale cyclone with two tornado vortexes and the damage path caused by the tornadoes. (After Snow and Agee, 1975)

several vortex suctions rotated around the axis of tornado.

5.1.5 Downburst

(1) The features of downburst

When the convective storm developed into mature stage, it can generates very strong cold downdraft and damaging winds. The wind speed on surface can be greater than 17.9 m · s^{-1} (~8 Beaufort wind scale). This kind of local severe downdraft outflow may be called as downburst.

According to Fujita (1981), downburst may be defined as a strong downdraft which induces an outburst of damaging winds on or near the ground. The damaging winds, either straight or curved, are highly divergent. Twisting downbursts characterized by curved streamlines are often induced by a rotating thunderstorm.

The downbursts can be further divided into mesoscale and small scale etc. The small downburst is also called microburst, which can be straight or curved airflow. Some microbursts are embedded inside a mesoscale downburst while others are induced by small convective storms with rain or virga.

Downbursts are the dangerous weather phenomena, especially for the aviation. When the aircraft take off or landing the downbursts may cause the serious accident events. In aviation, a localized downdraft with its downflow speed exceeding the climb or descent rate of an aircraft (3—4 m · s^{-1} at 100 m), and the crosswind or tail-

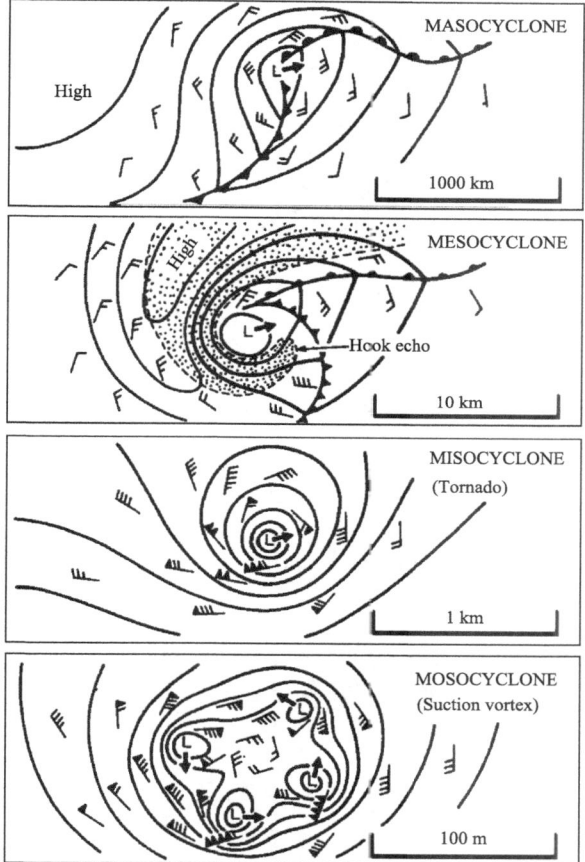

Fig. 5.17 Schematic drawings showing the features of maso-, meso-, miso-and moso-cyclones. (After Fujita, 1981)

wind shear in outburst winds of a strong microburst will endanger the aircraft operations near the ground.

The downdraft of the downburst is normally accompanied with rotation and will cause horizontal divergent flows. When the downdraft reached on surface ground, the airflows will be rolled upward to cause a horizontal axis vortex as shown in Fig. 5.18.

A microburst is a misoscale downburst with its horizontal dimension <4 km. A photograph of microburst is shown in Fig. 5.19a. From the picture it is seen clearly that the dust whirl which is whirled up by the horizontal vortex is in front of the downburst. Fig. 5.19b shows another ground microburst photographed in Kansas on

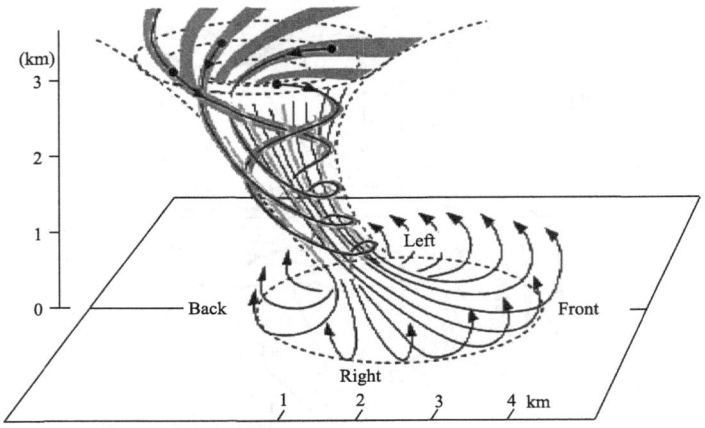

Fig. 5.18 A schematic diagram of the three dimensional structure of micro downburst. (After Fujita, 1985)

1 Jul., 1978. It is seen in the picture that the descending air spreads out violently on reaching near the ground.

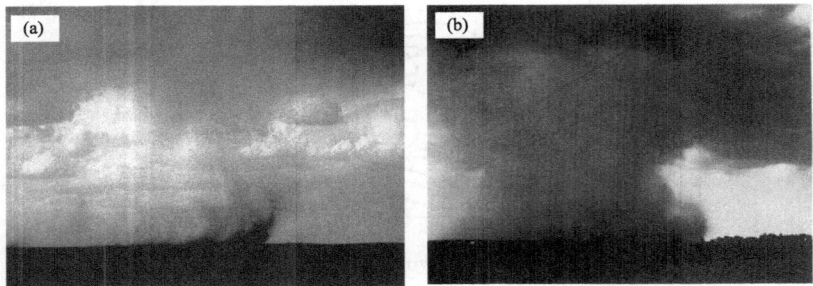

Fig. 5.19 (a) A photograph of microburst shows the dust whirl in front of the downburst (After Lemon et al., 1979); (b) A photograph of a microburst in action on 1 Jul., 1978, Courtesy of Mr. Mike Smith. (After Fujita, 1981)

According to the observations, the impacting area of a downburst can be about several tens meters to several hundred km. Fujita (1981) divided the downbursts into three types, i. e. micro scale, mesoscale and large scale (macro scale) and further divided them into 5 scales, i. e. the micro-β scale microburst, micro-α scale microburst, meso-β scale downburst belt, meso-α scale downburst and macro-β scale downburst family.

Where, the horizontal scale of the microburst is about 0.4—4 km, including micro-β scale microburst and micro-α scale microbursts, their downdrafts at the height of 100 m above surface ground can be as strong as 10—100 m·s^{-1} and the surface wind speed can be greater than 22 m·s^{-1}. The horizontal scale of the mesoscale downburst is about 10—100 km, including, meso-β scale downburst belt, macro-β scale downburst, their downdrafts at the height of 100 m above surface ground can be as strong as 1—10 m·s^{-1} and the surface wind speed can be greater than 18 m·s^{-1}. When a convective storm moving several hundred kilometers can form a macro-β scale downburst family. So that every macro-β scale downburst family may include several macro-β scale downburst; every macro-β scale downburst may include several meso-α scale downburst belts; every meso-α scale downburst belt may include several micro-α scale downbursts, and every micro-α scale downburst may include several micro-β scale downburst swath.

On the pressure field, different scale of the downbursts are corresponding to different high pressure systems. The straight line winds intersect with isobars. The front edges are the discontinuity lines in different scales. They are called as front, gust front, front of downburst and the front of the burst swatch respectively (see Fig. 5.20). When the different scale of fronts pass the station, they can bring local sudden strong surface wind and when they move forward they can successively trigger new convective cells in the front of themselves.

(2) The formation and monitoring of downburst

The formation of downburst is related with the collapse of the overshooting of convective cloud top. According to the satellite image analysis it can be seen that there is an overshooting dome at the cloud top when the convective storm developed into mature stage. This is due to the air particles are carried by strong updraft flow into the stable layer near the tropopause or above. During the overshooting process of the updraft, it will obtain the horizontal momentum from upper level atmosphere. The kinetic energy of the updraft will be changed into potential energy and stored at the top of cloud, on which air becomes heavy and cold. When the cloud top collapsed, the potential energy will be released and changed again into the kinetic energy of the downdraft.

The collapsing of the cloud top is related with the motion of the squall front underneath the storm cloud and the downburst front as well as the front of burst swatch. After the gust front formation it will move toward the front of the storm and

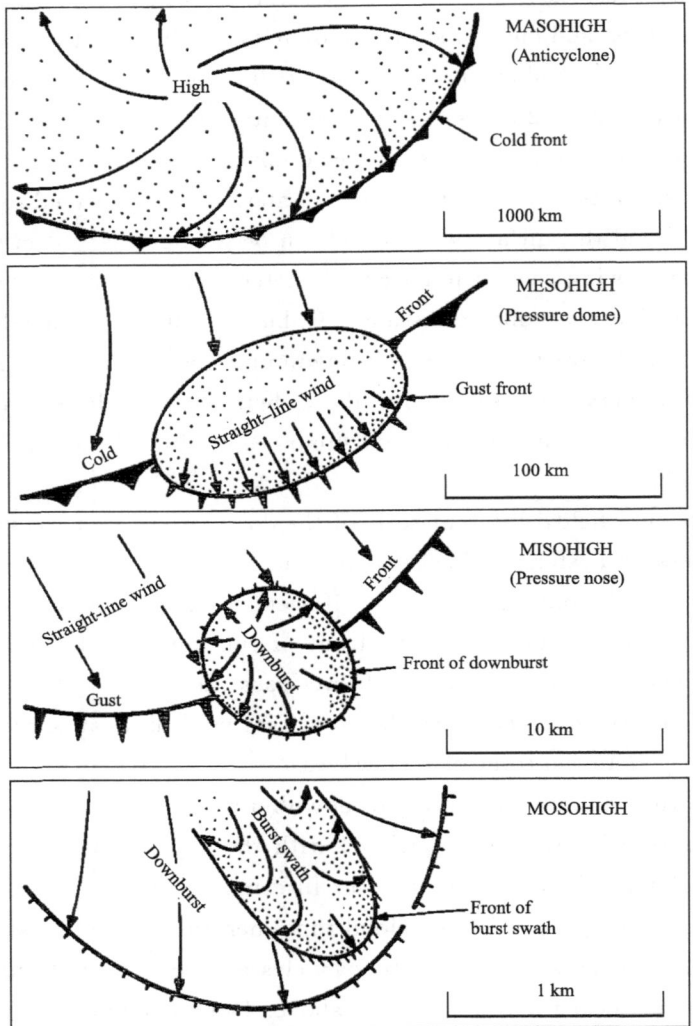

Fig. 5.20 Schematic drawings showing the airflow patterns accompanied by maso-, meso-, miso-and moso-high. (After Fujita, 1981)

gradually departure from the main part of the storm cloud, so that the supply of warm and moist air will be cut off by the gust front and to cause the updraft flows weaken and the cold cloud top collapse to form the downdraft (see Fig. 5.21). Due to the effect of entrainment the high momentum and low moisture air will enter the downdraft to enhance the downdraft and lead to form the downburst.

Chapter 5 Mesoscale Convective Systems • 145 •

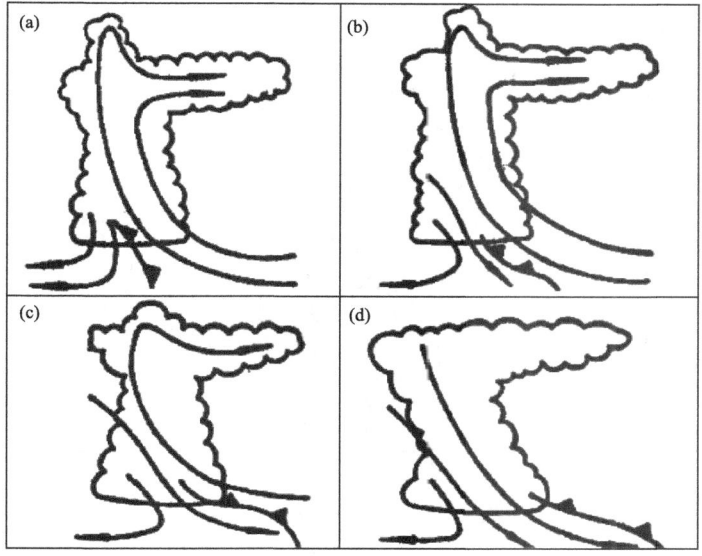

Fig. 5.21 The relationship between the collapsing of overshooting of cloud top and the motion of the gust front. (Adapted from McCanned, 1979)

The above process of the formation of downburst can be proved by the vertical section of the air flows in storm cloud based on the Doppler radar observational data as shown in Fig. 5.22. It can be seen clearly from the figure that the collapsing part of the cloud top is just corresponding to the downdraft, which stretched from the top of cloud to the surface ground so that to bring a downburst. Fig. 5.23 shows the distribution of the Doppler velocity in a travelling microburst in the warm sector. From the picture it is seen that the maximum horizontal velocity is located over the height about 50 m. An aircraft attempting to penetrate an outburst flow will encounter a tremendous increase in the tailwind as it flies out the high-wind core.

A stationary microburst descends to the ground in its contact stage. As the cold air brought down by the microburst accumulates on the ground, a cushion of the cold air prevents the descending air from reaching the surface. As a result, the height of the extreme wind above the ground increases with time. Meanwhile, the intensity of the extreme wind decreases as the spreading depth increases (Fig. 5.24).

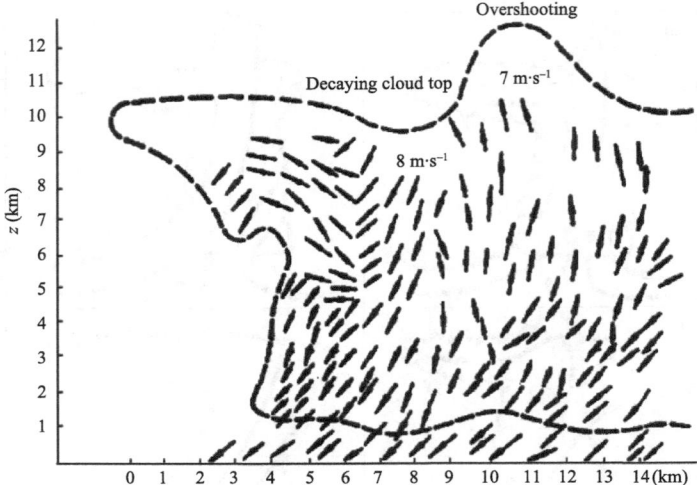

Fig. 5.22 The vertical section of the airflow in storm cloud measured by Doppler radar.

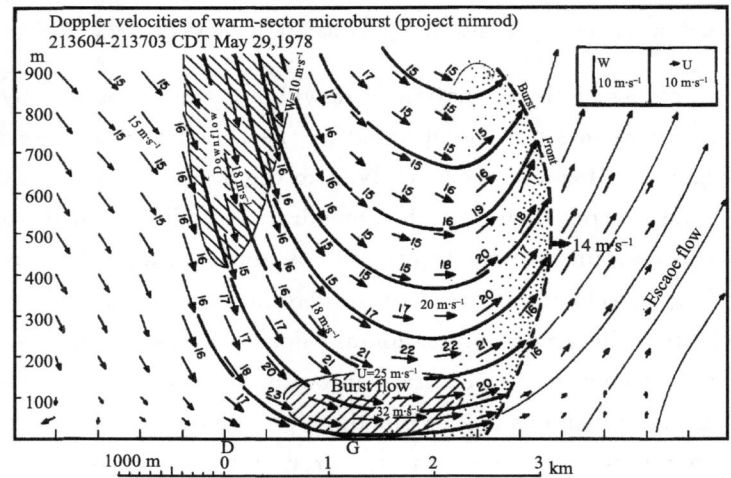

Fig. 5.23 Doppler velocities of a warm sector traveling microburst. The maximum wind of the horizontal outburst flow is located 50 m high above the ground. (After Fujita, 1981)

A traveling microburst, on the contrary, keeps descending in advance of the cold air which is left behind to form a mesoscale pressure dome. It is likely that the slanted surface of the cold air deflects the outburst winds forward (Fig. 5.25).

Chapter 5 Mesoscale Convective Systems • 147 •

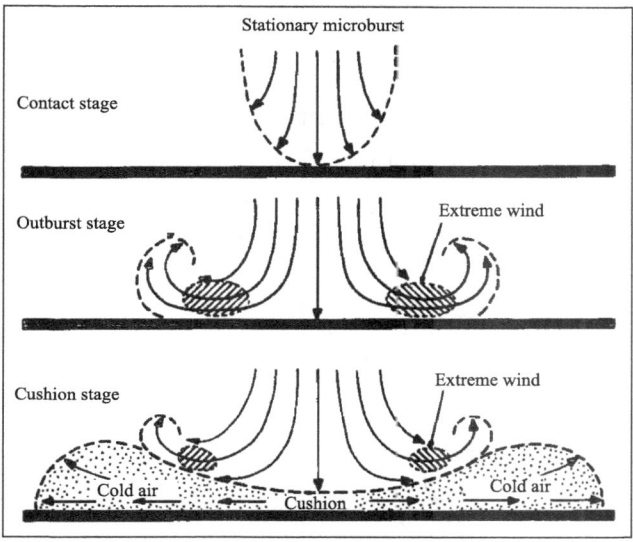

Fig. 5.24 Schematic diagram of a stationary microburst descending onto a cushion of cold air which weakens the outburst winds near the ground. (After Fujita, 1981)

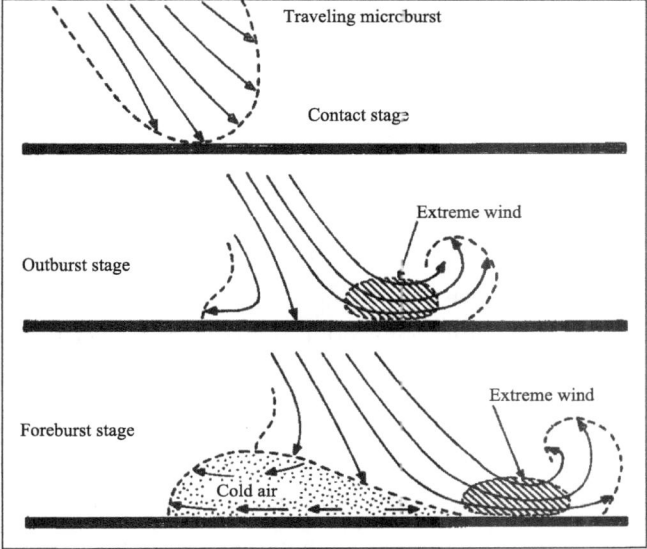

Fig. 5.25 Schematic diagram of a traveling microburst accompanied by an extreme wind near the ground. (After Fujita, 1981)

The downburst can be monitored by radar. There are two types of radar echoes of downbursts, i. e., the hock echo and the bow echo as shown in Fig. 5.26. Downbursts are normally found inner or around the hock echo. The hock echo can usually be found by the plan position indicator (PPI) of radar scanning the low level below 3 km. The hock echo reflects the location of the strong rotationally updraft in storm cloud. The strong downburst is normally found near the location of the bow echo forward, where the echo moves much faster than its two sides, so that the arch-shape presented.

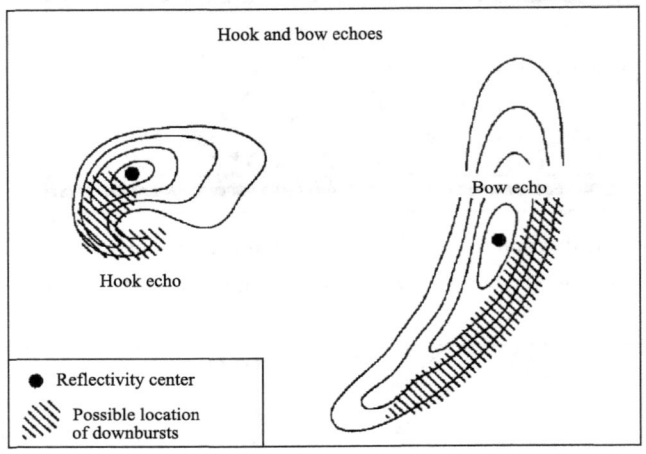

Fig. 5.26 Hook and bow echoes commonly observed during downbursts. (After Fujita, 1978)

A typical morphology of radar echoes associated with a bow echo, as envisioned by Fujita (1978), is presented in Fig. 5.27. It can be seen that the system usually begins as a single, large, and strong convective cell that may be either isolated or part of a more extensive squall line. As the strong surface winds develop, the initial cell evolves into a bow-shaped line segment of cells, with the strongest winds occurring at the apex of the bow. During its most intense phase (Fig. 5.27c), the center of the bow may form a spearhead echo. During the declining stage, the system often evolves into a comma-shaped echo (Fig. 5.27e). The stream lines show that at the north end of the bow echo, the air flows present cyclonic turning, while at the south end the flows are anti-cyclonic turning. The north end changes gradually into a comma shape, while the south end has no distinct changes.

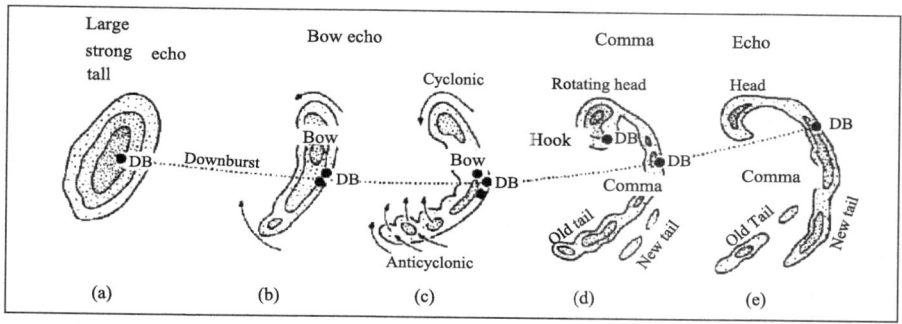

Fig. 5.27 A typical morphology of radar echoes associated with bow echoes that produce strong and extensive downbursts, labeled DB. (After Fujita, 1978)

According to the observations, Fujita (1979) pointed out that the bow echoes can be regarded as the prelude of the strong surface wind, but also the results brought by downbursts. In other words, when the bow echo occurs the process of the downburst has already been progressing. Furthermore, once the bow echo occurs its corresponding mesoscale wind field and convective structure feature will cause the damaging wind intensified and maintained.

The preliminary relationship between high wind and bow echo seems similar to the "chicken and egg" relationship. The model given by Fujita (1979) as shown in Fig. 5.28 shows that the bow evolution is caused by the downburst instead. Namely, a downburst is already in progress when a line echo takes the shape of a bow.

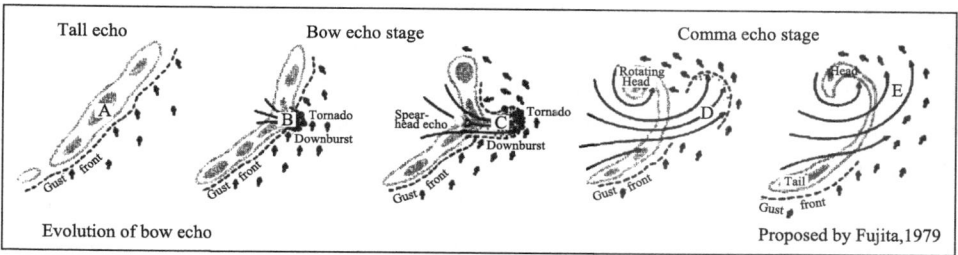

Fig. 5.28 Revised model of the evolution of bow echo. Bow echo in this new model is produced by a downburst thunderstorm as it snowballs, or cascades, down to the ground. This model is contrary to Fujita's (1978) first thought, "bow echo produces downbursts." (After Fujita, 1979)

According to the observation, the bow echoes related with the damaging downburst strong wind are usually accompanied with a strong rear inflow. When the downburst forms, the mid level air flows are accelerated and enter the convective body. The rear inflow leads to the quick motion of the convective body at the central part of the system. This is helpful to form the bow-shape echo. The rear inflows may play the roles to feed dry, higher momentum air into the downdraft, enhancing the strength of the resulting outflow at the surface through vertical momentum transport and increased evaporation. The rear inflows are presented on the radar echo by a rear inflow notch (in short RIN) or the weak echo channel. The bow echo with significant radar reflectivity feature is normally called as the significant bow echo (see Fig. 5. 29).

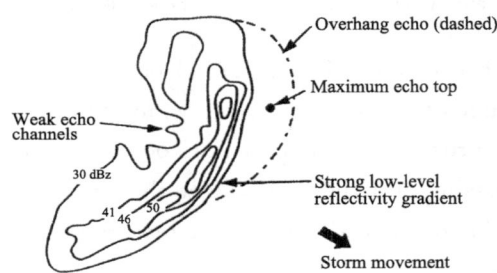

Fig. 5. 29 Radar reflectivity features of the significant bow echo.

The significant radar reflectivity features include: ① There is a high gradient region of radar reflectivity in front edge of the bow echo (at inflow side); ② at the inflow side there is a weak echo region (WER) (in early stage); ③ the top of the echo is located above WER or the high gradient region of the reflectivity; ④ to the rear side of the bow echo there is a weak echo channel or RIN, which means there is a strong inflow at the rear side. The distinct bow echoes imply they have more potential for producing the disasters.

The association between the "weak echo channel" and a rear-inflow jet flow has led to the identification of this feature as a "rear-inflow notch" (Przybylinski, 1995). Fig. 5. 30 highlights many of these attributes for a bow echo observed in central Minnesota on 19—20 Jul., 1983. Przybylinski and DeCaire (1985) also emphasized that while a bow echo is generally organized on a scale larger than a single convective cell, individual severe cells, sometimes supercells, may be contained within the larger-scale structure. However, the frequent observation of bow echoes without apparent supercells suggests that supercell processes are not crucial to the existence of bow echoes.

Chapter 5   Mesoscale Convective Systems   • 151 •

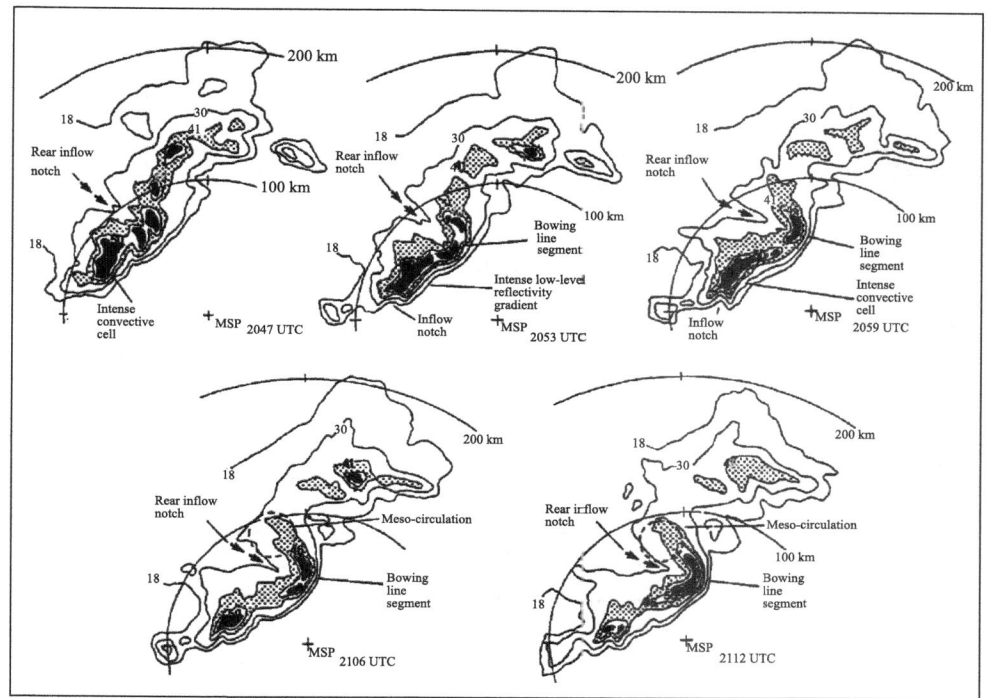

Fig. 5.30 Radar analysis of the central Minnesota derecho between 2047 and 2112 UTC from Minneapolis-St. Paul, MN (MSP). Reflectivity contours are 18, 30, 41, and 46 dBz. Shaded region represents reflectivity values greater than 50 dBz. (After Przybylinski, 1995)

Johns and Doswell (1992) pointed that the downburst may be generally found under the environmental conditions with stronger vertical wind shear. There are various types and scales of the convective storms for forming downbursts. The storms may include the downdraft in the front and back sides of the supercell storm, the downburst corresponding to a larger bow echo, the downdraft accompanied with the wavelike echoes, as well as the downdrafts of a long squall line contained bow echo and wavelike echoes (see Fig. 5.31).

Some downbursts are found companied with rainfall. They are called as moist downbursts. The features of the environment for forming moist downburst are higher convective instability (i. e., upper level dry and low level moist) and weaker synoptic scale forcing. The typical atmospheric stratification is shown in Fig. 5.32.

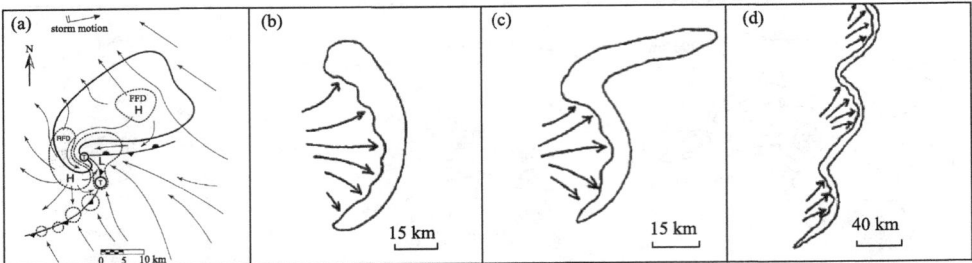

Fig. 5.31 The convective storms contribute to producing downbursts: (a) the downdraft at front and back sides of supercell. FFD = forward-flank downdraft, RFD = rear-flank downdraft, (Lemon and Doswell, 1979); (b) the downdraft corresponding to the larger bow echoes; (c) the downdraft accompanied with wavelike echoes; (d) the downdraft of a long squall line contained bow echoes and wavelike echoes. (After Johns and Doswell, 1992)

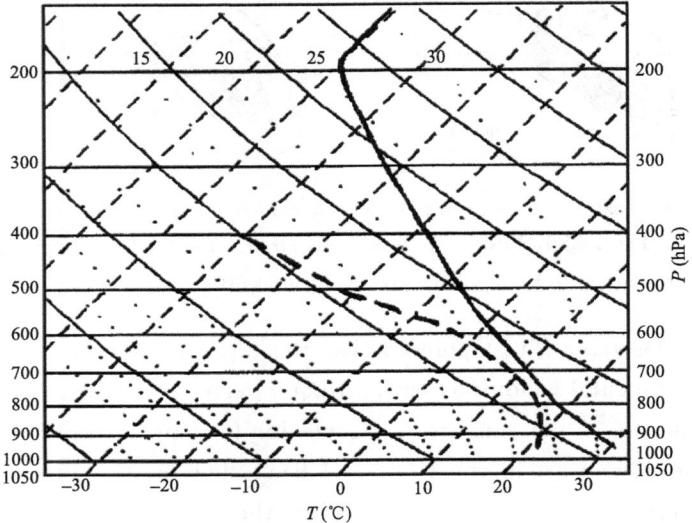

Fig. 5.32 The typical atmospheric stratification favorable to forming moist downburst. (After Atkins and Wakimoto, 1984)

(3) The differences between tornado and downburst

The damaging winds in generally may be divided into two types: tornadoes and straight line winds. But in recent years, the downbursts are regarded as the third type of damaging wind, The differences of the three types of damaging winds caused by severe thunderstorms are as follows: the tornadoes are highly convergent, rotationary damaging winds normally only affecting a narrow area; the straight winds occurred behind squall lines are non divergent straight line wind; the downbursts are highly divergent damaging winds blowing along a straight line or a curvature line.

The pictures in Fig. 5.33 show clearly the convergence and rotation as well as the narrow affecting area of the tornado winds. The pictures in Fig. 5.33 display the wind effects and the affecting ways of the downburst, which shows clearly the features of the downburst winds. Fig. 5.34 displays that tornadoes are companied with downbursts process. Fig. 5.35 shows the Doppler winds distribution, the feature of the vortex structure of the tornado embed in the downburst area.

Fig. 5.33 Examples of wind effects of downbursts and burst swaths. Location and dates of occurrences are: (B.1) northern Wisconsin downburst, 4 Jul., 1977; (B.2) Gessie, Indiana microburst, 30 Sept., 1977. (After Fujita, 1981)

(4) The wind intensity of the systems with different scales

The damaging winds may be induced by different scale of high and low pressures systems. The strong winds induced by high pressure systems include the gust

· 154 ·　　　　　　An Introduction to Mesoscale Meteorology

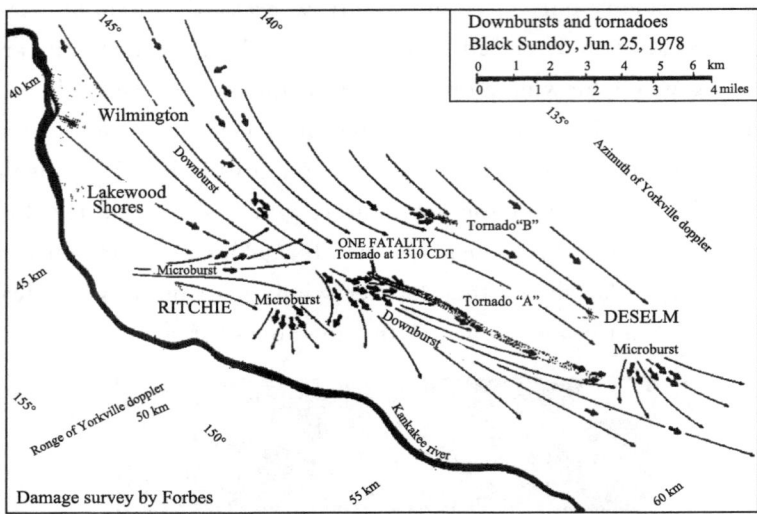

Fig. 5. 34　Downbursts and tornadoes to the southeast of Wilmington, Illinois, near 50 km range and 140° azimuth of The Yorkville (YKV) doppler radar. Damaging winds were on the ground between 1300 and 1320 CDT 25 Jun. , 1978. (After Fujita, 1981)

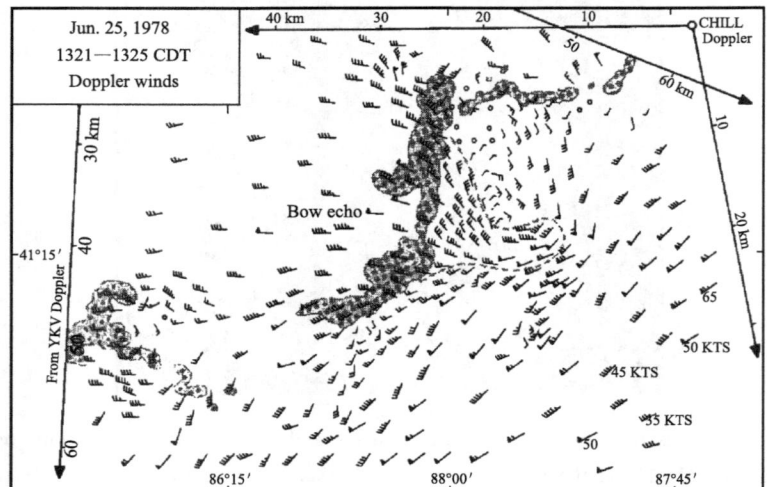

Fig. 5. 35　Dual-Doppler winds on 3. 5 ℃ scan surface of the Yorkville, Illinois, (YKV) radar between 1321 and 1325 CDT 25 Jun. , 1978. The AGL height of the Doppler beam in the high-wind areas is 3. 0—3. 5 km. (After Fujita, 1981)

front wind and the downburst wind. General speaking, when the horizontal scale of the high pressure system decrease the wind intensity will increase gradually. The intensity of the strongest damaging wind induced by micro downburst can be as great as about F3 scale. On other hand, the low pressure systems induced strong winds include the wind of hurricane, the mesoscale cyclonic wind, the winds induced by tornado and suction vortex etc. Generally, the wind intensity will increase gradually when the scale of the low pressure systems decreases. The most strong damaging wind induced by small scale suction vortex may be as great as F5 scale. As shown in Fig. 5.36.

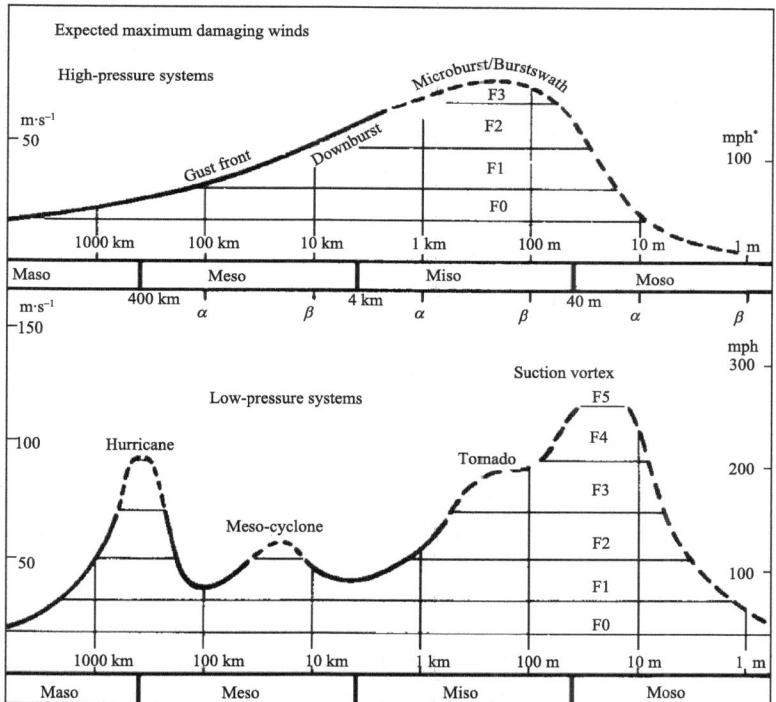

Fig. 5.36 Schematic distributions of expected maximum wind speeds induced by thunderstorms. The maximum winds associated with high pressure systems are seen predominately in misoscale outflow in microbursts and burst swaths. Maximum winds in low pressure systems show several peaks of maximum winds corresponding to hurricanes, mesocyclones, tornadoes and suction vortices. Dashed curves denote unconfirmed wind speeds. Localized high winds beneath mesocyclones are often divergent, associated with twisting downburst. (After Fujita, 1981)

---

* 1 mph = 1.609344 km·h^{-1}.

The F scale (the Fujita scale) mentioned above was defined by Fujita (1971). It may be divided into 6 scales (0—5). When F is 0, 1, 2, 3, 4, 5, the wind will be less than 18, 33, 50, 70, 93 and 117 (m • s^{-1}) respectively. According to the real observation records, the strongest tornado wind scale is $F=5$.

5.1.6 Tornado and downburst outbreak

Although the tornadoes and downbursts are the producers of isolated convective storms, under the certain synoptic situations and weather conditions sometimes the isolated convective storms will successively occur in a larger area, so that the phenomena such as tornadoes, downbursts and hails etc. will occur extensively and frequently in a large area and a long time duration. In the history of the United States used to happen many times of the severe weather events with several tens even several hundreds of tornadoes and downbursts or hailstorms simultaneously occurring in one or several days. These events with extensively and frequently occurring of tornadoes and downbursts etc. are normally called as outbreak of tornadoes and downbursts. The cases such as the outbreak in Apr. 11—12, 1965, the super outbreak in Apr. 2—5, 1974, the outbreaks in Apr. 19, 1996 and in May, 2003 etc. are the most outstanding examples of the outbreaks.

In the tornado outbreak process of Apr. 11—12, 1965 there were total 37 tornadoes occurring in 6 states with impacting track length of 1372 km; In the outbreak process of Apr. 19, 1996, there were 59 tornadoes and 128 reports of damaging winds as well as 239 reports of heavy hails in several states near Illinois State (see Fig. 5.37). During the process of May 3—11, 2003 there were 361 reports of tornadoes. During the super outburst process of Apr. 2—5, 1974 there were 148 tornadoes lasted 20 hours and impacted 13 states.

The above outburst processes are similar to each other in some degree. Here let's take the process of Apr. 19, 1996 as an example to show the background and features of the tornado outbreak.

In the tornado outbreak process of Apr. 19, 1996, most of the tornadoes are produced by the supercell storms, which are generally developed on the situations of dry line, warm front and the occluded front caused by dry line and warm front etc. Fig. 5.38a and Fig. 5.38b show the surface synoptic map at 2100 UTC Apr. 19, 1996 and the GOES-8 visible imagery at 2145 UTC 19 Apr., 1996 respectively. The figures show that the surface cyclone and its cold and warm fronts and dry line as well as the occluded front formed by warm front and dry line. There is a high level jet

Chapter 5 Mesoscale Convective Systems

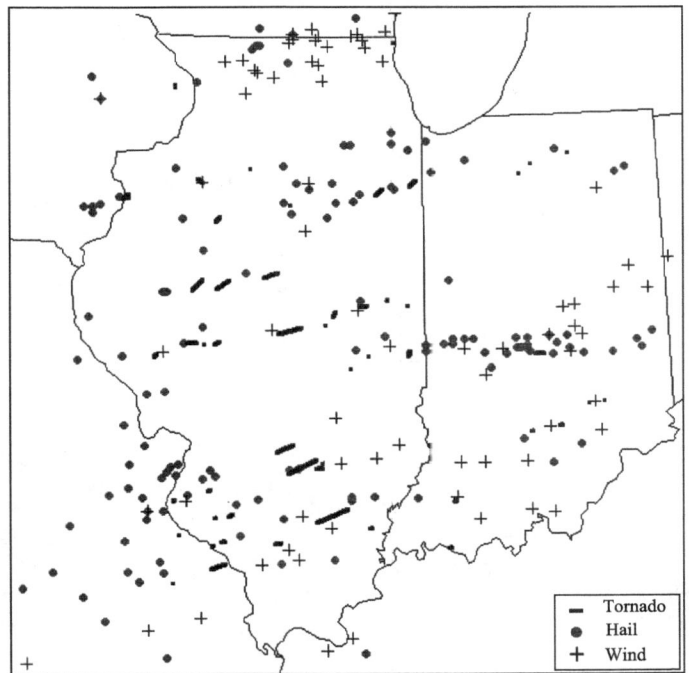

Fig. 5.37 Plot of archived Storm Prediction Center severe weather reports between 2000 UTC 19 Apr. and 0600 UTC 20 Apr., 1996. (After Lee et al., 2006)

stream over the surface cyclone. The severe convection region is located near the exit of the high level jet (see Fig. 5.38c). On the radar echo picture, the severe convective systems are corresponding to the strong supercell (Fig5.38d). Through tracing the 109 cells it is found that 85 original smaller cells are finally evolved into the supercells by merging. Some of the supercells are cyclonic, which have long lifecycle while some may be anticyclonic. The anticyclonic may be induced by splitting process of the storms. The sounding analysis shows that the atmospheric stratification is unstable (Fig. 5.39). Based on the case analyses of the large tornado outbreak, Hamill et al. (2005) proposed a conceptual model of synoptic conditions typically associated with a large tornado outbreak. The triangle area indicates region of expected tornadoes (Fig. 5.40).

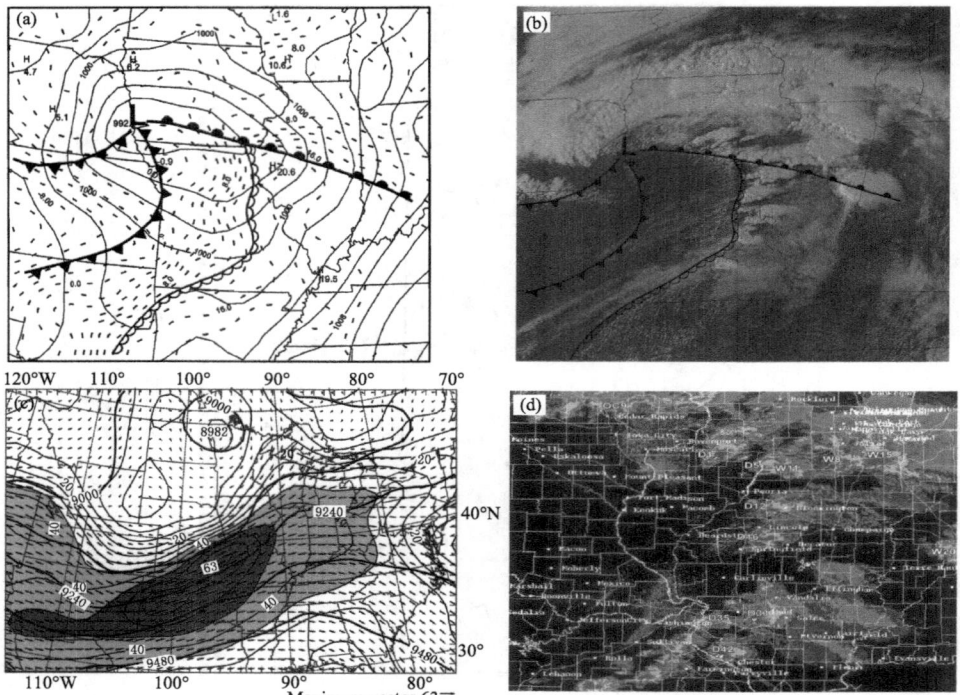

Fig. 5.38 (a) Objectively analyzed synoptic surface pressure and dew-point analysis and subjective frontal analysis for 2100 UTC 19 Apr., 1996. Isobars (isodrosotherms) are shown every 2 hPa (2 ℃); (b) GOES-8 visible imagery at 2145 UTC 19 Apr., 1996 with superposed frontal positions and surface low pressure center at 2100 UTC; (c) Geopotential height field (m, thick lines), isotachs (m · s^{-1}, thin lines), and vector winds at 300 hPa at 0000 UTC 20 Apr., 1996 (from MM5 RAWINS analysis). The light and dark gray shaded areas represent wind speeds greater than 35 and 45 m · s^{-1}, respectively. The maximum vector wind scale is shown in the lower right; (d) Radar reflectivity (0.5°) composite from KILX, KLSX, and KDVN at approximately 0000 UTC. (After Lee et al., 2006)(see the color illustrations)

§ 5.2 Belt-shaped convective systems

The belt-shaped convective systems refer to the mesoscale convective systems which consist of the convective cells arranged in a line. They can roughly be divided into two categories: the squall lines and the mesoscale rain-bands occurred near a frontal cyclone or a typhoon.

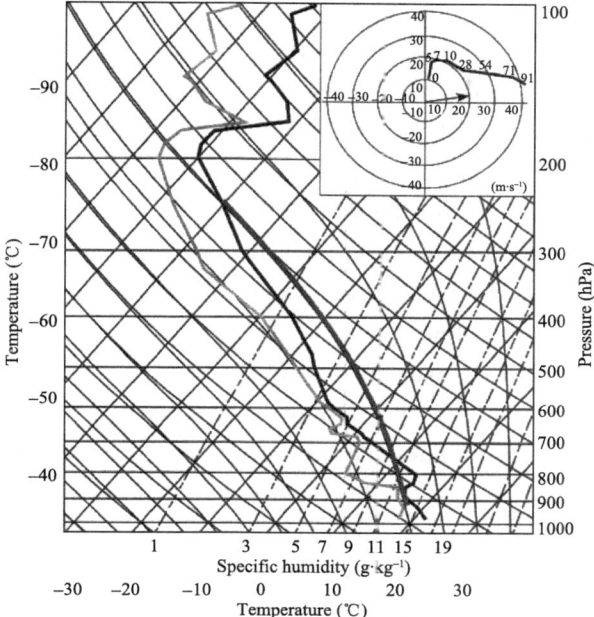

Fig. 5.39　Skew T-lg$p$ thermodynamic diagram of temperature (medium black line) and dew point (medium gray line) and hodograph for the 0000 UTC KILX sounding. The thick dark gray line represents the thermodynamic path for surface parcel ascent upon saturation. Altitudes are plotted on the hodograph (_100 m). The storm motion vector for the I-72 central IL supercell is indicated in the hodograph. (After Lee et al., 2006)

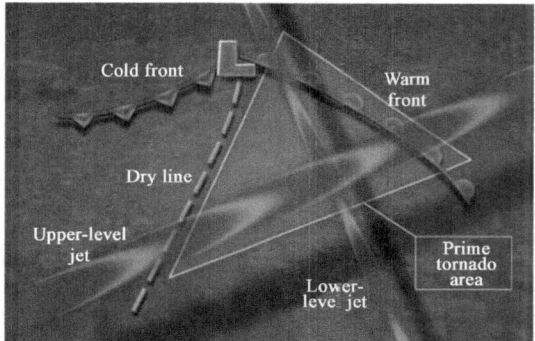

Fig. 5.40　Conceptual model of synoptic conditions typically associated with a large tornado outbreak. The triangle area indicates region of expected tornadoes. (After Hamill et al., 2005) (see the color illustrations)

5.2.1 Squall lines

The designation of squall line occurred as early as to the late of 19 century. In the early stage the squall line is defined as a gust line, which may include a front. For distinguishing the squall line and front, the squall line is defined as a non-frontal narrow active thunderstorm belt or an instability line since the 1950s. According to the configuration of the squall line it may be defined as a belt (or line)-shaped mesoscale convective system. The structures of the squall lines occurred in mid latitude extratropical region and in low latitude tropical region are discussed following respectively.

(1) Extratropical squall line

① Features in surface pressure and wind fields

In extratropical region the typical squall line usually occurs near a front and roughly parallel to the front. The length of squall line is normally in several hundred km and the width is about 50—100 km. The squall line has a four-stage life cycle, which includes the initiation stage, the development stage, the mature stage and the dissipation stage. The four stage's isobar patterns for the squall lines in the cold area, frontal zone and warm area are illustrated in Fig. 5.41 respectively. The structure at mature stage of the squall line in the warm area is illustrated in Fig. 5.42.

The squall line normally consists several segments. Every segment includes several large and isolated storms companied with the surface cold air dome and the horizontal outflow as well as the gust front in the mature stage of the storm. The smaller systems combined together may form a mesoscale thunderstorm high pressure. Beside the mesohigh, the other pressure and wind field systems in the mature stage of the storm also include the presquall mesolow and the wake low as shown in Fig. 5.43, as well as the mesoscale gust front, near which the gradients of temperature, pressure are great and the horizontal shear of the wind velocity and wind direction are strong similar to a front, so that the meteorological parameters will be sharply changed when the squall line passes as shown in Fig. 5.44.

② The formation of the mesoscale pressure systems

As mentioned above, the surface pressure systems of a squall line at mature stage include the mesoscale high, the pre-squall mesolow and the wake low. Following is an brief introduction about the formation of the mesoscale pressure systems based on the paper of Johnson et al. (1988).

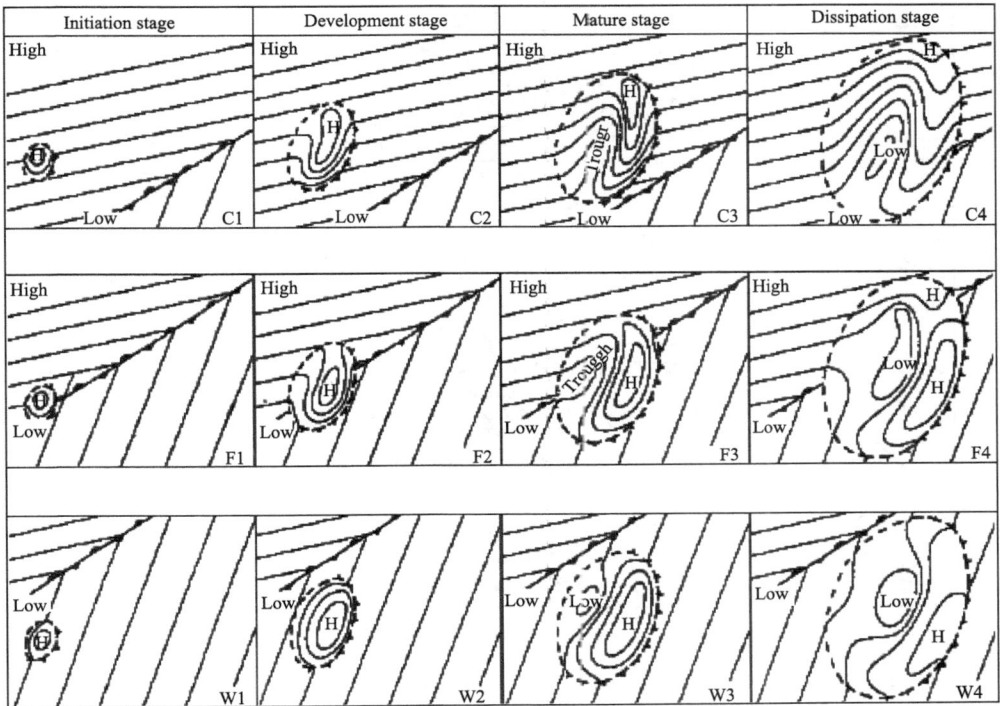

Fig. 5. 41 Typical isobar patterns of squall mesosystems obtained by combining the basic field and excess pressure patterns. Letters C, F, and W designate the basic fields: cold sector, on the front, and warm sector, respectively. The various systems go through stages 1—4 from left to right. (After Fujita, 1963)

About the mesohigh system, Fujita (1959) and many other meteorologists have convincingly demonstrated that the mesohigh owes its existence principally to evaporation in precipitation downdrafts. In this regard, the mesohigh can be considered primarily a hydrostatic phenomenon. However, subsequent studies have pointed out the role of specific nonhydrostatic effects. For example, Wakimoto (1982) schematically shows the relative contributions of hydrostatic and nonhydrostatic pressure to the mesohigh for an idealized thunderstorm downdraft. At the leading edge of the outflow, the gust front, there is a buildup of nonhydrostatic pressure as a result of converging air streams, which explains the surface pressure rises observed prior to the arrival of gust fronts.

• 162 • An Introduction to Mesoscale Meteorology

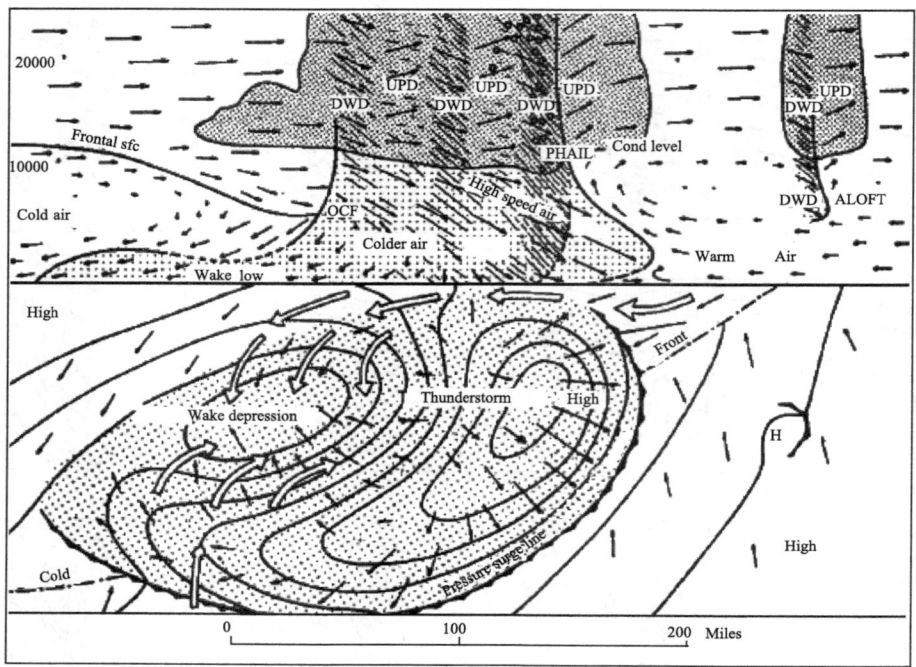

Fig. 5.42 Fujita's early model of squall-line circulation: DWD=downdraft, UPD=updraft. (After Fujita, 1962)

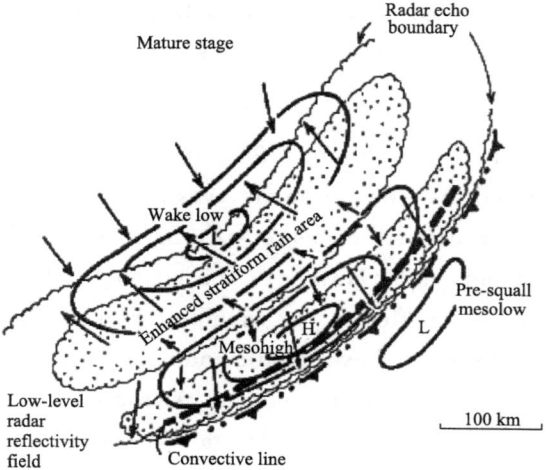

Fig. 5.43 Schematic surface pressure and wind fields and precipitation distribution during the squall-line mature stage. Winds in picture are system relative with the dashed line denoting zero relative wind. Arrows indicate streamlines, not trajectories, representing actual winds. (From Johnson and Hamilton, 1988)

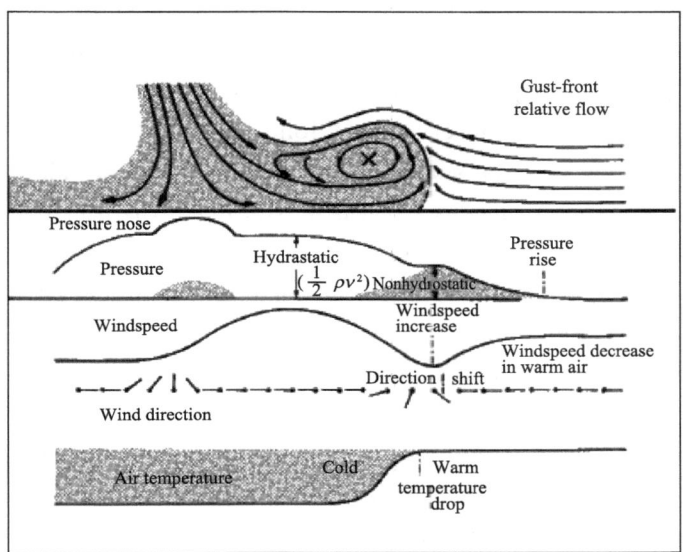

Fig. 5.44　Conceptual model of the surface observations during the passage of a gust front in its mature stage. (After Wakimoto, 1982)

About the presquall low, although not specifically mentioned it in Fujita's papers, a presquall trough or presquall low shows up in a number of his mesoanalyses. This feature (Fig. 5.43) has been attributed by Hoxit et al. (1976) to convectively induced subsidence warming in the mid-to-upper troposphere ahead of squall lines. Observational studies (e. g., Fankhauser, 1974; Sanders and Paine, 1975; Gamache and Houze, 1982; Gallus and Johnson, 1991) have confirmed the existence of presquall subsidence. The modeling study of Fritsch and Chappell (1980) provided strong evidence in support of the explanation of this feature by Hoxit et al. Warm advection may also play some role in the mesolow formation (Schaefer et al., 1985), although it is likely secondary.

Finally, as to the wake low, which has been clearly identified as a prominent feature of mature squall lines by Fujita (1955, 1963), while the dynamical mechanism for its formation has been elusive. Williams (1963) argued that it was a consequence of subsidence to the rear of convective lines, but the processes driving the subsidence were not explained. Zipser (1969, 1977) and Brown (1979) have shown that mesoscale downdrafts driven, in part, by precipitation evaporation can lead to adiabatic warming that exceeds evaporative cooling at low levels, hence producing a pressure fall at the surface. Miller and Betts (1977) argued that the subsidence is dynamically forced by spreading cool air at the surface. Rutledge et al. (1988) found the rear-inflow jet to be a prominent feature of PRE-STORM squall lines. Based on a dense network of surface and sounding observations, Johnson and Hamilton (1988) hypothesized that the wake low was the surface manifestation of a descending rear-inflow jet and that the warming due to the descent was maximized at the back edge of the precipitation area where there was insufficient evaporative cooling to offset adiabatic warming. Strong warming at low levels in the descending rear inflow jet suggests that air is overshooting its level of zero buoyancy.

③Environment conditions of the midlatitude squall line

The structure of the squall line is closely related with environment conditions. In midlatitude there are two common types of squall lines. One of them is formed in a distinct vertical wind direction shear environment. They often occurred in front of the aloft trough and consist of multicell and supercell storms. In the south end of the squall line the new convection is favorable to develop due to the vertical wind shear. So that the squall line will be elongated continuously. While at the north end of the squall line, the old cells are declined and evolved as the stratiform clouds gradually and then stretched toward northeast direction along the aloft wind direction to form a blockbuster of the anvil clouds. The characteristic of this type of the squall line is the anvil clouds stretch forward to the front of the squall line. This kind of squall line is the typical squall line in mid latitude. Fig. 5.45 is the vertical section structure of the squall line. From the diagram, some common features of the squall line such as the tilted updraft flow in mid level, the low level warm and moist inflow and the dry and cold air intrusion in mid level as well as the downdraft flow with low wet bulb potential temperature (θ_W) in the rear of the squall line.

In recent years, another kind of mid-latitude squall line has been noticed. This kind of squall lines are normally occurred in the environment with relative weaker vertical wind shear. In front of the storm there is an ascending inflow stretching from front to rear. while in the rear of the storm there is a descending inflow stretching from rear to front. A wide trailing stratiform cloud region formed due to the old cells declined in the ascending inflow from front to rear. Due to the ice crystals falling down from high level of the convective cloud into the trailing stratiform cloud region and the effect of the ascending motion caused by the secondary circulation in the trailing stratiform region, the precipitation under the stratiform cloud region is still obvi-

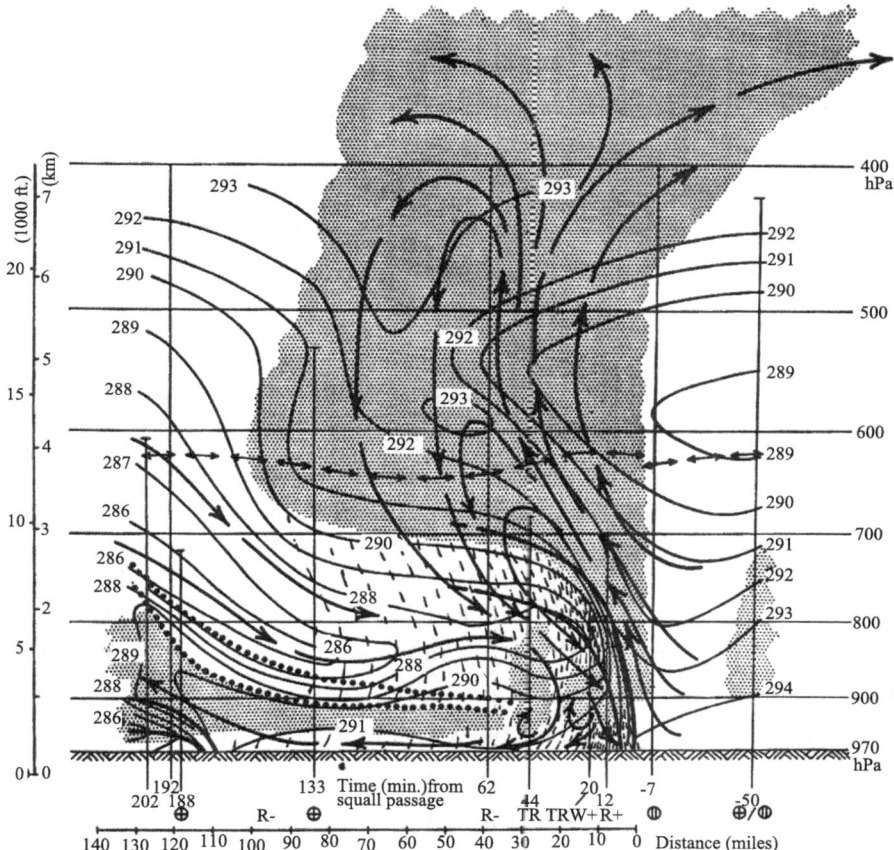

Fig. 5.45 The vertical section of the prefrontal squall line occurred in Ohio of the United States on May 29th, 1947. (The section is along the direction perpendicular to the squall line; the vector lines are the stream lines; the thin solid lines are the isolines of θ_w; the dot lines indicate the upper and lower bounds; the relative dry area is located above the stable layer; the shaded area is the cloud area.) (After Newton and Newton, 1959)

ously (Fig. 5.46).

(2) Tropical squall line system

The tropical squall line system also includes two parts: the convective region and the stratiform region. Where the convective region consists of the mature cumulonimbus belt. In the front of the convective cloud belt the new convective clouds are generated continually. While in the rear of the convective cloud belt the old clouds disap-

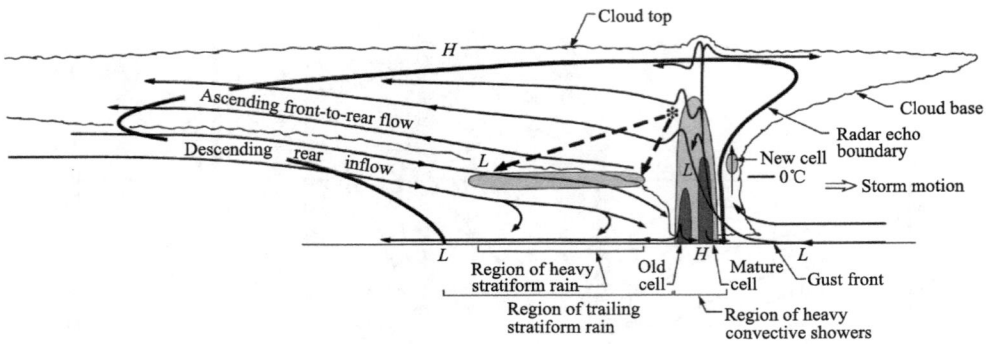

Fig. 5.46 A conceptual model of the kinematics and micro physics of the mesoscale convective system. The middle and heavy shadows indicate the middle and strong radar reflectivity respectively, H and L indicate the positive and negative pressure disturbance respectively, the chain line arrows present the trajectories of the scattering ice crystals passing through the melting level. (After Houze et al., 1989)

pear and a wide trailing stratiform cloud region is formed and presented the special structure with the leading line and trailing stratiform region. The warm and moist air flow in ahead of the leading line flows into the convective clouds of the squall line system. The height of the convective cloud top can be as high as 16—17 km and the range of the stratiform region can be 200 km or more. The precipitation in the stratiform cloud region is horizontally uniform. The major precipitation substances at the upper level are the ice crystal particles, which are from the convective cells on the leading line of the squall line system. When the squall line moves forward the particles carried by the a mid level jet stream stretching from front to rear move backward relatively into the trailing stratiform cloud region. The ice crystal particles drop down and melted to form a melting level, which presents a bright belt in the radar echo picture. Fig. 5.47 shows the illustration of structure of the tropical squall line given by Houze (1977).

The tropical squall line is different from extratropical squall line, it has no steering level. Whole troposphere is an inflow layer. Thus the tropical squall line system has a wide trailing stratiform region. The most strong echo of the tropical squall line system is found in the low part of the convective cell. Moncrieff and Klinker(1997) revealed the characters of the mesoscale convective system such as the curving convective updraft area and the mesoscale inflow under the wide anvil cloud in the rear part of the system etc (Fig. 5.48).

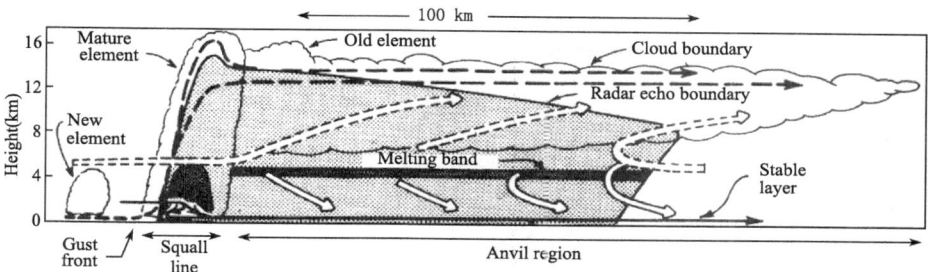

Fig. 5.47 An illustration of the structure of a tropical squall line. (In the picture the dashed single stream lines indicate the convective scale updraft; the dashed double stream lines indicate the mesoscale ascending circulation; the solid double stream lines indicate the mesoscale descending circulation; the light shadow area indicates the weak radar echo; the dark shadow area indicate the strong radar echo in the melting belt and in the strong precipitation area in the mature convective cloud; the corrugated lines indicate the exterior boundary of the cloud.) (After Houze, 1977)

Zipser (1977) proposed an ideal model of tropical mesoscale convective system with the leading squall line and trailing stratiform structure as shown in Fig. 5.49. The air rises in the boundary under the cloud and forms updraft of the convective system.

The ambient air is entrained into the updraft. The air parcels rise in the updraft continuously until they lost Buoyancy force produced by the entrainment or met a stable layer in the environment. The updraft weakened by the entrainment of ambient low θ_e air and the convective scale downdraft is formed and reaches to the surface in the convective precipitation belt area. It should be noted that the system is in three dimension, thus the trajectories of the updraft and downdraft flows are not coincident, and contains a "crossover zone" coexisting the convective scale updraft and downdraft. The conceptual model given by Zipser shows that the air entering the mesoscale convective system from middle layer has very low θ_e, so that it must be subsidence. According to the ideal of the crossover zone, the updraft layer should be allowed let the low θ_e downdraft to pass through.

(3) Three modes of linear mesoscale convective system

Before the 1970s the researches only paid attention to the convection part of the squall line. After that Houze (1977) and Zipser (1977) et al. pointed out that as a mesoscale system the squall line should include two parts, i.e. the convective region

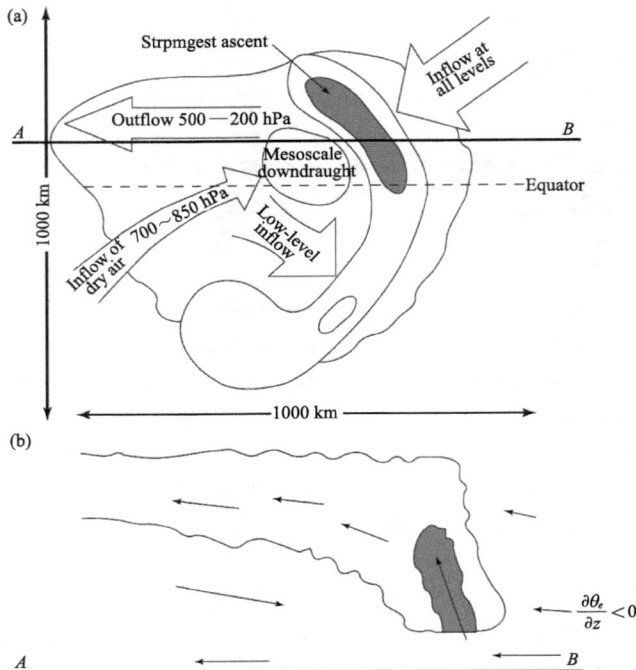

Fig. 5.48 A conceptual model of a large mesoscale convective system cloud cluster occurred over the west Pacific ocean tropical area. (After Moncrieff and Klinker, 1997)

and the stratiform region in radar echo picture as shown in Fig. 5.50.

Parker and Johnson (2004) classified the 88 linear MCSs from May 1996 and May 1997 into three basic categories (i. e. LS, TS and PS). They considered the relative extent of stratiform precipitation ahead of, behind, and parallel to the convective line as the types LS (leading stratiform), TS (trailing stratiform) and PS (parallel stratiform) respectively (Fig. 5.51).

① Trailing stratiform

The TS MCS archetype (Fig. 5.51a) was described by HSD90. The definition of the TS MCS include a convective line, "convex toward the leading edge," with "a series of intense reflectivity cells solidly connected by echo of more moderate intensity." The line has a "very strong reflectivity gradient at the leading edge (i. e. gradient much stronger at the leading edge than the back edge of the convective region)," and a large trailing stratiform precipitation region, often exhibiting a "sec-

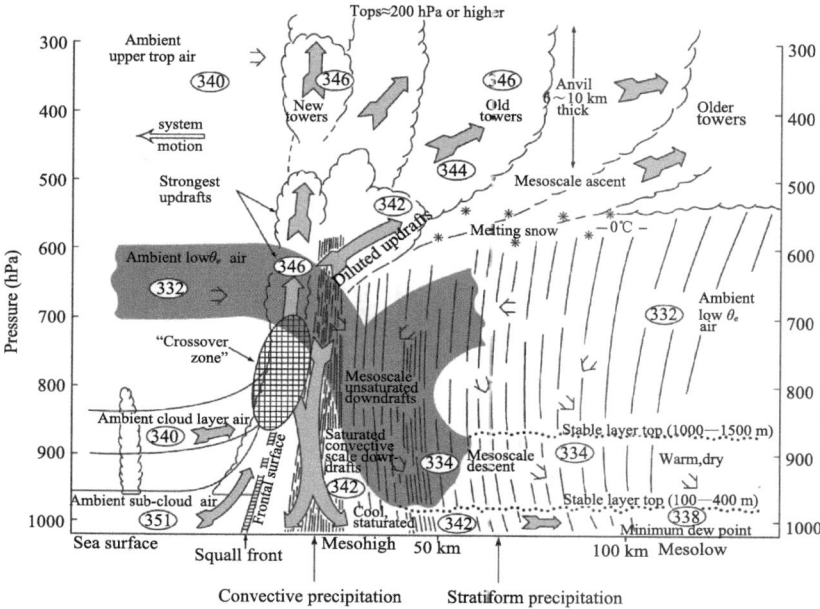

Fig. 5.49 An ideal model of the tropical ocean mesoscale convective system with the leading squall line and the trail stratiform region. (After Zipser, 1977)

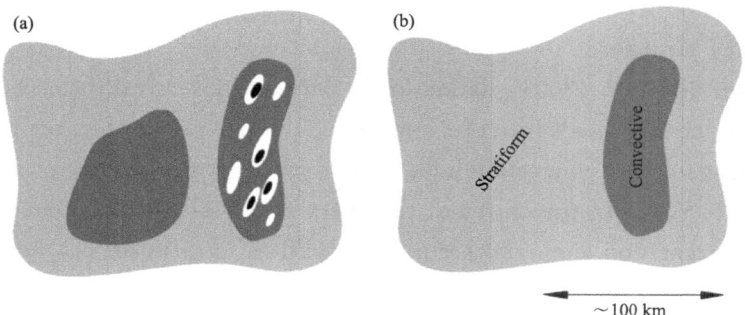

Fig. 5.50 (a) An ideal horizontal diagram of the radar reflectivity; and (b) the convective region and the stratiform region. (After Houze, 1997)

ondary maximum of reflectivity separated from the convective line by a narrow channel of lower reflectivity" (the transition zone). It is relevant to note that the TS archetype depicted by HSD90 exhibited very little leading stratiform precipitation.

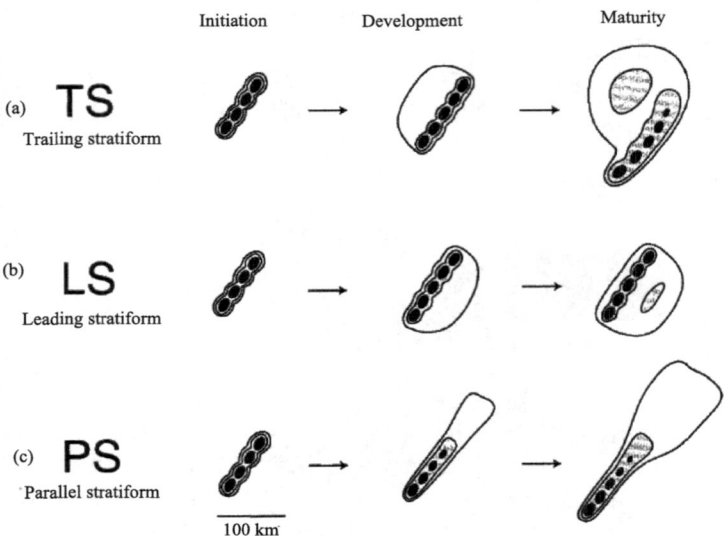

Fig. 5.51 Schematic reflectivity drawing of idealized life cycles for three linear MCS archetypes from Parker and Johnson (2000): (a) leading-line TS; (b) convective line with LS; and (c) convective line with parallel stratiform (PS). Approximate time interval between phases: for TS 3—4 h; for LS 2—3 h; for PS 2—3 h. Levels of shading roughly correspond to 20, 40, and 50 dBz. (After Parker and Johnson, 2004)

② Leading stratiform

Parker and Johnson (2004) define linear MCSs whose stratiform precipitation is predominantly located in advance of a convective line (i.e., in the area toward which the MCS propagates) as convective lines with leading stratiform precipitation (or simply LS) (Fig. 5.51b). In the extreme, members of this class exhibit a convective line preceded by a transition zone and secondary swath of stratiform precipitation with a reflectivity maximum. More frequently, LS MCSs exhibit moderate regions of leading stratiform precipitation without transition zones and secondary bands. The tendency of a MCS to generate predominantly preline precipitation is a criterion for LS classification even if postline precipitation is present. Cases with extensive preline and postline precipitation obviously represent a transition between the two extreme archetypes.

③ Parallel stratiform

Parker and Johnson (2004) deem a linear MCS to have a convective line with parallel stratiform precipitation (or simply PS) if most or all of the stratiform precipitation region associated with the convective line moves parallel to the line itself (in a storm-relative framework) and to the left of the line's motion vector throughout its life cycle (Fig. 5.51c). Very little stratiform precipitation surrounds the convective lines of PS cases. More formally, the reflectivity gradient is relatively large on both sides of the convective line. The stratiform regions' movements (as determined from animations) in PS cases generally deviate less than 308 from the convective lines' (azimuthal) orientations. In some cases, PS convective lines backbuild to the right of their motion vectors. Such behavior may be accompanied by the general decay of convective cells to the left of a line's motion vector, yielding a progressively larger region of lower reflectivity echoes to the left of and parallel to the convective line. In other cases, PS convective lines do not appear to backbuild substantially. Rather, a location of persistent deep convection appears to give rise to the PS echoes via line-parallel advection. Probably both processes are at work to some degree in most PS MCSs.

(4) Formation mechanism of the line-shaped convective system

Squall line is a line-shaped convective system. The formation is very often closely related with line-shaped atmospheric disturbances. When the line-shaped disturbance, for instance a front, approaches an instable area and moves faster than the unstable area, then in the across places that the disturbance line intersects with the unstable boundary, the line occurring sources of severe convection, it is possibly to form thunderstorms. If the storm moves faster than the cold front, the squall line may be formed in front of the cold front as shown in Fig. 5.52.

Except front the triggering mechanisms of squall line in atmosphere still include sea breeze front, dry line, gravity waves, topographic lifting, thermal lifting, low level jet stream, old thunderstorm outflow (arc-like cloud line), mesoscale systems and symmetric instability of the atmosphere etc. The dry line is a discontinuity line of the moisture in horizontal direction. In its one side the air is dry and warm and another side is cold and moist. Across the dry line, the horizontal gradient of dew point temperature can be as high as 5 ℃ $\cdot$ km^{-1}. Over the cold-moist side of the dry line there is usually a capping inversion, also called as warm-dry lid, which may play a role for storing instability energy. Observations show that near the dry line cumulus belts will occur and spread outward very often. This means the dry line is a kind of disturbance source for triggering convections. The dry line can be a triggering mechanism is due to the different radiation situation in the two sides of the dry line. When there is a upper level jet stream over the dry side of the dry line, in the day time the low level wind velocity will increase since the momentum downward transferring.

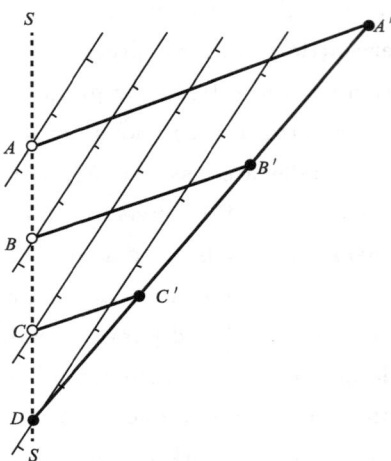

Fig. 5.52 Illustration of the line source SS, which may be west boundary of the unstable tongue, and the formation process of the prefrontal squall line. (After Newton, 1950)

Thus the convergence zone will be formed near the dry line and to cause the dry warm-dry lid lifted and the the convection break out.

Early as to 1950s, Tepper explained how the gravity waves can play the role for triggering the squall line. He thought when the front advanced to be accelerated, the density wave will be formed on the frontal inversion layer due to the effect of the gravity force. While the front velocity decreased the rare factional wave will be formed subsequently and spreads forward along the inversion layer. The wave is the gravity wave. Since the action of the ascending of the gravity wave the inversion layer will break and the moist unstable air will be lifted. Therefore the convection will break out and a prefrontal convective cloud belt will be formed.

Recent years, many people regard the roller shape circulation caused by the symmetric instability as a triggering mechanism of the squall line. So called symmetric instability means the instability caused by the slantwise rising motion under the original condition with gravity static stable and inertial stable. The instability will accelerate the slantwise rising flow and to cause the roller shape circulation and to form a convective cloud belt. In general speaking the symmetric instability is usually to be found under conditions with smaller absolute vorticity and stronger vertical wind shear as well as the weaker static instability. The symmetric instability may be indicated by the index S, which may be expressed as follows

$$S = \frac{\zeta_a}{f} - \frac{1}{R_i} \qquad (5.2.1)$$

where ζ_a is absolute vorticity, f is Coriolis parameter, R_i is Richardson number. When $S<0$, the atmosphere is under symmetric unstable. More criteria of symmetric instability will be discussed in Chapter 6. Emanuel (1982) pointed out based on the case study that the severe squall line was just located in the area where S is a small or negative value in most part of the troposphere. This means the symmetric instability is really a favorable mechanism for triggering squall line.

5.2.2 The mesoscale rain bands near the frontal cyclone

The early frontal cyclone model given by the Norwegian school described the simple type of the precipitation distribution near front. However, the recent observation and investigations show that the distribution of the precipitation and vertical motion is much more complicated. In general speaking, a mid-latitude frontal cyclone normally includes synoptic scale precipitation area, mesoscale precipitation area (or belt) and small scale convective cells (see Fig. 5.53). The synoptic precipitation area may be as large as 10000 km² or more and may last for 12 hours or longer. In the synoptic precipitation area (PA) there are several larger mesoscale PAs with the area of 1000—2000 km², while each larger mesoscale PA may include 3—6 smaller mesoscale PAs with the area of 250—400 km². Every smaller PA usually includes 1—7 convective cells. Each of them has an area about 5—10 km². In general speaking the vertical velocity and precipitation rate is inversely proportional to the horizontal scale of the precipitation system. The dynamic mechanisms for different scale precipitation system are different. The synoptic scale precipitation system majorly is caused by the large scale slantwise motions. while the mesoscale and convective scale systems are caused by mesoscale circulations and small scale convective cells respectively. The convective cloud near the front sometimes may run through whole troposphere, they are called deep convection or D-type convection. Under the more common frontal situation, the convections near the front are usually limited in a shallow layer. If they occur in high or mid level of the troposphere, they are called U-type convections. While if they occur in a low level of the troposphere or near the boundary layer, they will be called L-type convections. The line-convection is a special L-type convection. In the line-convection updraft is only concentrated in a very narrow area, which is called convective element with width of several kilometers and length of several ten kilometers.

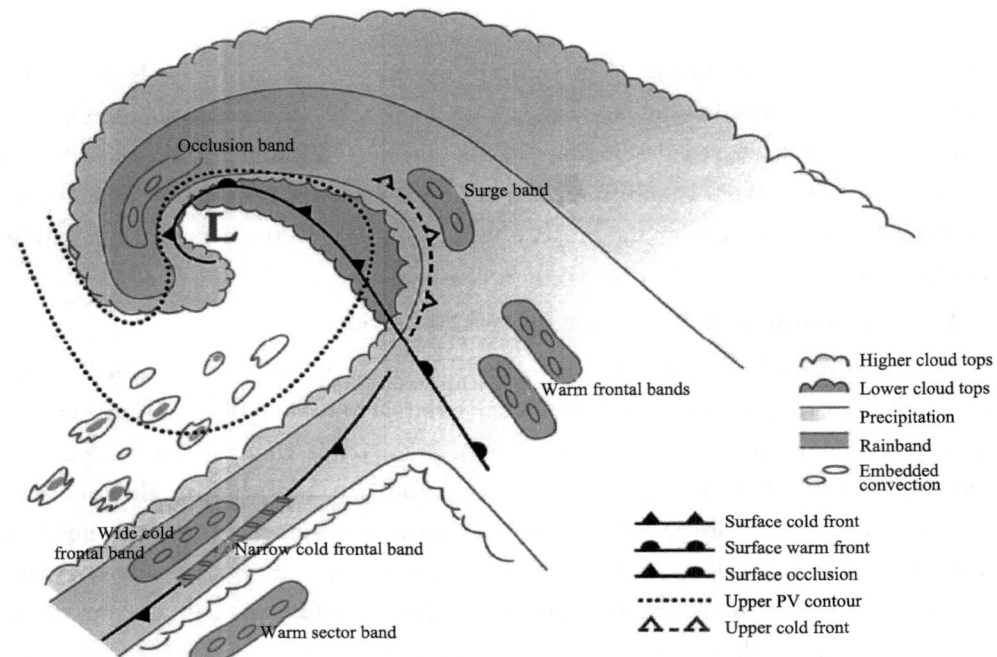

Fig. 5.53 Idealization of the cloud and precipitation pattern associated with a mature extratropical cyclone. (After Houze, 2014)(see the color illustrations)

Following we will discuss the precipitation systems in different scale near the extratropical cyclone in some detail.

(1) The mechanisms of synoptic scale precipitation

Here we will majorly discuss the warm conveyor belt and cold conveyor belt (see Fig. 5.54). In the boundary of the convergence zone in front of trough there is usually a narrow and elongated cloud belt. This cloud belt is formed by the boundary warm air, when it moves northward and upward to the mid and upper level of the troposphere originally from low level of low latitude. The narrow and long air flow plays the role to transport thermal energy, moisture and momentum poleward and upward and therefore it is called the warm conveyor belt (WCB). Generally, the WCB has the following features.

① It locates in the front of the cold front and its west boundary is clear, but the east boundary is not very clear. WCB is usually corresponding to a low level jet stream, since there is a great pressure gradient to the west side of the WCB and there-

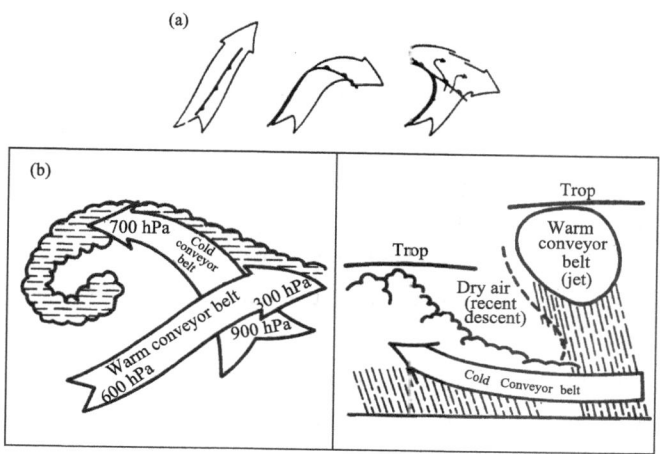

Fig. 5.54 (a) Configuration of the warm conveyor belt (WCB) undergoing modification from a baroclinic cloud leaf to a comma (left to right) and from a rearward-sloping to warm front type conveyor belt; (b) Schematic model of a mature cold conveyor belt (CCB) or instant occlusion in plan view (left) and cross section (right). (Adapted from Carlson, 1998)

fore there is a strong southerly wind there. Generally, the west side is warmer than the east side of the WCB, thus there is a northerly thermal wind, which causes the wind decreases with height. Additionally, the surface frictional effect can also decrease the wind velocity near surface, so that the vertical distribution of the wind velocity will present the feature of low level jet stream, i. e. the wind velocity will be weak at upper and low levels, while strong at middle level.

② The WCB is usually as long as several kilometers. This kind of low level jet stream has no distinct diurnal variation. But in the mid part of the United States some low level jet streams have distinct diurnal variation. Their lengths are relatively shorter, normally only few hundreds kilometers long.

Except WCB, there is another important air flow called cold conveyor belt (CCB), which is originally from ambient of the high pressure located to the northeast of the cyclone. It is a anti-cyclonic low level air flow. It plays the role to transport the cold air from north to south, that is the reason why it is called CCB.

In a mature cyclone, the precipitation and cloud large scale distribution may be explained by the above two moist air flows (WCB and CCB) originated from boundary layer.

Firstly, let's look at WCB, which comes from the boundary layer of the warm sector. It rises along the cold front and arrives at the upper level of troposphere to form high cloud systems. When it moves over the cold air in front of the surface warm front, it will change direction clock-wisely. When it enters the northwest flow in the front of upper level ridge, it will sink and evaporate and dissipate obviously.

Secondly, let's look at CCB, which originates from the ambient of the high pressure to the northeast of the cyclone. It is an anticyclonic low level jet stream. It moves westward relative to the advancing cyclone. It just located in front of the surface warm front and under the WCB. Near the surface warm front the low level air at the edge of CCB rises due to frictional convergence. Then it moves westward continuously and arrives the place near the apex of the warm area at mid level of the troposphere. Gradually, thus a cloud belt is formed. When the CCB occurs to the west side of the WCB, it will possibly be anti-cyclonic veering around the low pressure center, so that the cloud belt will dissipate distinctly. Its figure may look like as shown in Fig. 5.55. The superposition of the cloud belts formed by WCB and CCB respectively will form a large belt-shaped cloud type feature.

The WCB and CCB sometimes may exist as an independent producing system of cloud and precipitation. Some cold air vortex may be regarded as an expressive form of CCB. Their corresponding cloud systems are normally some small comma clouds. When the main cloud belt, which is equal to a WCB, meets the small comma clouds the north edge of the main cloud belt will move over the pre-existing comma cloud and curl up over the cold airt vortex cloud, and thus to form the cloud pattern similar to the classical occlusion cyclone cloud type. During this "occlusion" process, the WCB provides the appearances of a cold front and a warm front, while the cold air vortex then provides the appearances of the low pressure center and occlusion part of the system.

The WCB may be divided into two general types, i.e. the backward slantwise ascent and the forward slantwise ascent. The former has a component of the relative air flow pointing to the back of the cold front; while the latter has a component of the relative air flow pointing to the front of the warm front as shown in Fig. 5.56.

(2) The mesoscale rainbelts near the frontal cyclone

The mesoscale rainbelts near the frontal cyclone have different types. Fig. 5.57 is a schematic diagram of the airflow and the precipitation types in an occluded extratropical cyclone given by Browning and Monk (1982). In the diagram the numbers represent as follows: ① warm front precipitation, ② convective precipitation-genera-

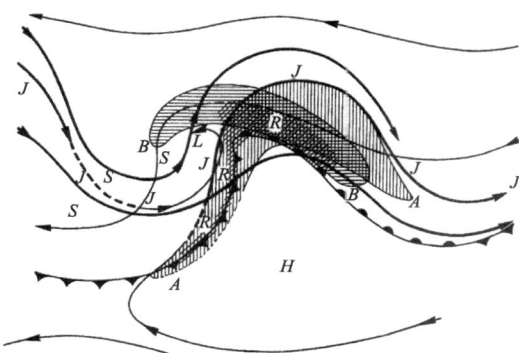

Fig. 5.55 The model of the basic large scale cloud system and major trajectories of the air in a occlusion wave cyclone. The arrow lines is the air trajectories relative to the center of the cyclone. The letter L indicates the location of the cyclone center at surface. The thin arrow lines are trajectories of the relative air flows majorly flow from east to west, which are at low level of the troposphere. The thick arrow lines indicate the relative air flows majorly flow from west to east on the high level of troposphere. The letter J marks the position of the axis of the upper level jet stream. The areas of maximum ascending and descending areas are showed by the dashed lines and marked by R and S respectively. Two basic cloud systems are showed in the diagram. One of the cloud belts is the WCB caused by the rising warmer air showed by the vertical shadow lines and marked by the letter B. Another cloud belt is CCB formed by the rising colder air showed by horizontal shadow lines and marked by the letter B. The surface wide precipitation area is indicated by the diagonal line shadow area and the position of the surface front is indicated by using the conventional method. (After Ludlam, 1980)

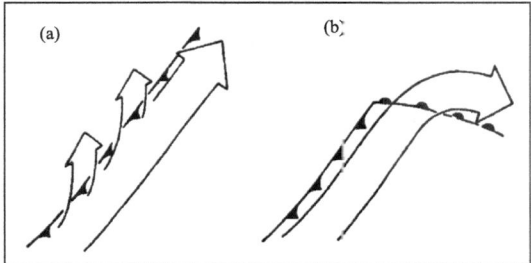

Fig. 5.56 Illustration of the backward slantwise ascent (a) and the forward slantwise ascent (b). (Adapted from Browning, 1983)

ting cells associating with dry air intrusion aloft, ③ precipitation from upper cold frontal convection descending through an area of warm advection, ④ shallow moist zone characterized by warm advection and scatted outbreaks of mainly light rain and drizzle, and ⑤ precipitation at the surface cold front.

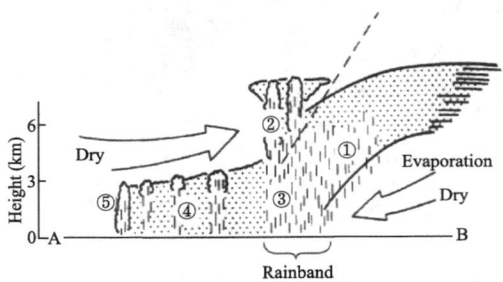

Fig. 5.57 Schematic of the airflow in an occluded extratropical cyclone. Numbers represent the precipitation types (see the text). Dashed line represents the leading edge of the dry air. (Adapted from Browning and Monk, 1982)

Based on the observational studies the mesoscale rain belts near the frontal cyclone may roughly be divided into three types: U-type, L-type and D-type according to the height that the convective clouds occurring. The shallow layer convection occurred in the mid to upper level of the troposphere is called U-type; the shallow layer convection occurred in the lower level of the troposphere is called L-type; and the vertical deep convection is called D-type (Fig. 5.58 and Fig. 5.59). Hobbs (1978) suggested that the above rain-belts may be furtherly divided. For instance, the U-type may be further divided into the cold front rain-belt, the warm front rain-belt and the pre-cold frontal cold surge rain-belt (as shown in Fig. 5.60); the L-type may be further divided into the narrow cold rain belt and the warm sector small rain belt (transverse and longitudinal rain belts); and the D-type may be further divided into the rain belt in warm sector and the rain belt behind front etc.

The structures of the three types of the U-type rain belts are shown in Fig. 5.60. The warm front rain belt is presented in Fig. 5.60a. It occurs under the forward slantwise rising situation and is located near the warm front and is parallel to the warm front. And sometimes it companied with the L-type rain belt. The prefrontal cold surge rain belt is presented in Fig. 5.60b. It occurs also under the forward slantwise rising situation and is located just in the ahead of the upper level cold front and is parallel to the cold front. The cold front rain belt presented in Fig. 5.60c occurs under

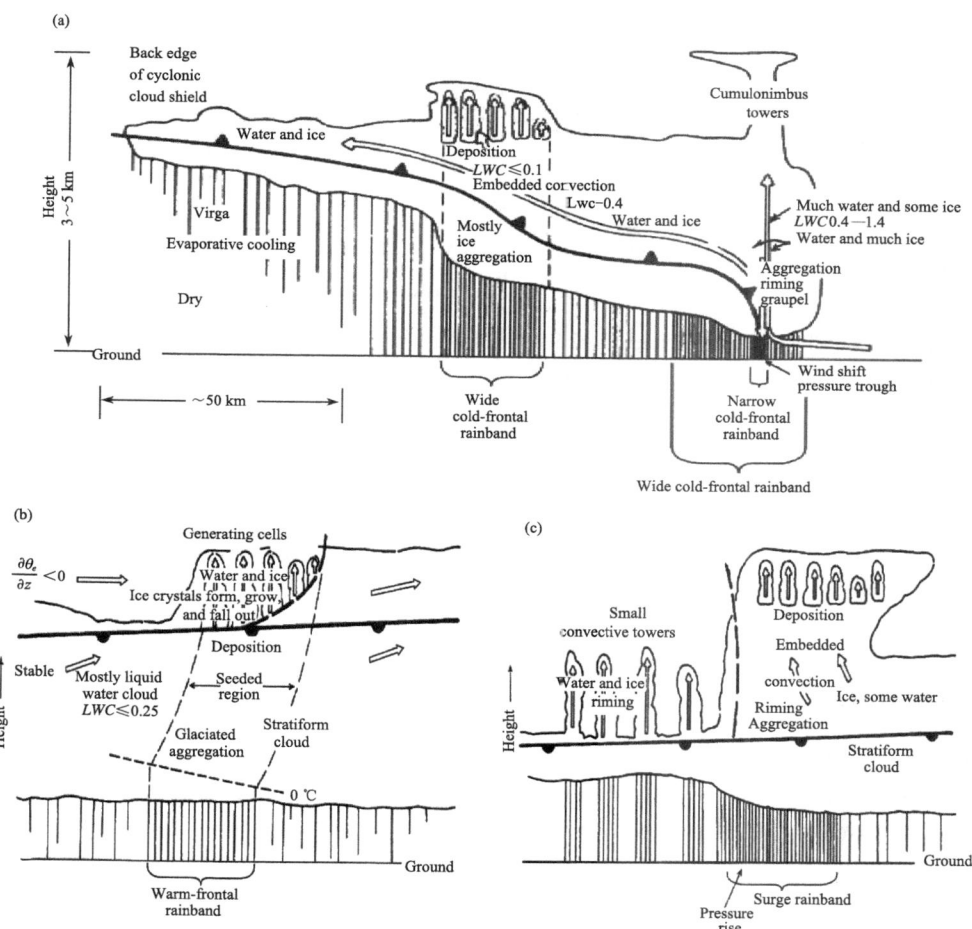

Fig. 5.58 The cloud structure, air motion and precipitation mechanisms across a cold front (a), a warm front (b) and a surge rainband (c) of a extratropical cyclone as revealed by instrumented aircraft, radar, sounding and other observations of frontal cyclones passing over Washington State of the United States. Vertical hatching below cloud bases represents precipitation, the density of the hatching corresponds quantitatively to the precipitation rate, Open arrows depict airflow relative to the front: a strong convective updraft and downdraft above the surface front and pressure trough and a broader ascent over the cold front aloft. Cloud liquid water contents (lwc) are in $g \cdot m^{-3}$. The motion of the rainband is from left to right. (Adapted from Matejka, 1980).

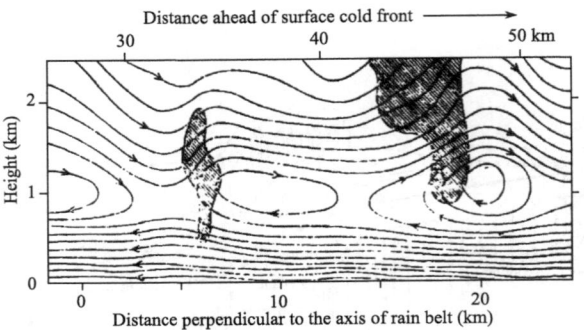

Fig. 5.59 The airflow pattern in the vertical section direct across with the warm sector small rain belts arranged along the geostrophic wind direction. (After Browning et al., 1975)

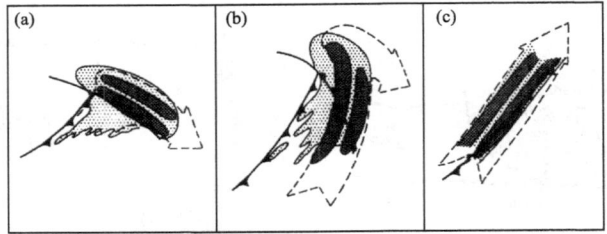

Fig. 5.60 The idea structures of the U-type rainbelts (The light and heavy dot shadow areas are the light rain area and the mid-heavy rain area respectively). The wide dashed arrow represented the warm conveyor belt in mid level of the troposphere. (Adapted from Browning, 1983)
(a) Warm front rainbelt; (b) Prefrontal cold surge rainbelt; (c) Cold front

the backward slantwise rising situation and is parallel to the cold front. Near the surface cold front there is narrow cold front rain belt. However, it should be noted that in the real situation the rain belts are normally irregularly. As shown in Fig. 5.61 the radar observational data displayed the irregularity of the warm front and cold front rain belts respectively.

Normally the U-type rain belts have some common features as follows: ① they are all related with the rising part of the warm front conveyor belt. Normally they occur in the mid-upper level of the troposphere, typically in the level between 700 hPa. ② Their widths are normally about 50 km, and the typical length is about several hundred km. Their orientations are normally parallel to the baroclinic belt in the height they located on. ③ They consist of the convective cells in mid and upper level and are usually in groups. They formed in

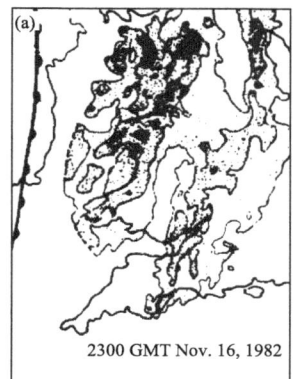 

Fig. 5.61 The cases of rain belts observed by the meteorological radar network in England. (a) two wide warm front rain belts; (b) two wide cold front rain belts. The light, medium and heavy dot shadow areas indicate the radar echo areas corresponding to the light, medium and heavy precipitational intensities respectively. (After Browning, 1983)

a shallow potential instability layer ($\frac{\partial \theta_w}{\partial z} < 0$).

As mentioned above the L-type rain belts may be further divided into two types: the warm sector small rain belt and the narrow cold front rain belt (line element). These low level convection in WCB are the warm sector phenomena. In the warm sector PBL, the air sometimes is nearly saturated and the vertical lapse rate is very small ($\frac{\partial \theta_w}{\partial z} \cong 0$). When the air is disturbed mechanically and organized, the convective circulation will occur in the scroll vortices form arranged along the geostrophic flow. This is a possible explanation of the formation of the warm sector small rain belts.

Except the warm sector small rain belts, the narrow cold front rain belt is another kind of the L-type rain belts. This kind of rain belt occurs in the west edge of the WCB in a line convection form. The width of the line-convective elements are normally only several kilometers, while their vertical velocity may be as strong as several meters per second.

As mentioned above, the layer that the low level jet located is a nearly saturated and neutral stable atmospheric layer. It is usually covered by a capping stable layer on 700 hPa. The vertical ascending motion will be caused by the convergence between the airflow in low part of the low level jet and the airflow in the back of the front. A steep updraft area will be formed after the moist air lifted and the latent heat is released as shown in Fig. 5.62 and Fig. 5.63.

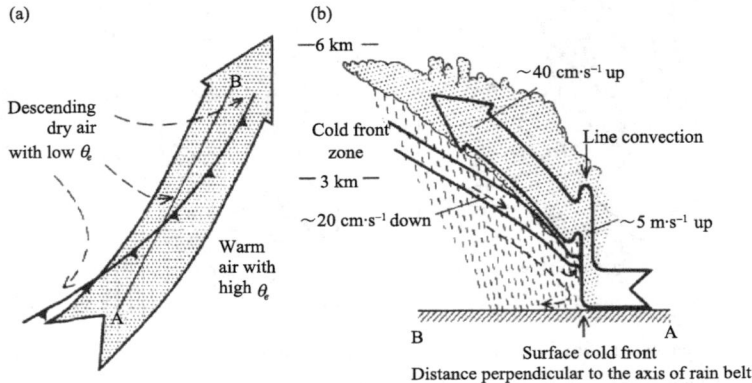

Fig. 5.62 The vertical section structure of a line convection. (After Browning and Harrold, 1970)

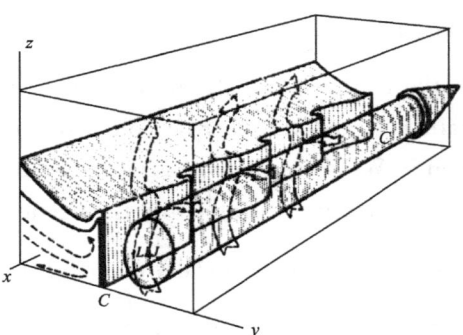

Fig. 5.63 Illustration of the airflow related with the line convection to the left side of the low level jet. *CC* indicates the front edge of the curved cold outflow, *LLJ* indicates the low level jet. (After Browning, 1983)

The line-convection is usually broken as a series of line elements, each of which has a length about ten kilometers to several ten kilometers. Very often they are the narrow heavy rain belts and sometimes accompanied with small hails and tornadoes. The small rain belts have only about 3 km wide and with clear boundary. They are the shallow systems mainly located in the low atmosphere below 700 hPa. Since they are usually embed in the deep stratiform precipitation area, so that it is difficult to find them in the satellite cloud pictures. Between two sides of the line convection element the wind changes distinctly. Most frequently the line convection will curl up and sometimes may form a clear vortex and probably to cause tornadoes (see Fig. 5. 64, Fig. 5. 65. Fig. 5. 66).

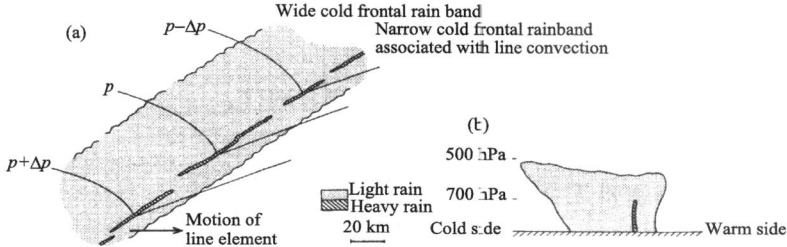

Fig. 5. 64 Illustration of the precipitation distribution on a steep cold front. (After Browning, 1983)
(a) in the horizontal plane; (b) in the vertical section perpendicular to the front

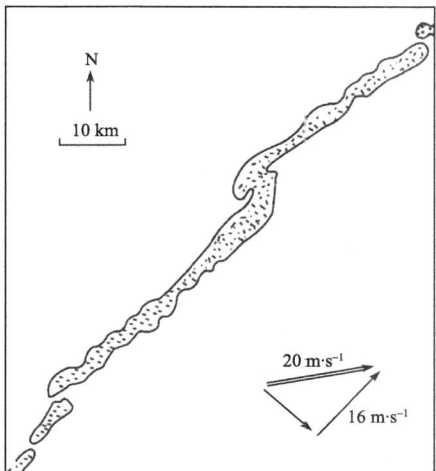

Fig. 5. 65 The contour lines of the radar PPI echo corresponding to the precipitation of the line convection. (After Browning and Harrold, 1970)

5.2.3 The mesoscale rain belts in a typhoon

(1) The general structure of typhoon

Typhoon is the tropical cyclonic vortex with a nearly round shape and warm center structure. The observations based on satellite and radar showed that there is a clear air area in the center of typhoon called typhoon eye. Around the eye there is a

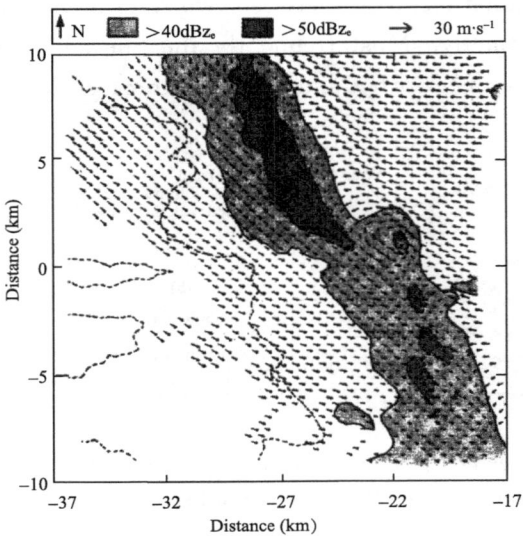

Fig. 5.66 The vortex on the line element. The dot area represents the radar echo area related with the severe precipitation in the line element. (Adapted from Carbone, 1982)

deep convective cloud ring called eye-wall of typhoon. Its inner diameter is about 15—80 km. The mesoscale spiral rain belts located outside of the eye wall of typhoon. The rain belts occurred inside of the typhoon cyclonic circulation are called inner rain belts, while that occurred outside of cyclonic circulation are called outside rain belts or the squall lines before the typhoon. The distribution of the mesoscale rain belts near typhoon is shown in Fig. 5.67, Fig. 5.68 and Fig. 5.69.

(2) The structure of the rain belts in the eye wall of typhoon

Jorgensen (1984) studied the structure of the typhoon eyewall rain belts by using the composite analysis method and presented the radial direction vertical section of the typhoon eyewall as shown in Fig. 5.70. It can be seen in Fig. 5.70 that the eye wall cloud region is tilted with height. The vertical motion is updraft in the eye wall and is downdraft in the typhoon eye region respectively. In two sides of the eye wall at low level (from surface to 1.5 km above the ground) the horizontal air flows flow into the eye wall. The maximum wind velocity occurs in the eye wall. The contours of the 10 dBz radar reflectivity and the maximum tangential wind velocity (v_θ) are all tilted with height. And the amounts of the radar reflectivity and v_θ are all increasing with radius. Above the frozen level both the content of the liquid water and the radar re-

Chapter 5 Mesoscale Convective Systems   • 185 •

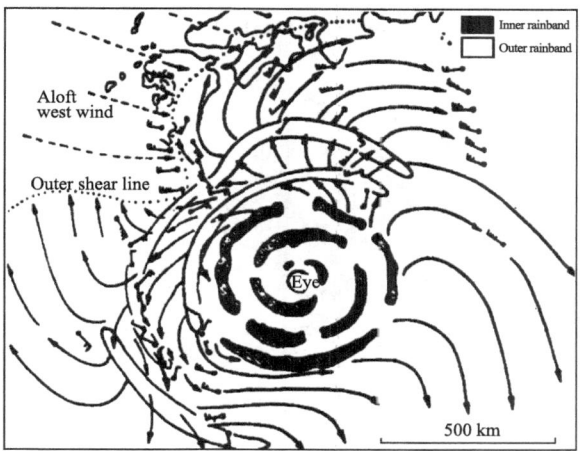

Fig. 5.67 Illustration of the rain belts near a typhoon. The solid lines are the stream lines of the upper level outflow over the typhoon. (After Fujita, 1976)

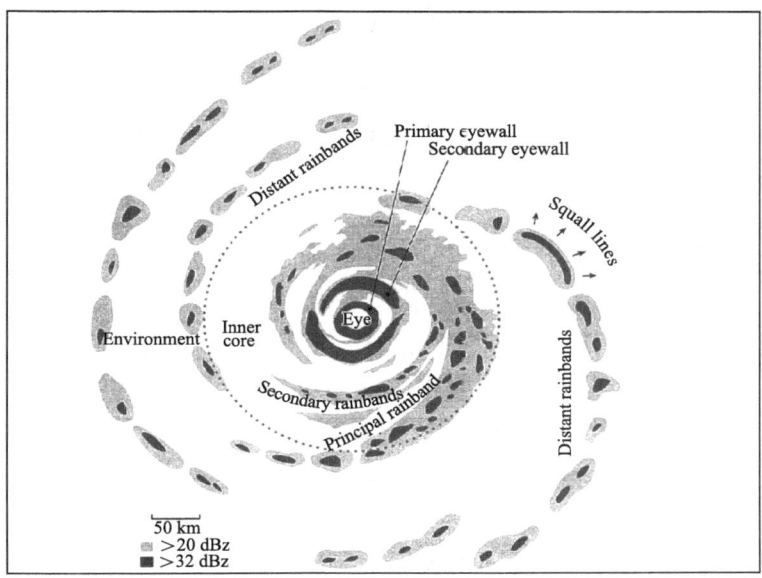

Fig. 5.68 Schematic illustration of the radar reflectivity in a Northern Hemisphere tropical cyclone with a double eye wall. (After Houze, 2010)

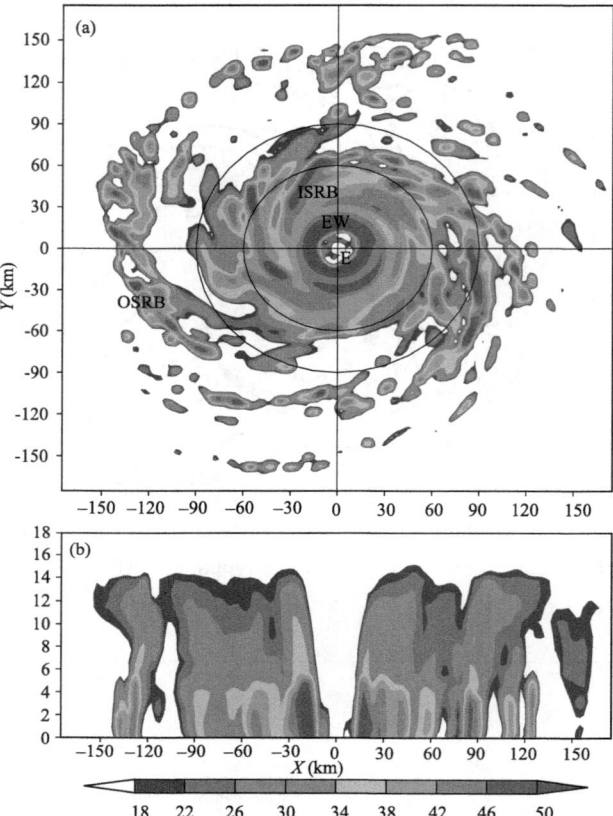

Fig. 5.69 plan view (a) and cross section (b) of radar reflectivity in a typhoon.
(E=eye, EW=eye wall, ISRB=inner spiral rainband, OSRB=outer spiral rainband)
(see the color illustrations)

flectivity are quickly decrease with height. Near the frozen level there is a severe reflectivity band (the radar echo bright band). This feature shows the structure of the eye wall rainbelt is distinctly similar to the tropical squall line.

(3) Features of the mesoscale circulation in typhoon cyclone

Based on the numerical simulation results, Liu and Zhang (1997) proposed a conceptual model that describes the axisymmetric structures in the inner core of a mature hurricane as shown in Fig. 5.71. In the diagram the light-shading areas indicate the regions with cloud and precipitation; the dark regions represent the convective eyewall and spiral rain bands; the slash-hatched area represents the eye inversion layer

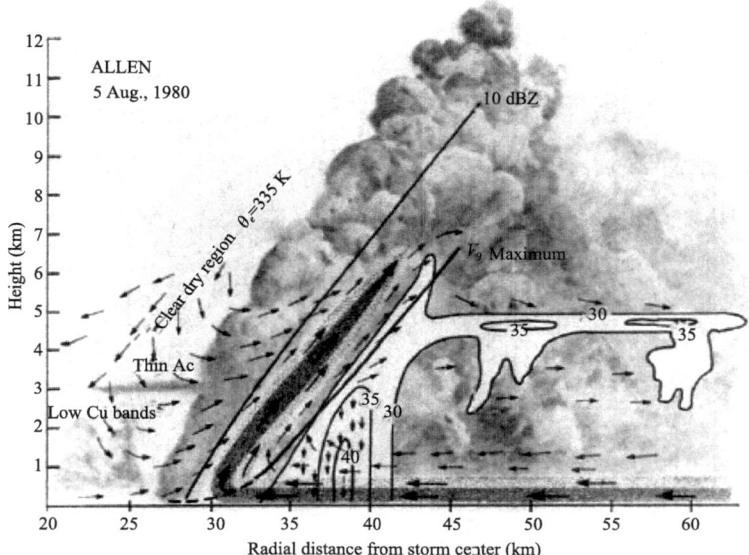

Fig. 5.70 Illustration of the vertical section of the eye wall of typhoon Allen occurred on Aug. 5, 1980. The arrow lines indicate the radial and vertical flows, and the thin solid lines indicate the radar reflectivity and maximum wind boundaries. (After Jorgensen, 1984)

(EIL); the cross-hatched regions is the occluded eye air (OEA) with low θ_e; the freezing level is marked with a dashed line; MTD and SR stand for moist downdrafts between the convective bands and spiral rain band updrafts, respectively.

The main ingredients of the conceptual model consist of the basic and local circulations, the inertial-gravity wave oscillations, and the process of lateral mixing. As shown in Fig. 5.71, the basic circulations include a layer of main inflow (MI) below 1.5 km in the MBL, a sloping updraft (SU) in the eyewall, a main outflow (MO) in the upper troposphere, and a mean descent (MD) in the eye. The main inflow (MI) originates from the far outer regions, it intensifies mostly in the inner core as a result of the deepening of the storm and the transfer of angular momentum. Its speed reaches a maximum slightly outside the radius of maximum wind (RMW) and then decelerates rapidly toward the center of the eye. The MI plays an important role in feeding the high θ_e air to the eyewall convection. The SU has its roots in the MBL and it is fed by the high-air from the MI and the eye. In the midtroposphere, the SU is characterized by a weak, divergent, radial outflow. The local vertical circulations include a ra-

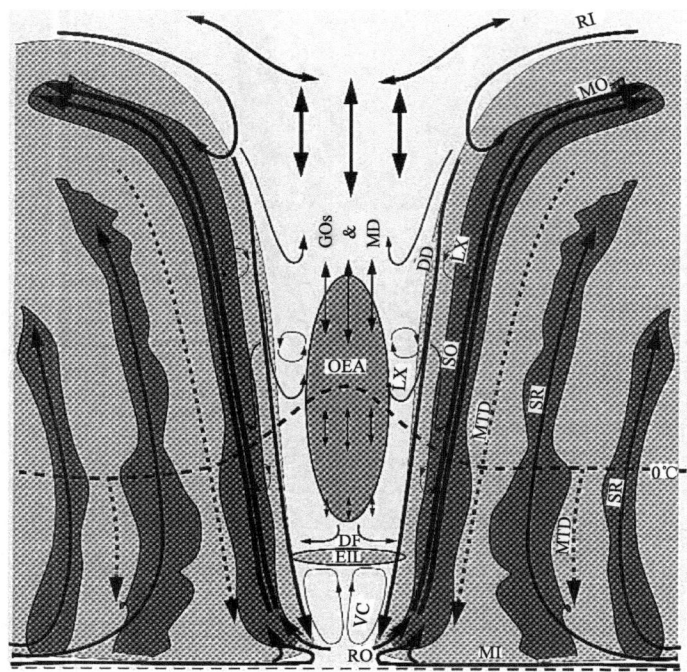

Fig. 5.71 A schematic (radial-height) conceptual model of a mature hurricane in the inner-core region. Refer to the text for the meaning of the symbol in the diagram. (After Liu and Zhang, 1997)

dially narrow zone of dry downdraft (DD) along the inner edge of the eyewall, a return inflow (RI) near the tropopause, the MD advects the warm core and the minimum θ_e layers downward, thereby strengthening the inversion below. The altitude of the inversion remains nearly constant but its intensity increases with time as a result of subsidence aloft and ascent below. These local circulations are mainly forced by the basic circulations, but they also contribute significantly to the development of the hurricane. The RI, a branch of the local upper-eye vertical circulation, is generated by the release of latent heat and momentum sources in the upper eyewall. It streams inward from above the eyewall, descends after entering the eye, and finally intrudes into the eyewall. The RI is the basic mass source that initiates and feeds the strong subsidence in the eye, particularly the penetrative DD. It induces convergence toward the eye and appears to be responsible for the generation of inertial-gravity waves. While the DD is initiated by the RI, its further downward development to the MBL is believed to be due to the sublimative/evaporative cooling of detrained condensates from the eyewall and the compensating subsidence associated with the local heating in the eyewall. The VC is frictionally forced (Holton, 1992) and influenced by penetrative DDs; it induces maximum ascent at the center of the eye and recycles the mass into the eyewall through the low-level RO. The other two kinematic elements, inertial-gravity oscillations (GOs) and lateral mixing (LX), also play a role in determining the time evolution of the inner-core structures. The GOs are generated in the upper eye by the RI; they weaken (intensify) outward (inward) from the eyewall and also diminish downward. The GOs appear to account primarily for the generation of the fluctuating W in the eye. The symbol LX denotes both entrainment and detrainment of individual flow elements described above as well as any diffusive effects. The upper level RI and part of the DD can be considered as detrained masses from the eyewall that help drive or modify the local circulations.

(4) Structure and motion of the spiral rain-belts

The spiral rain-belts are consist of convective clouds and stratiform clouds. Normally the clouds at the up-wind and the down-wind sides of the spiral rain-belts are convective clouds, stratiform clouds respectively. While at the middle part of the rain-belts the clouds are usually the transitional clouds between convective and stratiform. The mesoscale convective zone sometimes moves along the rain-belts toward the down-wind side and to form more convective clouds at the down-wind side of the rain-belt. There are no strong and well organized updrafts and radar reflectivity core in the spiral rain-belts as in the eyewall rain-belts. The precipitation may be formed in a wide area below the melting level (about at the height of 4.5—5.0 km) characterized by the radar reflectivity bright belt. According to the estimation by Jorgensen (1984), the stratiform precipitation area will be 10 times of the convective precipitation area.

The motion of the spiral rain-belt is depending on the motion of the convective cells formed the rain-belt. Tatebira (1962) pointed out that the single convective cell normally forms at the outside end of the rain-belt and passes over the rain-belt and then dissipates at the inside end of the rain-belt.

The mesoscale rain-belts in typhoon are sometimes relatively stationary and sometimes propagate around a center. In general speaking, the quasi-stationary rain-belts are normally corresponding to the smooth track of typhoon. When the rain-belts spread around the typhoon center, the typhoon track will be swing and oscillation very often.

(5) Mesoscale distribution of the weather near typhoon

According to the observational study by Parrish et al. (1982), the maximum surface wind in typhoon occurs in the inner boundary of the typhoon eyewall. The heavy rainfall area is normally consistent with the strong wind area. Tornadoes are usually found in the area to the east or northeast of the typhoon center or in any quadrant within the 100 km scope near the typhoon center as shown in Fig. 5.72. Most of the tornadoes occur in the severe convective zone near the outer rain bands of typhoon.

The stronger low level wind shear is an important factor to form the tornadoes in typhoon. The low level wind shear may be caused by the thermal wind effect due to the development of the surface cold core when the typhoon quickly is filled. The wind shear may be also related with the seacoast effect. Near the coast the wind at the low level of typhoon decreases due the frictional effect, while the upper level wind velocity is still strong, therefore the strong vertical wind shear is caused. According to the statistics by Novlan and Gray (1974), most of the tornadoes are formed in the regions departure from coast less than 200 km.

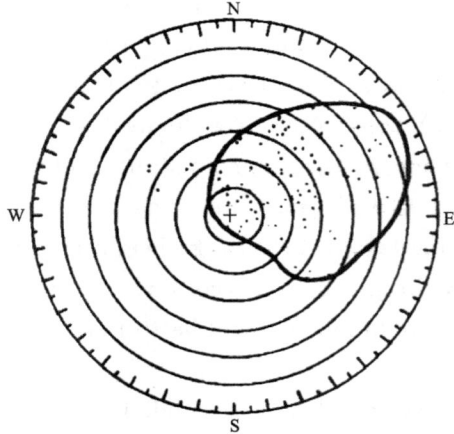

Fig. 5.72 The positions of the tornadoes relative to the typhoon center during 1973—1980. The thick line indicates the area where 95% of tornadoes are formed. The concentric circles are 50 km apart each other. (After Gentry, 1983)

§5.3 Mesoscale convective complex (MCC)

5.3.1 Definition and characteristics of MCC

MCC is a meso-α scale convective system which may be recognized from the enhanced satellite image. It consists of many smaller convective systems such as the tower cumulus, convective clusters or the meso-β scale squall lines etc. and is organized as a convective complex. It is a large, persistent and nearly a round-shaped cloud cluster (see Fig. 5.73). For recognizing the systems in conventional weather analysis and forecasting Maddox (1981) proposed the definition and physical characteristics of the mesoscale convective complex (MCC) in its mature stage as shown in Table 5.1.

Table 5.1 Definition and physical characteristics of mesoscale convective complex (MCC).

Size	A. Cloud shield with IR temperature $\leqslant -32$ ℃, must have an area $\geqslant 100000$ km²
	B. Interior cold cloud region with temperature $\leqslant -52$ ℃, must have an area $\geqslant 50000$ km²
Initiate	Size definitions A and B are first satisfied
Duration	Size definitions A and B must be met for a period $\geqslant 6$ h
Maximum extent	Contiguous cold cloud shield (IR temperature $\leqslant -32$ ℃) reaches maximum size
Shape	Eccentricity (minor axis/major axis) $\geqslant 0.7$ at time of maximum extent
Terminate	Size definitions A and B on longer satisfied

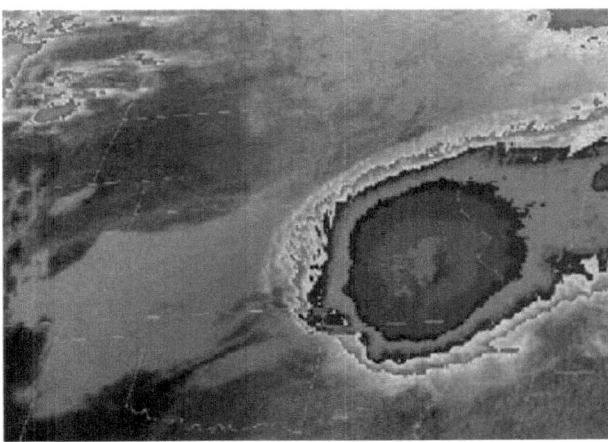

Fig. 5.73 Infrared satellite image of a mesoscale convective system over Missouri. Courtesy of J. Moore, St. Louis University, St. Louis, Missouri. (After Houze, 2014)

(see the color illustrations)

Maddox (1980a) found that many MCCs produced locally intense rainfalls and flash flooding, and some MCCs may also produce a variety other severe convective weather phenomena including tornadoes, hails, damage winds, and severe electric storms etc. All these indicate that the mesoscale systems (MCCs) are truly significant events.

5.3.2 Life cycle of MCC

The life cycle of MCC includes four developing stages. First is the genesis stage. In the stage some scattered convective systems begin to develop under the favorable situation such as the conditional unstable stratification, low level convergence and ascending, thermal heating and mechanical lifting effects of the topography etc. Second is the developing stage. During the stage the outflows of the thunderstorms converge gradually into a strong mesoscale cold high pressure and an outflow boundary line to press the warm and moist flows flow into the system and to cause the cloud cluster getting enhanced gradually. The third stage is the mature stage. In this stage the low level convergence, upper level divergence and the mesoscale ascending motion developed vigorously and the characteristics on satellite image meet all the conditions listed in the table 5.1. The fourth stage is the dissipating stage, In this stage the cloud system of storm on the satellite image becomes scattering and in disorder but may still see a nearly successive stratiform anvil clouds.

5.3.3 Structure of MCC

Maddox (1981) made a composite analysis for 10 MCCs. The result showed that in mature stage the MCC has a relative steady mesoscale circulation as shown in Fig. 5.74 with following characteristics: at the upper level of troposphere there is a cold core anticyclone; at the mid-upper level of troposphere there is a warm core mesoscale cyclonic vortex; at the low level there is the upward mass flux caused by convergence.

The similar descriptions are shown in Fig. 5.75 given by Fritsch et al. (1994). Fig. 5.75 is a conceptual diagram of the structure and development mechanism of a mesoscale warm core vortex associated with an MCS. It can be seen from the diagram that there are distinct vertical shears of environmental winds at high and low levels respectively. At the low level there are the vorticity components perpendicular to the plane of the cross section produced by the cold pool and by the environmental vertical wind shear. The system is propagating left to right and is being overtaken by air of high equivalent potential temperature in the low level jet. Air overtaking the vortex

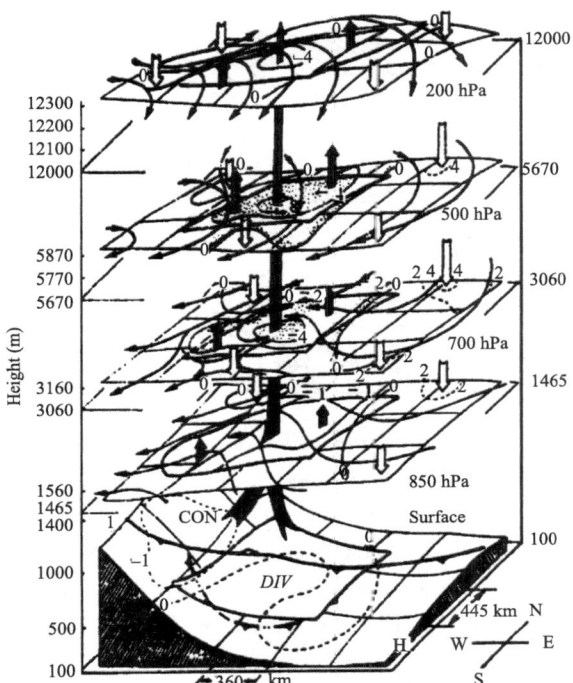

Fig. 5.74 Illustration of the mature MCC and its vicinal environment. (After Maddox, 1981)

ascends isentropic surfaces, reaches its level of free convection (LFC), and thereby initiates deep convection.

5.3.4 Precipitation distribution in MCC

In general speaking, in the central part of MCC the precipitation possibility and rainfall are higher than other parts of the MCC as shown in Fig. 5.76.

Cotton and McAnelly (1984) analyzed the composite distribution of the hourly precipitation intensity in MCC and got the results similar to the above. The maximum mean area of precipitation occurs at the time about 1—2 hours later than when the area of MCC reaches maximum. At the early stage of MCC, the convective precipitation (rainfall 25.4—50.8 mm) has bigger rate in total rainfall. This shows the important effect of convection for forcing the circulation of MCC.

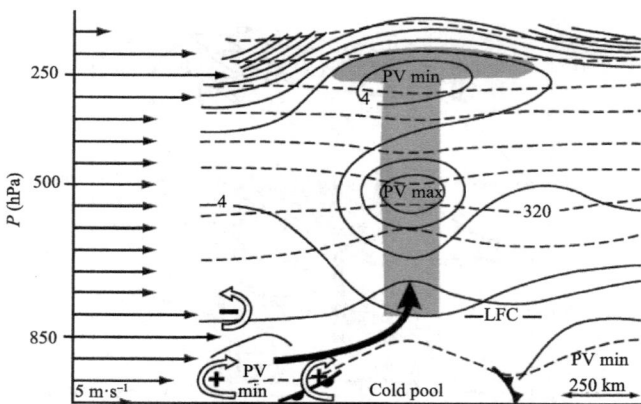

Fig. 5.75 Conceptual diagram of the structure and development mechanism of a mesoscale warm core vortex associated with an MCS. Thin arrows along the ordinate indicate the vertical profile of the environmental wind. Open arrows with plus or minus signs indicate the sense of the vorticity component perpendicular to the plane of the cross section produced by the cold pool and by the environmental vertical wind shear. The bold solid arrow indicate the updraft axis created by the vorticity distribution. Frontal symbols indicate outflow boundaries. Dashed lines are potential temperature (5 K intervals), and solid lines are potential vorticity ($2 \times 10^{-7} \, m^2 \cdot s^{-1} \cdot K \cdot kg^{-1}$ intervals). The system is propagating left to right at about 5—8 m·s^{-1} and is being overtaken by air of high equivalent potential temperature in the low level jet. Air overtaking the vortex ascends isentropic surfaces, reaches its level of free convection (LFC), and thereby initiates deep convection. Shading indicates cloud. (After Fritsch et al.,1994)

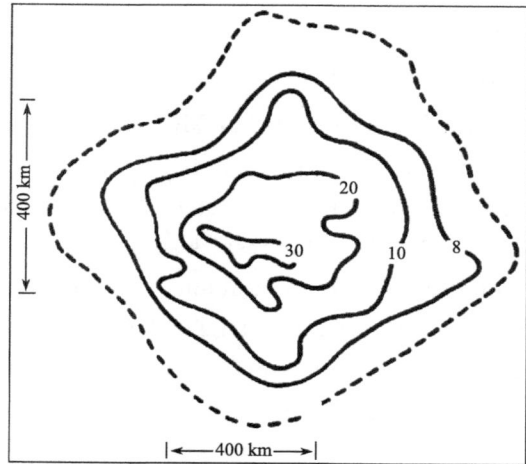

Fig. 5.76 The composite rainfall distribution of the eight MCCs.
(unit:mm)(After Fritsch et al., 1981)

5.3.5 Synoptic environment of MCC development

Maddox pointed out the features of the synoptic environment in different stage of MCC's life cycle. The features prior to MCC development are given in Fig. 5.77.

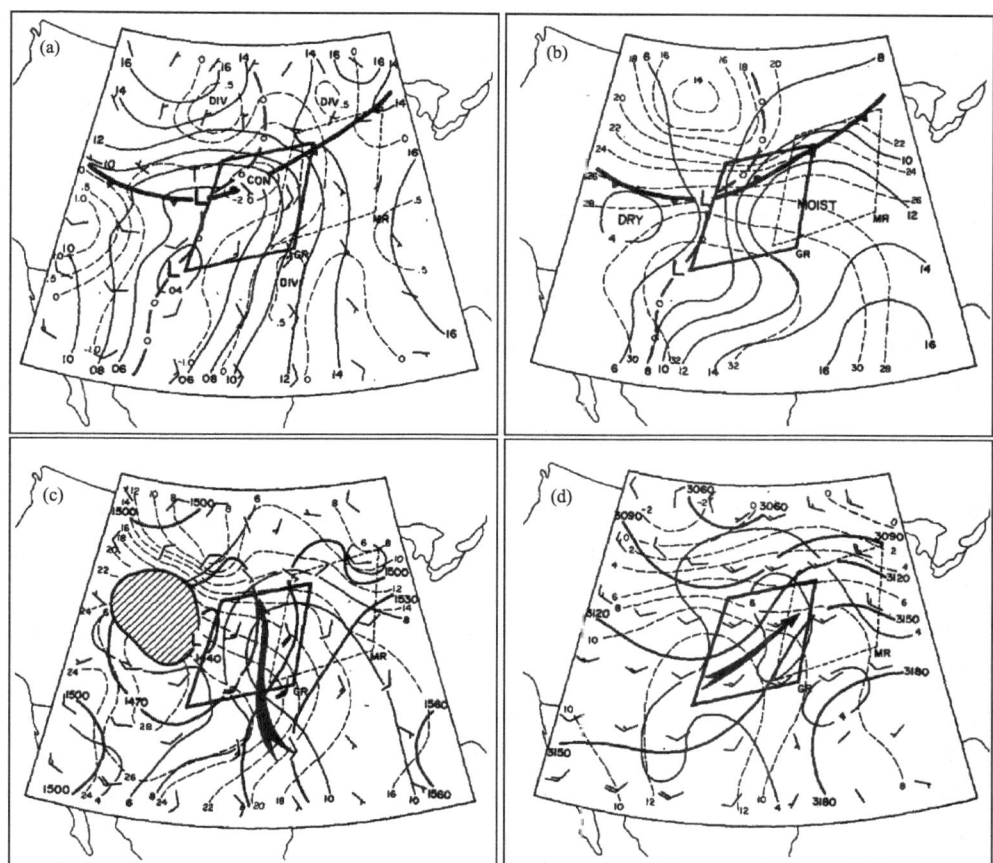

Fig. 5.77 The features prior to MCC development (a) Surface winds (full barb=5 m · s^{-1}) and isobars (solid lines) and surface divergence (dashed lines, $\times 10^{-5}$ s^{-1}); (b) Isopleths of mixing ratio (g · kg^{-1}) are solid lines and isotherms (℃) are dashed; (c) Analysis of the 850 hPa level: heights (m) (heavy solid lines), isotherms (℃) (dashed lines), mixing ratio (g · kg^{-1}) (light solid lines), winds (full barb=5 m · s^{-1}), the dark arrow shows the axis of maximum winds, the cross-hatched region indicates terrain elevations above 850 hPa level; (d) As (c) but for the 700 hPa level. (After Maddox, 1983)

In the MCC's genesis region (GR) there is a low level (850 hPa) strong wind belt (velocity$>$10 m $\cdot$ s^{-1}) and warm advection and a short wave trough is approaching. At 200 hPa there is a weak jet stream (velocity$\sim$32 m $\cdot$ s^{-1}). MCC is formed to the right side of the 200 hPa jet stream.

In the MCC's mature region (MR), there is a SW wind jet flow and a moist tongue on 850 hPa to the southwest of MR. On the constant pressure surface of 500 hPa the mixing ratio is over 3 g $\cdot$ kg^{-1}. A distinct warm ridge located over MR region. There are cold advection to the west and northwest sides of MR and warm advection to the east and northeast sides. On the 200 hPa constant pressure surface along the north and northeast boundary of MCC there are a jet stream with anticyclonic curvature and maximum wind velocity over 50 m $\cdot$ s^{-1} and a distinct cold core. There is a extreme unstable area in the southwest part of the MR region.

5.3.6 Features of the quasi-stationary MCC

Some MCCs move slowly or present the quasi-stationary state. These MCCs may cause local heavy rainfall and flash flood very often. The place where the new cells generate and develop is the most active part in the storm. The area where the convections are most violent in MCC is called the meso-β scale element (MBE). Most of the severe convective weather phenomena induced by MCC are related with the area of MBE.

Merritt et al. (1984) investigated the motion rules for more than hundreds of MBEs and pointed out that there is no steering level to control the motion of MBE. While they found that the mean vertical wind shear vector of the cloud layer may be used for estimating the motion of the MBE in MCC.

In general speaking, MBE moves along the direction of vector of the vertical wind shear between 850 hPa and 300 hPa or moves along the isoline of the thickness between 850 hPa and 300 hPa as shown in Fig. 5.78. Meanwhile the storm will normally propagate toward mesoscale convergence zone. If the propagation direction is just against with the motion direction of the MBE with similar velocity, then the MBE will keep quasi stationary state.

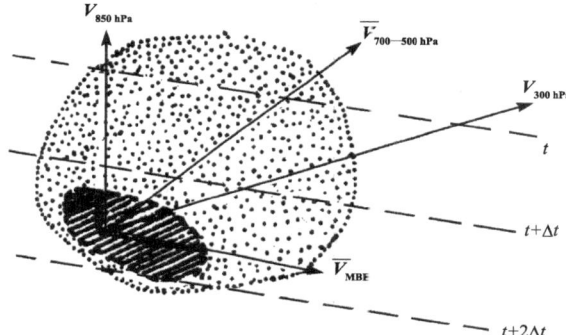

Fig. 5.78 Illustration of the motion of the MBE relative to the winds of 850 hPa and 300 hPa or the mean wind of the mid troposphere (700—500 hPa). (the dashed lines are the isolines of the thickness between 850—300 hPa; the light shadow area indicates the cold cloud shield; the dark shadow area indicates the MBE region) (After Merritt and Fritsch, 1984)

References

BLUESTEIN H, 1984. Dynamics of mesoscale weather systems[C]. NCAR Summer Colloquium Lecture Notes, 11 June—6 July, 497-516.
BROWNING K A, 1983. Mesoscale Structure and Mechanisms of Frontal Precipitation Systems. Mesoscale Meteorology[M]. Sweden: SMHI.
BROWNING K A, FANKHAUSER J C, CHALON J-P, et al, 1976. Structure of an evolving hailstorm. Part V: Synthesis and implications for hail growth and hail suppression[J]. Mon Wea Rev, 104:603-610.
BROWNING K A, HARDMAN M E, HARROLD T W, et al, 1973. The structure of rainbands within a mid-latitude depression[J]. Q J Roy Meteor Soc, 99:215-231.
BROWNING K A, HILL F F, PARDOE C W, et al, 1974. Structure and mechanism of precipitation and effect of orography in a wintertime warm sector[J]. Q J Roy Meteor Soc, 100:309-330.
BURGESS D W, WOOD V T, BROWN R A, 1982. Mesoscale Evolution Statistics[C]. Preprints, 12th Conference on Severe Local Storms, San Antonio, TX, 12—15 Jan, 422-424.
BYERS H R, BRAHAM R R Jr, 1949. The Thunderstorm[M]. Washington D C: U S Government Printing Office:287.
CARLSON T N, 1998. Mid-latitude Weather Systems[M]. AMS:137.
CHAPPELL C F, 1986. Quasi-Stationary Convective Events[M]. Ray P S. Mesoscale Meteorology and Forecasting. Boston: Am Meteor Soc.
CHISHOLM A J, ENGLISH M, 1973. Alberta hailstorms[J]. AMS Met Monographs, 14:101.

CHISHOLM A J, RENICK J H, 1972. The kinematics of multicell and supercell Alberta hailstorms [C]. Alberta Hail Studies, Research Council of Alberta Hail Studies Rep, Canada, 72(2): 24-31.

COTTON W R, 1983. Upscale Development of Moist Convective Systems. Mesoscale Meteorology [M]. Sweden: SMHI.

COTTON W R, LIN M S, MCANELLY R L, et al, 1989. A composite model of mesoscale convective complexes[J]. Mon Wea Rev, 117:765-783.

DOSWELL C, CHURCH C, BURGESS D, et al, 1993. The Tornado: Its Structure, Dynamics, Prediction, and Hazards[M]. Church C, et al. Geophysical Monograph 79, Amer Geophys Union: 161-172.

FANKHAUSER J C, MOHR C G, 1977. Some correlations between various sounding parameters and hailstorm characteristics in Northeast Colorado[C]. Preprints, 10th Conference on Severe Local Storms, Oct 18—21, Omaha, Nebraska, 218-225.

FRITSCH J M, BROWN J M, 1982. On the generation of convectively driven mesohighs aloft[J]. Mon Wea Rev, 110:1554-1563.

FRITSCH J M, FORBES G S, 2001. Mesoscale convective systems[J]. Meteorol Monogr, 28: 323-357.

FRITSCH J M, MADDOX R A, 1981a. Convectively driven mesoscale weather systems aloft. Part I: Observations[J]. J Appl Meteor, 20:9-19.

FRITSCH J M, MADDOX R A, 1981b. Convectively driven mesoscale weather systems aloft. Part II: Numerical simulations[J]. J Appl Meteor, 20:20-26.

FRITSCH J M, MURPHY J D, KAIN J S, 1994. Warm core vortex amplification over land[J]. J Atmos Sci, 51:1781-1806.

FUJITA T T, 1971. Proposed mechanism of suction spots accompanied by tornadoes[C]. Preprints, Seventh Conf on Severe Local Storms, Kansas City, MO, Am Meteor Soc:208-213.

FUJITA T T, 1978. Manual of downburst identification for project NIMROD[R]. SMRP Research Paper, No 156, Dept of Geophysical Sci Chicago University.

FUJITA T T, 1981. Tornadoes and downbursts in the context of generalized planetary scales[J]. J Atmos Sci, 38(8):1511-1524.

FUJITA T T, 1986. Review of the History of Mesoscale Meteorology and Forecasting. Mesoscale Meteorology and Forecasting[M]. Am Moteor Soc.

HAMAN K E, 1976. On the airflow and motion of quasi-steady convective storms[J]. Mon Wea Rev, 104:49-56.

HAMAN K E, 1978. On the motion of a three dimensional quasi-steady convective storm in shear[J]. Mon Wea Rev, 106:1622-1626.

HAYASHI Y, 1970. A theory of large-scale equatorial waves generated by condensation heat and accelerating the zonal wind[J]. J Met Soc Japan, 48:140-160.

HOBBS P V, 1978. Organization and structure of clouds and precipitation on mesoscale and microscale in cyclonic storms[J]. Rev Geophys Space Phys, 16:741-755.

HODGES K I, THORNCROFT C D, 1997. Distribution and statistics of African mesoscale convective weather systems based on the ISCCP Meteosat imagery[J]. Mon Wea Rev, 125:2821-2837.

HOUZE R A Jr, 1977. Structure and dynamics of a tropical squall-line system[J]. Mon Wea Rev, 105:1540-1567.

HOUZE R A Jr, 1982. Cloud clusters and large-scale vertical motions in the tropics[J]. J Met Soc Japan, 60:396-410.

HOUZE R A Jr, 1993. Cloud Dynamics[M]. San Diego:Academic: 573.

HOUZE R A Jr, 1997. Stratiform precipitation in regions of convection: A meteorological paradox? [J]. Bull Am Meteor Soc, 78:2179-2196.

HOUZE R A Jr, 2004. Mesoscale convective systems[J]. Rev Geophy, 42: 2004RG, 000150.

HOUZE R A Jr, 2014. Cloud Dynamics[M]. 2nd ed. International Geophysics Series, Volume 104, Academic Press.

HOUZE R A Jr, BETTS A K, 1981. Convection in GATE[J]. Rev Geophys, 19:541-576.

HOUZE R A Jr, HOBBS P V, 1982. Organization and structure of precipitation cloud systems[J]. Advances in Geophysics, 24:225-300.

HOUZE R A Jr, RAPPAPORT E N, 1984. Air motions and precipitation structure of an early summer squall line over the eastern tropical Atlantic[J]. J Atmos Sci, 41: 553-574.

HOUZE R A Jr, SMULL B F, DODGE P, 1990. Mesoscale organization of springtime rainstorms in Oklahoma[J]. Mon Wea Rev, 118:613-654.

JOHNSON R H, HAMILTON P J, 1988. The relationship of surface pressure features to the precipitation and airflow structure of an intense mid-latitude squall line[J]. Mon Wea Rev, 116: 1444-1472.

KLEMP J B, WEISMAN M L, 1983. The dependence of convective precipitation patterns on vertical wind shear[C]. Preprints, 21st Conference on Radar Meteorology, Edmonton Alberta, Canada, Sept 19—23, 44-49.

LEMON L R, DOSWELL C A Ⅲ, 1979. Severe thunderstorms evolution and mesocyclone structure as related to tornadogenesis[J]. Mon Wea Rev, 107:1184-1197.

MADDOX R A, 1980. Mesoscale convective complexes[J]. Bull Am Meteor Soc, 61:1374-1387.

MADDOX R A, 1983. Large scale meteorological conditions associated with midlatitude mesoscale convective complexes[J]. Mon Wea Rev, 111:1475-1493.

MADDOX R A, 1986. Mesoscale Convective Complexes in the Midlatitudes. Mesoscale Meteorology and Forecasting[M]. Am Meteor Soc.

MADDOX R A, CHAPPELL C F, HOXIT L R, 1979. Synoptic and meso-α scale aspects of flash flood events[J]. Bull Am Meteor Soc, 60:115-123.

MADDOX R A, PERKEY D J, FRITSCH J M, 1981. Evolution of upper tropospheric features during the development of a mesoscale convective complex[J]. J Atmos Sci, 38(8):1664-1674.

MADDOX R A, RODGERS D M, HOWARD K W, 1982. Mesoscale Convective Complexes over the United States during 1981-Annual summary[J]. Mon Wea Rev, 110(10):1501-1514.

MARKOWSKI P, RICHARDSON Y, 2010. Mesoscale Meteorology in Midlatitudes[M]. UK: Willey-Blackwell.

MARWITZ J D, 1972a. The structure and motion of severe hailstorms. Part I: Supercell storms[J]. J Appl Meteor, 11:166-179.

MARWITZ J D, 1972b. The structure and motion of severe hailstorms. Part II: Multicell storms[J]. J Appl Meteor, 11:180-188.

MATEJKA T J, 1980. Mesoscale organization of cloud processes in extratropical cyclone[D]. Ph D dissertation, Dept of Atmospheric Sci, University of Washington.

MONCRIEFF M W, GREEN J S A, 1972. The Propagation and transfer properties of steady convective overturning in shear[J]. Q J Roy Meteor Soc, 98:336-353.

MONCRIEFF M W, MILLER M J, 1976. The dynamics and simulation of tropical cumulonimbus and squall lines[J]. Q J Roy Meteor Soc, 102:373-394.

OGURA Y, LIOU M T, 1980. The structure of a midlatitude squall line: A case study[J]. J Atmos Sci, 37:553-567.

ORLANSKI I, 1975. A rational subdivision of scale for atmospheric processes[J]. Bull Am Meteor Soc, 56:527-530.

PARKER M D, JOHNSON R H, 2000. Organization modes of midlatitude Mesoscale Convective Systems[J]. Mon Wea Rev, 128:3413-3436.

PARKER M D, JOHNSON R H, 2004a. Simulated convective lines with leading precipitation. Part I: Governing dynamics[J]. J Atmos Sci, 61(14):1637-1655.

PARKER M D, JOHNSON R H, 2004b. Structures and dynamics of quasi-2D mesoscale convective systems[J]. J Atmos Sci, 61(5):545-567.

SHOU S W, CHEN X R, LIN J R, 1982. Formation and structure of a severe squall line in eastern China[C]. 12th Conference on Severe Local Storms AMS.

SHOU S W, HOUZE R A Jr, 1990. The mesoscale structure of a mid-latitude squall line with a wide trailing stratiform region[R]. Research Report of NIM, 1987—1989.

SHOU S W, LI S S, 1991. Diagnoses of kinetic energy of a decaying onland typhoon[J]. Adv Atmos Sci, (8):4.

SNOW J T, AGEE E M, 1975. Vortex splitting in the mesocyclone and occurrence of tornado families [C]. 9th Conference on Severe Local Storms.

WALLACE J M, 1975. Diurnal variations in precipitation and thunderstorm frequency over the conterminous United States[J]. Mon Wea Rev, 103:406-419.

WALLACE J M, HOBBS P V, 2006. Atmospheric Science: An Introductory Survey[M]. Pittsburgh: Academic Press:504.

WEISMAN M L, KLEMP J B, 1982. The dependence of numerically simulated convective storms on vertical wind shear and buoyancy[J]. Mon Wea Rev, 110: 504-520.

WETZEL P J, COTTON W R, MCANELLY R L, 1983. A Long lived mesoscale convective complex. Part II: Evolution and structure of the mature complex[J]. Mon Wea Rev, 111:1919-1937.

WILHELMSON R B, KLEMP J B, 1983. Numerical simulation of severe storms within lines[C]. Preprint 13th Conf on Severe Local storms AMS, 231-234.

Chapter 6
Atmospheric Instabilities

The atmospheric instability (or stability) implies the evolution tendency of the disturbance of the air parcel, which is originally under a sort of balance state. If the disturbance will be increased or decreased or not changed with time, then the situation is called instable or stable or neutral respectively. If an air parcel is lifted up to a certain height and, if left to itself, it will return to its original location, however, if the air parcel is lifted beyond a certain height (i. e., the level of free convection), and is then left to itself, it will continue rising. This situation is called conditional instability. An analog situation of stable is shown in Fig. 6.1a, a sort of balance state where a ball is originally located at the lowest point in a valley. If the ball is displaced in any direction and is then left itself, it will return to its original location at the base of the valley. An analog situation of unstable is shown in Fig. 6.1b, where a ball is initially on the top of a hill. If the ball is displaced in any direction, and is then left itself, it will roll down the hill. An analog situation of stable is shown in Fig. 6.1c, where a ball is on a flat surface. If the ball is displaced, and then left to itself, it will not move. An analog situation of conditionally instability is shown in Fig. 6.1d where a displacement of a ball to a point A, which lies to the left to the hillock, will result in the ball rolling back to its original position. However, if the displacement takes the ball to a point B on the other side of the hillock, the ball will not return to its original position but will roll down the right-hand side of the hillock. The formations of many atmospheric convective phenomena are closely related with various atmospheric instabilities. For example, the common small perturbations or the cumulus convections with the horizontal scale about ten or less kilometers are normally related with the static (or gravity) instability or shearing (or Kelvin-Helmholtz) instability; while the mesoscale cloud clusters and rain bands may related with CISK (the conditional instability of second kind) and the inertial-buoyancy instability, i. e. the symmetric instability. In this chapter we will introduce the concepts and the criteria of the atmospheric instabilities closely related with mesoscale atmospheric disturbances.

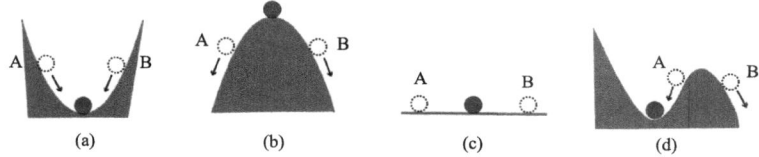

Fig. 6.1 Analogs for stable (a); unstable (b); neutral(c); and conditional instability (d). The red point is the original position of the ball, and the white circles are displaced positions. Arrows indicate the direction the ball will move from a displaced position if the force that produced the displacement is removed. (After Wallace, 2006)

§ 6.1 Conditional instability

6.1.1 Static instability

Now let us consider the stability of an air parcel originally in static balance state after suffering a vertical disturbance in the stratification atmosphere, which can be characterized by the vertical gradient of the potential temperature.

Suppose the vertical velocity, the vertical acceleration, the vertical displacement, the potential temperature, the vertical gradient of the potential temperature and the environmental potential temperature as well as the environmental vertical gradient of the potential temperature of the air parcel are $w, \frac{dw}{dt}, \delta z, \vartheta, \frac{d\vartheta}{dz}, \bar{\vartheta}, \frac{d\bar{\vartheta}}{dz}$ respectively, and suppose there are no exchanges of the heat, moisture, mass and momentum as well as no friction between the air parcel and environment. Meanwhile the quasi-static condition is satisfied, in other words, the pressure of the air parcel is always same with that of the environment. Based on the above assumptions, the following equation can be derived from the equation (1.3.12):

$$\frac{dw}{dt} = \frac{d^2}{dt^2}(\delta z) \cong g\left(\frac{\theta - \bar{\theta}}{\bar{\theta}}\right) = \frac{g}{\bar{\theta}}[\theta(z) - \bar{\theta}(z)] \qquad (6.1.1)$$

Suppose at the initial height $z_0 = 0$, $\bar{\theta}_{z_0} = \theta_{z_0} = \theta_0$. hence at the height $z = z_0 + \delta z$, the environmental potential temperature will be:

$$\bar{\theta}(z) = \bar{\theta}(\delta z) = \bar{\theta}_{z_0} + \frac{d\bar{\theta}}{dz}\delta z \qquad (6.1.2)$$

Under the dry adiabatic condition the potential temperature is conservation, i.e., $\frac{d\theta}{dz} = 0$, thus

$$\theta(z)=\theta(\delta z)=\theta_0+\frac{d\bar\theta}{dz}\delta z=\theta_0 \tag{6.1.3}$$

Substituting the above two formulas into the formula (6.1.1), we can obtain

$$\frac{d^2}{dt^2}(\delta z)=-N^2\delta z \tag{6.1.4}$$

where

$$N^2=\frac{g}{\bar\theta}\frac{d\bar\theta}{dz} \tag{6.1.5}$$

The equation (6.1.4) is a simple harmonic oscillation equation, it has solution as follows:

$$\delta z=A\exp(iNt) \tag{6.1.6}$$

where A is the initial displacement of the air parcel; N is the frequency of the buoyancy oscillation, called as the buoyancy or Brunt-Väisälä frequency.

It can be seen from equation (6.1.6) that when $N^2>0$ the air parcel will oscillate around its initial locating level with a period T, $(T=2/N)$. Under the mean circumstances of the troposphere $N\cong1.2\times10^{-2}$ s^{-1}, thus $T\cong8$ minutes. When $N=0$, the acceleration equals zero. When it arrives to a new position it will reach balance at the new position. When $N^2<0$, the vertical displacement of the air parcel will increase with time according to an exponential form. The above three circumstances denote the atmospheric stratification are in static stable, neutral and static instable respectively. According to the formula (6.1.5), we know that the criteria of the instabilities expressed by using N^2 can also be expressed by using the stratification parameter $\frac{d\bar\theta}{dz}$ as shown in (6.1.7).

$$\frac{d\bar\theta}{dz}\begin{cases}>0 & \text{static stable}\\ =0 & \text{neutral}\\ <0 & \text{static instable}\end{cases} \tag{6.1.7}$$

6.1.2 Conditional instability and convective instability

The static instable is favorable to form convection. However, the real atmosphere is usually static stable because for the dry air the potential temperature is usually increasing with height. Thus we have to consider the effect of the water vapor in the moist air. When the moist air parcel is lifted adiabatically the water vapor will be cooling to generate condensation. While the condensation will cause the releasing of the latent heat to increase the buoyancy suffered by the air parcel and then to cause it accelerating upward, in other word to change it into unstable. Commonly we call the circumstance that is stable for dry or unsaturated air while is unstable for the saturated moist air as the conditional instable.

For the moist adiabatic motions, we take the equivalent potential temperature θ_e, which is conserved quantity in the moist adiabatic process and may be expressed as

$$\theta_e \cong \theta \exp(L_c q_s / C_p T) \qquad (6.1.8)$$

where L_c, q_s, C_p and T are condensation latent heat, saturation specific humidity, constant pressure specific heat and temperature respectively, θ is potential temperature. Taking the logarithm and differential for the formula (6.1.8), and to obtain the following express:

$$d\ln\theta = d\ln\theta_e - d(L_c q_s / C_p T) \qquad (6.1.9)$$

For the moist adiabatic motion, $d\ln\theta_e \cong 0$, thus the above formula may be rewritten as follows:

$$\frac{d\theta}{\theta} \cong -d\left(\frac{L_c q_s}{C_p T}\right) \cong -\frac{\partial}{\partial z}\left(\frac{L_c q_s}{C_p T}\right)$$

or

$$\frac{d\theta}{dz} \cong -\theta \frac{\partial}{\partial z}\left(\frac{L_c q_s}{C_p T}\right) \qquad (6.1.10)$$

This implies the potential temperature θ of the air parcel is no longer conservation when it is rising, but it has a lapse rate of the potential temperature. Thus, when we discuss the formula (6.1.3), $\theta(\delta z) \neq \theta_0$, but should be

$$\theta(\delta z) = \theta_0 + \frac{d\theta}{dz}\delta z$$

i. e.

$$\theta_0 = \theta(\delta z) - \frac{d\theta}{dz}\delta z \qquad (6.1.11)$$

substituting the equation (6.1.11) into (6.1.2), and to combine equation (6.1.10) then we can get

$$\bar{\theta}(\delta z) - \theta(\delta z) = \theta \frac{\partial}{\partial z}\left(\frac{L_c q_s}{C_p T}\right)\delta z + \frac{d\bar{\theta}}{dz}\delta z$$

Substituting the above formula into the equation (6.1.1) to obtain the following

$$\frac{d^2}{dt^2}(\delta z) = g\left(\frac{\theta - \bar{\theta}}{\theta}\right) \cong -g\frac{d}{dz}\left[\ln\bar{\theta} + \left(\frac{L_c q_s}{C_p T}\right)\right]\delta z$$

$$= -g\frac{d\ln\bar{\theta}_e}{dz}\delta z = -\frac{g}{\bar{\theta}_e}\frac{d\bar{\theta}_e}{dz}\delta z = -N_w^2 \delta z \qquad (6.1.12)$$

The equation (6.1.12) has the same form with (6.1.4), the only difference is the parameter N^2 is changed into N_w^2. Where $N_w^2 = \frac{g}{\bar{\theta}_e}\frac{d\bar{\theta}_e}{dz}$ is called as the moist buoyancy frequency, in which $\bar{\theta}_e$ is the equivalent potential temperature of the environment.

Similar to the above discussion, now we can get the solution of the equation (6.1.12) as follows:

$$\delta z = A\exp(iN_w t) \tag{6.1.13}$$

And we can obtain the following criteria:

$$\frac{\partial \bar{\theta}_e}{\partial z} \begin{cases} >0 & \text{conditional stable} \\ =0 & \text{neutral} \\ <0 & \text{conditional unstable} \end{cases} \tag{6.1.14}$$

The convective weather are usually occurred under the conditional unstable situation. While sometimes even under the original conditional stable situation with the stratification that the upper level is dry and lower level is moist may also possibly cause the convective weather if there is a strong lifting motion to cause the whole stable layer lifted. The originally stable layer with aboard dry and under wet may be changed into an instable layer after the layer is lifted entirely, this sort of the instability is called the convective instability, the conception of which is illustrated in Fig. 6.2. The criteria of the convective instability are same with that of the conditional instability as shown in formula (6.1.15).

$$\frac{\partial \bar{\theta}_{sw}}{\partial z} \left(\text{or } \frac{\partial \bar{\theta}_{se}}{\partial z}, \frac{\partial \bar{\theta}_e}{\partial z} \right) \begin{cases} >0 & \text{convective stable} \\ =0 & \text{neytral} \\ <0 & \text{convective unstable} \end{cases} \tag{6.1.15}$$

where $\bar{\theta}_{sw}$, $\bar{\theta}_{se}$ and $\bar{\theta}_e$ are the pseudo wet bulb potential temperature, the pseudo equivalent potential temperature and the equivalent potential temperature of the environment respectively.

Both the conditional instability and the convective instability are the potential instability. Most of the severe convective weather occurred under the potential instable situations. For instance, the strong squall line process occurred in Nanjing Jiangsu Province on Jun. 17, 1974 and the hailstorm process occurred in Linbi, Anhui Province on Jun. 6, 1975 were formed both under the strong potential instability condition previous the storms. The weather of the former was much stronger than the later since the instability of the former was much stronger than the later (Fig. 6.3).

6.1.3 The scale of storm induced by potential instability

The thunderstorms are caused by the overturning of the potential unstable atmospheric layer. Now let us analyze the features of the atmospheric layers, which are restrained between two horizontal surfaces with the interval of h as shown in Fig. 6.4, when the convective overturning occur.

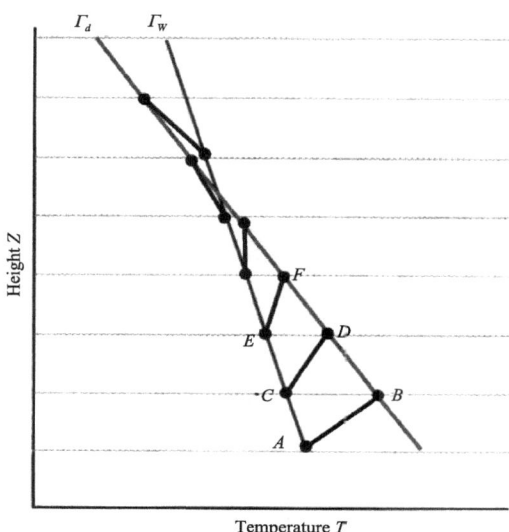

Fig. 6.2　Illustration of the increase in the lapse rate $-dT/dz$ within an inversion layer is lifted. The black line segment AB represents the temperature profile within the inversion layer before it is lifted; CD represents the temperature profile in the same layer after it is lifted one height increment, EF after it is lifted two height increments, etc. It is assumed that the bottom of the layer is saturated with water vapor and cools at the saturated adiabatic lapse rate as the layer is lifted, while the top of the layer is unsaturated and cools at the dry adiabatic lapse rate. The steepening of the lapse rate due to differential rate of cooling is partially compensated by the expansion of the air within the layer as it rises. This effect is not represented in the diagram. (After Wallace, 2006)

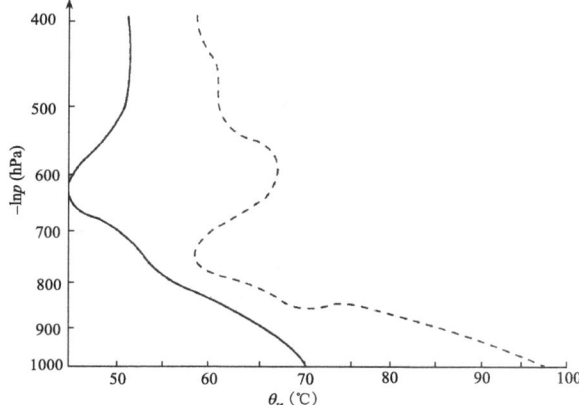

Fig. 6.3　The vertical profiles of $\bar{\theta}_{se}$ in Linbi, Anhui Province on Jun. 6, 1975 (solid line) and in Nanjing, Jiangsu Province on Jun. 17, 1974 (dashed line).

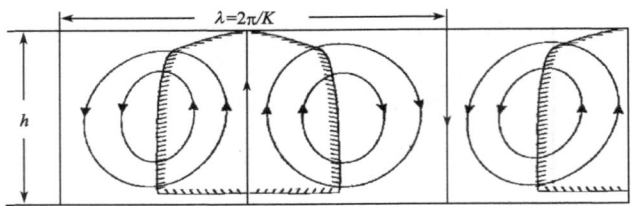

Fig. 6.4 The stream line distribution derived from a simple linear model of the convection between two horizontal planes with interval of h. The shaded area denotes cumulus. (After Kessler, 1987)

Assume the atmosphere is incompressible and non-viscous fluid. The basic atmospheric state is stationary, and the disturbance is the two dimensional motions in (x, z) plane. Thus the atmospheric state can be expressed as follows:
$$u=u', \quad w=w', \quad p=P(z)+p', \quad \rho=\bar{\rho}(z)+\rho'$$
Considering the effect of the Earth's rotation and neglecting the second order small quantities, then we get the equations:

$$\frac{\partial u}{\partial t}=-\frac{1}{\bar{\rho}}\frac{\partial p'}{\partial x} \tag{6.1.16}$$

$$\frac{\partial w}{\partial t}=-\frac{1}{\bar{\rho}}\frac{\partial p'}{\partial z}-\frac{\rho'}{\bar{\rho}}g \tag{6.1.17}$$

$$\frac{\partial u}{\partial x}+\frac{\partial w}{\partial z}=0 \tag{6.1.18}$$

Assume the density satisfying the continuity equation:

$$\frac{\partial \rho'}{\partial t}+w\frac{\partial \bar{\rho}}{\partial z}=0 \tag{6.1.19}$$

Setting the equations (6.1.16)—(6.1.19) have the solution in the following form:
$$(u,w,p',\rho') \propto \exp(ikx+\sigma t) \tag{6.1.20}$$
Substituting the equation (6.1.20) into the equations (6.1.16)—(6.1.19), then we can obtain four equations for four variables as follows:

$$\sigma\bar{\rho}u+ikp'=0 \tag{6.1.21}$$

$$\sigma\bar{\rho}w+\frac{\partial p'}{\partial z}+g\rho'=0 \tag{6.1.22}$$

$$iku+\frac{\partial w}{\partial z}=0 \tag{6.1.23}$$

$$\sigma\rho'+w\frac{\partial \bar{\rho}}{\partial z}=0 \tag{6.1.24}$$

Suppose $\bar{\rho}$ is a constant number, the only exception of the supposing is in the equation (6.1.24), in which $\dfrac{\partial \bar{\rho}}{\partial z} \ne 0$, unless $h\left|\dfrac{\partial \bar{\rho}}{\partial z}\right| \ll \bar{\rho}$. The equations (6.1.21)—(6.1.24) can be changed into an equation about the only variable w, i. e. the vertical velocity, as follows:

$$\frac{\partial^2 w}{\partial z^2} + k^2\left(\frac{g}{\sigma^2 \bar{\rho}}\frac{\partial \bar{\rho}}{\partial z} - 1\right)w = 0 \qquad (6.1.25)$$

When deriving the equation (6.1.25) from the equations (6.1.21)—(6.1.24), the term $\left(\dfrac{1}{\rho}\right)\left(\dfrac{\partial \bar{\rho}}{\partial z}\right)\left(\dfrac{\partial w}{\partial z}\right)$ was neglected. If we set $\sigma_0^2 = (g/\bar{\rho})(\partial \bar{\rho}/\partial z)$, and take σ_0 as a constant number, then the function

$$w = w_0 \sin(\pi z/h) \qquad (6.1.26)$$

where w_0 is a constant number. The equation (6.1.26) will satisfy the equation (6.1.25). Meanwhile as long as

$$\sigma^2 = k^2 \sigma_0^2/(k^2 + \pi^2/h^2) = \sigma_0^2/[1 + (\pi^2/h^2)/k^2] \qquad (6.1.27)$$

it will be considered that $w = 0$ at the upper and lower planes. Thus, when $\partial \bar{\rho}/\partial z \propto \sigma_0^2 > 0$, w will be exponentially increased.

From the above analysis, we can furtherly learn the maximum growth rate of the disturbance and the horizontal scale of the storm. It can be seen from Fig. 6.4 derived from the equations (6.1.21)—(6.1.24) that the cell's width of the convective cells is $\lambda = 2\pi/k$, and the depth is h. It can be seen that for given h, if $k^2 \gg \dfrac{\pi^2}{h^2}$, in other words the wave number k is bigger, the growth rate σ^2 will be equal to σ_0^2, i. e. $\sigma^2 = \sigma_0^2$; when the wave number k is smaller, then the growth rate σ^2 will be linear increased with k. When $k^2/\pi^2/h^2 \geqslant 1$, i. e. when $k^2 \geqslant \dfrac{\pi^2}{h^2}$, the growth rate σ^2 will be the largest. Since $k = \dfrac{2\pi}{\lambda}$, thus $\lambda \leqslant 2h$. This shows that when the horizontal scale is equal to or less than two times of the vertical scale, the disturbance growth rate will be the greatest. In general speaking, in spite of the patterns of the initial disturbance, the pattern of its final result must be in the type that determined by the greatest growth rate. Since the horizontal scale of the thunderstorm is $\lambda/2$, thus the horizontal scale of the thunderstorm is normally equivalent to the depth of the convective instable layer, roughly say is about 10 kilometers. It is also easy to understand the reason why the shorter waves grow faster, because for the given temperature amplitude (T'), under the circumstance of short wave the horizontal gradient of density is much greater than that under the longer wave situation, so that the buoyancy force and the vertical acceleration will be much greater.

§ 6.2 Conditional instability of second kind (CISK)

The pure conditional instability, we may call it the first kind of the conditional instability, could not well explain the formation of the cloud clusters which are organized, with large horizontal scales and long temporal scale existed in the tropical and mid-latitude areas.

Firstly this is because that the conditional instability requires not only to satisfy the condition $\partial \theta_e/\partial z < 0$, but also the moist air should be saturated. For the unsaturated air, it is necessary to have low level convergence to force the air to be lifted and causing saturation and convection. For the tropical atmosphere, the condition $\partial \theta_e/\partial z < 0$ is commonly satisfied at the low atmosphere, but the mean relative humidity in tropical area is lower than 100% as shown in Fig. 6.5. Hence in the tropical area the cumulus convection development is not always vigorous. The vigorous convections are only be founded under the situation cooperated with convergence and ascending motions. This shows that the conditions for the formation of the vigorous convective activities in the tropical area should include not only the pure convective instability but also need the cooperation of the large scale fluid field for causing the convergence and ascending motion.

Secondary, the analysis from § 6.1 has showed that the greatest growth rate of the unstable waves induced by the conditional instability of the first kind is corresponding to the motion system with the scale of the single cumulus cloud system. Thus it is difficult to explain the formation of the huge convective cloud clusters by using the pure conditional instability. This problem then promotes the peoples to consider the effects of the small scale convective heating for increasing the large scale fluid field after the convection occurring. Charney and Eliassen (1964) and Ooyama (1964) et al. studied the interaction between the small scale convection and large scale fluid filed and reduced it as the following process. At first the large scale fluid field supplies the necessary water vapor convergence and ascending motion for the developing of the cumulus convections by means of the Ekman Pumping effect in the frictional boundary layer. And then the latent heat released by the cumulus convective condensation turns to be the energy source necessary for driving the large scale disturbance. Hence both the small scale convection and the large scale fluid field are co-developed by means of the interaction and cooperation between them. This kind of the unstable increasing physical mechanism of the convection and large scale fluid field through interaction each other may be called as "the conditional instability of the second kind", or CISK for short.

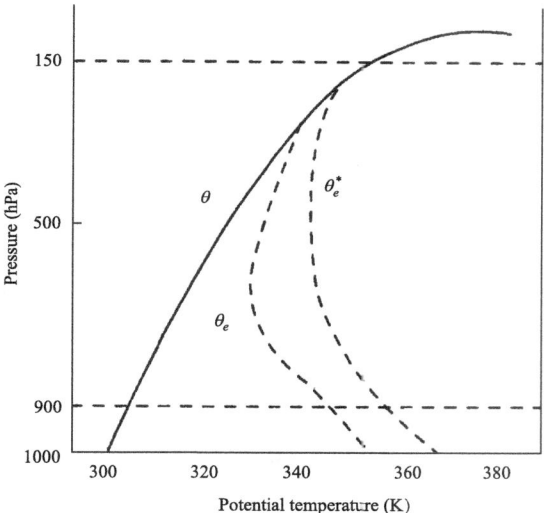

Fig. 6.5 The typical sounding curves of the tropical atmosphere. (θ is potential temperature; θ_e is equivalent potential temperature; θ_e^* is the equivalent potential temperature at the layer with same temperature and saturated water vapor)

Now let us use the mathematic analysis method to explain furtherly the CISK process. As normally, we will take the parameterization method for describing the cumulus convective heating problem, in other words, we will use the parameters of the large scale field to express the effect of the small scale cumulus convective heating. In order to simplify the problem, a most simple cumulus convective parameterization scheme is taken, i. e. , to assume all the water vapor drawing from the boundary layer will totally provide for the cumulus development, and all the water vapor will be condense and release latent heat. The quantity of the condensation is proportional to the vertical velocity in the frictional layer. The mathematic form can be expressed as follows:

$$Q_c = F\eta\left\{\frac{1}{2}(|\omega_E| - \omega_E)\right\} \tag{6.2.1}$$

where Q_c is heating of latent heat; η is a parameter determined by experiences or the assumed conditions; F is the vertical distribution function of the condensation latent heat, which is commonly taken as the following forms approximately:

$$F=1 \quad \text{or} \quad Q_c=-\eta\omega_E \quad \text{(when } \omega_E<0\text{)}$$
$$F=0 \quad \text{or} \quad Q_c=0 \quad \text{(when } \omega_E\geqslant 0\text{)} \tag{6.2.2}$$

where ω_E is the pressure coordinate vertical velocity at the top of the frictional boundary layer (FBL) indicating the pumping effect of FBL. Charney and Eliassen (1949) derived the following relational expression:

$$\omega_E=-\rho_E g(K/2f)^{\frac{1}{2}}\zeta_g \tag{6.2.3}$$

where K is the turbulence eddy coefficient, f is Coriolis parameter, ρ_E is the air density at the top of boundary layer, ζ_g is the vorticity of geostrophic wind. The formula (6.2.3) shows that the vertical velocity at the top of the boundary layer is proportional to the geostrophic wind vorticity of the large scale fluid field. To combine the formulas (6.2.2) and (6.2.3) and therefore the parameterization of the small scale cumulus convective heating may be expressed by using the large scale field parameters.

Based on the above discussions, we can discuss the CISK problem now. Taking the two layers model as shown in Fig. 6.6, the vorticity equations and the thermodynamic equation can be written as follows:

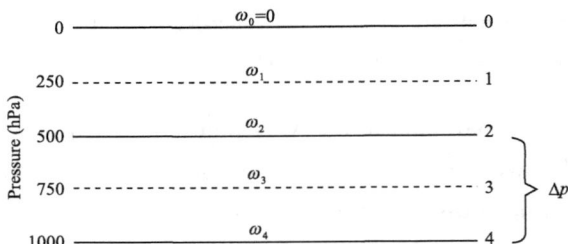

Fig. 6.6 The arrangement of variables in vertical direction in the two layers model.

$$\frac{\partial \zeta_1}{\partial t}+\mathbf{V}_1\cdot\nabla(\zeta_1+f)=\frac{f_0}{\Delta p}\omega_2 \tag{6.2.4}$$

$$\frac{\partial \zeta_3}{\partial t}+\mathbf{V}_3\cdot\nabla(\zeta_3+f)=-\frac{f_0}{\Delta p}(\omega_2-\omega_4) \tag{6.2.5}$$

$$\frac{\partial T_2}{\partial t}+\mathbf{V}_2\cdot\nabla T_2-\sigma\omega_2=\frac{Q}{C_p} \tag{6.2.6}$$

where $\mathbf{V}_{1,3}=\mathbf{k}\times\nabla\psi_{1,3}$ are the wind velocities at the level 1 and 3 respectively, the corresponding vorticities are

$$\zeta_{1,3} = \nabla^2 \psi_{1,3}, \quad V_2 = \frac{1}{2}(V_1 + V_3)$$

where V_2 is the wind velocity at midlevel, which is assumed as the mean value of the wind velocities at upper level and low level; ψ is stream function; $\sigma = -\frac{T}{\theta}\frac{\partial \theta}{\partial p}$ is the static stability parameter; Δp is the pressure difference between upper level and low level; f_0 is taken as a constant number.

If in the right side of the equation (6.2.6) Q is only the sum of condensation heating caused by the large scale motion and the cumulus convection, that is

$$Q = -L_c \frac{\partial q_s}{\partial p}\omega_2 - \eta \omega_E \tag{6.2.7}$$

From the static equation we know that

$$-\frac{\partial \phi}{\partial p} = \alpha = \frac{R}{P}T$$

where $\phi = f_0 \psi$ is the geopotential of the constant pressure surface. Thus in the two layers model, the temperature can be expressed as

$$T_2 = \frac{P}{R}f_0\left(-\frac{\partial \psi}{\partial p}\right) \simeq \frac{P\,f_0}{R\,\Delta p}(\psi_1 - \psi_3) = \frac{P\,f_0}{R\,\Delta p}h \tag{6.2.8}$$

where $h = (\psi_1 - \psi_3)$, corresponding to the thickness between 1—3 level. Substituting the equation (6.2.7), (6.2.8) and (6.2.3) into the equation (6.2.6), we can obtain

$$\frac{\partial h}{\partial t} + V_2 \cdot \nabla h - \sigma_m \omega_2 = \eta \lambda \zeta_3 \tag{6.2.9}$$

where $\sigma_m = -\frac{R}{f_0}\left(\frac{T\partial \theta_*}{\theta_* \partial p}\right)$ is the moist static stability parameter; $\lambda = \chi \rho_E g (K f_0^{-3}/2)^{1/2}$, $\chi = R/C_p$ and $\Delta p = p_2$; and the vorticity at the third level ζ_3 is set as the vorticity at the top of the boundary layer approximately. In the right side of the equation (6.2.9), there is only the cumulus convection heating η term, while the original large scale ascending condensation heating effect is already included in the moist static stability parameter σ_m in the left side of the equation. To solute ω_2 from equation (6.2.9), and substitute it into (6.2.4) and (6.2.5) and only the linear terms are considered, then we have

$$\frac{\partial \zeta_1}{\partial t} = \frac{f_0}{\sigma_m \Delta p}\left(\frac{\partial h}{\partial t}\right) - \eta \frac{\lambda f_0}{\sigma_m \Delta p}\zeta_3 \tag{6.2.10}$$

$$\frac{\partial \zeta_3}{\partial t} = -\frac{f_0}{\sigma_m \Delta p}\left(\frac{\partial h}{\partial t}\right) + \eta \frac{\lambda f_0}{\sigma_m \Delta p}\zeta_3 - \frac{f_0 \lambda'}{\Delta p}\zeta_3 \tag{6.2.11}$$

In the derivation process of the formula (6.2.11) the assumptions that $\omega_4 \cong \omega_E = -\lambda' \zeta_3$, $\lambda' = \rho_E g \left(\dfrac{K}{2f_0}\right)^{\frac{1}{2}}$ have already been made. For highlighting the effect of the cumulus convective heating, the fluid field changes induced by the term of η in the linear equations (6.2.10) and (6.2.11) are taken into account as the only factor. That is

$$\left(\frac{\partial \zeta_1}{\partial t}\right)_\eta = -\eta \frac{\lambda f_0}{\sigma_m \Delta p} \zeta_3 \qquad (6.2.12)$$

$$\left(\frac{\partial \zeta_3}{\partial t}\right)_\eta = \eta \frac{\lambda f_0}{\sigma_m \Delta p} \zeta_3 \qquad (6.2.13)$$

It can be seen from the above two formulas that, when there is the cyclonic vorticity in the low level atmospheric fluid field (namely, $\zeta_3 > 0$), the ascending motion will be induced due to the frictional convergence in the boundary layer ($\omega_E < 0$, see the formula 6.2.3). Due to the ascending motion the small scale cumulus convection heating will be induced ($Q_c = -\eta, \omega_E > 0$). This heating effect will in turn promote the low level large scale atmospheric fluid field to be further intensified ($\dfrac{\partial \zeta_3}{\partial t} > 0$, see the formula (6.2.13)), and at the meantime the upper level anticyclone circulation increased, $\dfrac{\partial \zeta_1}{\partial t} < 0$. Thus the large scale fluid field and the small scale cumulus convection heating will be interacted as both the cause and results and developing together. Fig. 6.7 is the schematic diagram to denote the CISK process.

The above analysis and Fig. 6.7 show that CISK is actually a self-excited instability of the large scale fluid field and is also one of the forms to present the feedback effect of the latent heat.

CISK may be used for explain the development of the tropical disturbances. Some people also use it to explain the development of the mesoscale convective systems (MCSs) in mid-high latitude area. For instance, Rasmussen (1979) explained the development of the polar low by use CISK mechanism. In a low level convergence area induced by some kind of initial disturbances and in a conditional unstable unsaturated atmosphere, when air parcel is lifted to the condensation level the cumulus convection will be formed. The latent heating will cause divergence at high level and convergence at low level. The low level convergence will increase the water vapor convergence in boundary and also the positive vorticity of the initial disturbance. The layer will cause increasing the vertical velocity at the top of the Ekman layer. Thus the surface air will be lifted to the condensation level and much more water vapor will condensate to cause

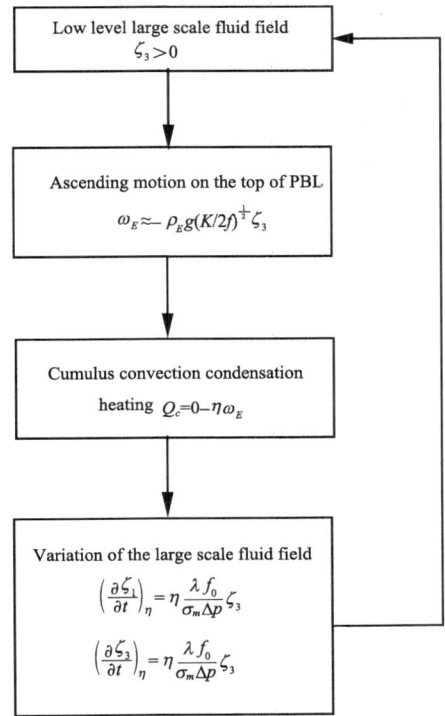

Fig. 6.7 The schematic diagram of the CISK process.

new convection development. This will form a positive feedback. Once the process begins, the scale of both the convection and the disturbance (the depression) will be increased. Rasmussen employed a linear quasi-geostrophic model, of which the pressure at 0 level is 400 (or 650) hPa, the vertical velocity is 0, the pressure at the 1st, 2nd, 3rd, 3.5th and 4th are 500, 600, 700, 800 and 900 (or 700, 750, 800, 850 and 900) hPa respectively. The 4th level is coincided with the top of boundary layer, on which the vertical velocity $\omega_4 = -K\nabla^2 \phi_4$, where K is the coefficient of turbulence eddy, and ϕ is the geopotential of the constant pressure surface. The sensible heat and latent heat pumped from Ekman layer will be transported upward through the top of Ekman layer and distributed linearly between the 1st and the 3rd level. Applying the vorticity equation on the 2nd and 3.5th levels and assuming the basic flow is constant, and the solutions are as following:

$$\phi, \omega \sim \exp(ikx)\exp(\sigma t)$$

where k is wave number, σ is the growth rate of the disturbance. The results from the original equation show that for example, when the wavelength is 200 km, $\sigma = -3.61 \times 10^{-6}$ s^{-1}; when the wavelength is 400 km, $\sigma = 1.99 \times 10^{-5}$ s^{-1}; when the wavelength is 600 km and 800 km, σ will be 1.68×10^{-5} s^{-1} and 1.16×10^{-5} s^{-1} respectively. The relationship between the growth rate and the wave length of the disturbance is denoted in Fig. 6.8. From the figure it can be seen that when the wavelength is about 400 km, the growth rate is the biggest.

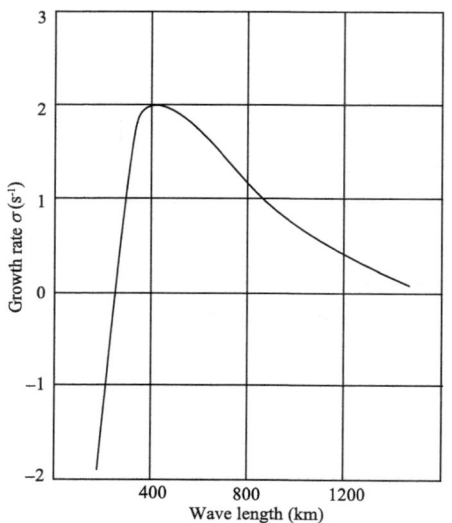

Fig. 6.8   The relationship between the growth rate and the wave length.
(After Rasmussen, 1979)

The CISK mechanism discussed above emphasized the frictional convergence effect in the boundary layer. Bates (1973) considered both the effects of the frictional convergence in boundary layer and of the convergence induced by allobaric wind at meantime. This kind of CISK mechanism is called as "Generalization of CISK".

§ 6.3   Wave-CISK

In the mechanism of CISK, the key point is to introduce the pumping effect of the frictional boundary, namely to require the formula (6.2.3) is set up. However, in practical situation, the frictional convergence in boundary layer is not the only factor to cause the vertical motion to initiate convection. There are still many other factors to cause the low level vertical motions to lift the air to the free convective level (LFC) to cause the moist convection take place. For instance the low level allobaric wind convergence mentioned above is just about one of the factors. In recent years, many people noted that the waves in the atmosphere, especially the gravity inner waves may generate strong low level horizontal convergence, thus in this situation the CISK process may be realized dispense with the Ekman pumping effect. Lindzen (1974) called the CISK process created by the gravity inner waves the wave-type of CISK (wave-CISK). Raymond (1975, 1976, 1983) et al. think that the Wave-CISK mechanism seems particularly suitable for mesoscale phenomena. For instance, the observations show that under the situation without external forcing the life cycle of the most of isolated moist convective cells, except the rotational supper cell, are quite short, while the life cycles of the clustering of the convective cells or the mesoscale cloud masses are relatively much longer, their development may be regarded as a convective self-excited process. To be specific, when the convection occurs, the gravity waves will be generated in the anvil, namely in the outflow layer, at the top of the convective cloud. In the area that the gravity wave intersected with surface will occur convergence (divergence). The convergence generated by gravity waves will cause the low level air to be lifted and promote the convection develop furtherly. In turn, more gravity waves will be created. Fig. 6.9 is a schematic diagram to show the relationship between the gravity waves and the convective storms. In the diagram we can see the wave pocket consisted of a convergence zone and a divergence zone, their wave length are corresponding to the diameter of the mesoscale storms. The convective plumes develop in the convergence area and move to the divergence area.

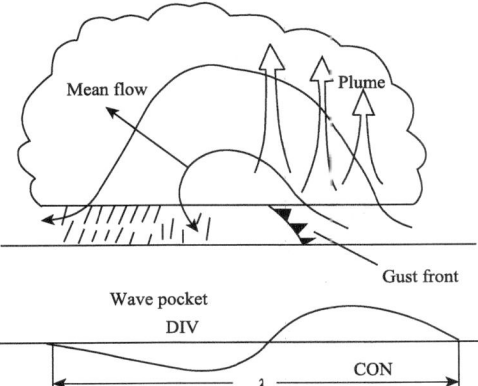

Fig. 6.9 A schematic diagram of the relationship between the gravity wave pocket and the convective storm. (After Raymond, 1957)

In order to explain the wave-CISK mechanism furtherly, Lindzen (1974) introduced a simple Boussinesq model. Considering the 2-dimensional wave in the stratified, irrotational, Boussinesq fluid, assuming the temporal scale is O(1h), so that the hydrostatic assumption is valid. Thus the disturbance field equations are as follows:

$$\frac{\partial u}{\partial t}=-\frac{1}{\rho_0}\frac{\partial p}{\partial x} \qquad (6.3.1)$$

$$\frac{\partial p}{\partial z}=-\rho g \qquad (6.3.2)$$

$$\frac{\partial \rho}{\partial t}+w\frac{\partial \rho_0}{\partial z}=-Q \qquad (6.3.3)$$

$$\frac{\partial u}{\partial x}+\frac{\partial w}{\partial z}=0 \qquad (6.3.4)$$

where u is the disturbance horizontal velocity, w is the disturbance vertical velocity, ρ is the disturbance density, others are the common symbols. Meantime,

$$\rho=\rho_0 e^{-\frac{z}{H}} \quad (H\cong 7.5 \text{ km}) \qquad (6.3.5)$$

where ρ_0 is the basic density distribution, Q is the thermal forcing created due to the releasing of the latent heat, and assuming

$$Q=\hat{Q}(z)e^{ik(x-ct)} \qquad (6.3.6)$$

Similarly, all the disturbance fields have the same form with above formula, so that the equations (6.3.1)—(6.3.4) can be changed into the following forms:

$$-ikc\hat{u}=-\frac{1}{\rho_0}ik\hat{p} \qquad (6.3.7)$$

$$\frac{d\hat{p}}{dz}=-\hat{\rho}g \qquad (6.3.8)$$

$$-ikc\hat{\rho}-\frac{\hat{W}}{H}\rho_0=-\hat{Q} \qquad (6.3.9)$$

$$-ik\hat{u}+\frac{d\hat{W}}{dz}=0 \qquad (6.3.10)$$

From the equation (6.3.10), $\hat{u}$ can be obtained. Substituting it into (6.3.7) we can obtain $\hat{p}$, substituting $\hat{p}$ into (6.3.8) to obtain $\hat{\rho}$, and then substituting it into (6.3.9), and assuming

$$\widetilde{W}=\rho_0^{\frac{1}{2}}\hat{W} \qquad (6.3.11)$$

To get

$$\frac{d^2 \widetilde{W}}{dz^2}+\lambda^2 \widetilde{W}=\frac{g}{C^2}\hat{Q}\rho_0^{-\frac{1}{2}} \qquad (6.3.12)$$

where
$$\lambda^2 = \frac{g}{HC^2} - \frac{1}{4H^2} \qquad (6.3.13)$$

Taking the boundary condition
$$\widetilde{W} = 0 \quad \text{at} \quad Z = 0 \qquad (6.3.14a)$$
$$\widetilde{W} \propto e^{-\bar{a}z} \quad \text{for} \quad Z > Z_t \qquad (6.3.14b)$$

(Z_t is the height of the cloud top), then we can obtain the solutions of the equation (6.3.12) as follows:

$$\widetilde{W} = -\frac{\sin\lambda Z}{\lambda} \int_{Z_c}^{Z_t} f e^{-\bar{a}z} dz, \quad 0 \leqslant Z \leqslant Z_c \qquad (6.3.15a)$$

$$\widetilde{W} = -\frac{\sin\lambda Z}{\lambda} \int_{Z_c}^{Z_t} f e^{-\bar{a}z} dz - \frac{e^{-\bar{a}z}}{\lambda} \int_{Z_c}^{Z} f \sin\lambda z \, dz, \quad Z_c \leqslant Z \leqslant Z_t \qquad (6.3.15b)$$

$$\widetilde{W} = -\frac{e^{-\bar{a}z}}{\lambda} \int_{Z_c}^{Z_t} f \sin\lambda z \, dz, \quad Z \geqslant Z_t \qquad (6.3.15c)$$

where
$$f \equiv \frac{g}{C^2} \hat{Q} \rho_0^{-\frac{1}{2}} \qquad (6.3.16)$$

and Z_c is the height that the surface air is changed into convective instability after lifted.

By using the mechanism of wave-CISK, the mesoscale convective systems can be well simulated. For example, Raymond (1984) successfully simulated a squall line process occurred in Oklahoma on May 22, 1976. For this process, Ogura and Liou (1980) used to do a detail mesoscale analysis. The results showed that preceding the passing of the squall line the stratification over the observational station presented conditional instability. From the formative stage to the mature stage of the squall line, the convective cells will go through a concentration process. In the mature stage, a mean structure of the air flow inner the squall line as shown in Fig. 6.10 will be formed. After given the buoyancy (N), the moist static energy (S), and the wind velocity vertical distribution etc. Raymond obtained the result of the simulation as shown in Fig. 6.11. It can be seen clearly that the simulation result is very similar to the real situation.

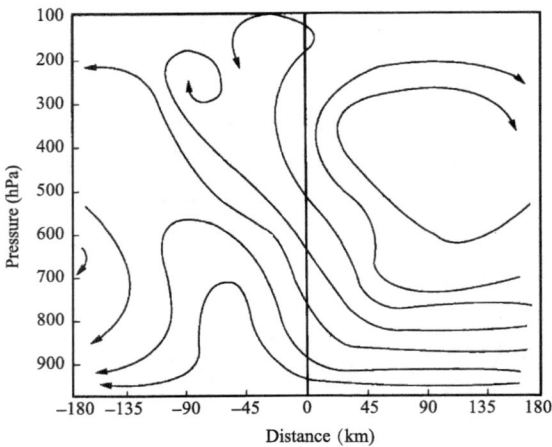

Fig. 6.10 The stream lines determined by u and v components in the vertical section traversed the squall line. (After Ogura and Liou, 1980)

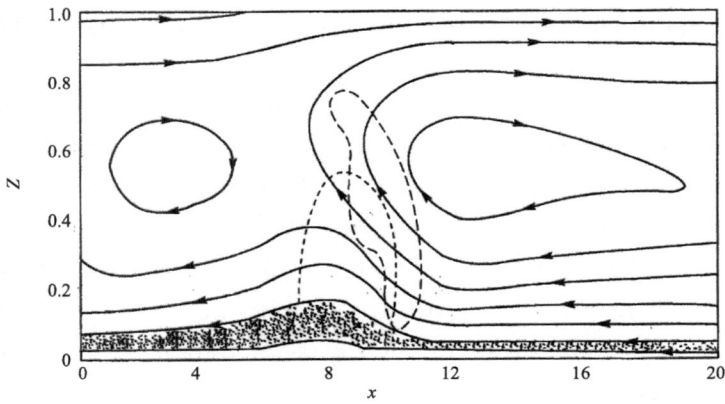

Fig. 6.11 The vertical section of the squall line based on the simulation results. (The stream line denoted by the arrow line; the upward convective mass flux area enclosed by dashed lines; the downward convective mass flux area enclosed by dot lines; and the stream function, of which the quantity is proportional to the grey level of the shaded area. (After Raymond, 1984).

§ 6.4 Inertial instability

6.4.1 Concept of inertial instability

When the basic air flow, which is under the geostrophic balance in a horizontal surface, suffered a transversal disturbance, if the disturbance is accelerated then the situation is called inertial instable, while the situation on the contrary is called as inertial stable.

Assuming the considered basic air flow is the geostrophic westerly wind along the x direction, with the wind velocity u_g. At the moment the pressure gradient is $\frac{\partial p}{\partial y} = -\rho f u_g$. Suppose there is a small air parcel pass laterally through the basic air flow, and suppose the disturbance displacement in south-north direction does not change the pressure field. Now introducing a physical quantity M, which means the absolute momentum for a unit mass air.

$$M = u - fy \tag{6.4.1}$$

Under the condition neglecting friction, M is a conservational quantity, namely,

$$\frac{dM}{dt} = \frac{d}{dt}(u - fy) = \frac{du}{dt} - fv = 0 \tag{6.4.2}$$

Thus

$$\frac{du}{dt} = fv = f\frac{dy}{dt} \tag{6.4.3}$$

The above equation shows that when air parcel moves northward, the zonal velocity will increase. Its increment will be f time of the moving speed.

Now let us furtherly introduce the second motion equation:

$$\frac{dv}{dt} = -\frac{1}{\rho}\frac{\partial p}{\partial y} - fu = f(u_g - u) \tag{6.4.4}$$

It can be seen that when u increase, the geostrophic departure will cause the acceleration in south-north direction. Assuming at the initial time the air parcel is located at the place $y = y_0$. If it moves northward a distance y, then by means of integrating the equation (6.4.3) the zonal velocity at the place $y = y_0 + \delta y$ will be

$$u(y_0 + \delta y) = u_g(y_0) + f\delta y \tag{6.4.5}$$

At the place $y_0 + \delta y$, the environmental geostrophic wind is

$$u_g(y_0 + \delta y) = u_g(y_0) - \frac{\partial u_g}{\partial y}\delta y \tag{6.4.6}$$

Substituting (6.4.5) and (6.4.6) into (6.4.4), we get

$$\frac{dv}{dt} = \frac{d^2(\delta y)}{dt^2} = -f\left(f - \frac{\partial u_g}{\partial y}\right)\delta y = -F^2\delta y \tag{6.4.7}$$

where $F^2 = f(f - \frac{\partial u_g}{\partial y})$.

6.4.2 Criteria of inertial instability

The equations (6.4.7) and (6.1.4) have the similar forms. so that it has the so-

lution similar to form of (6.1.6), namely
$$\delta y = A\exp(iFt) \quad (6.4.8)$$
Obviously, when $f(f-\frac{\partial u_g}{\partial y})<0$, δy will increase with time by exponent, that is the inertial unstable situation. Conversely, when $f(f-\frac{\partial u_g}{\partial y})>0$, δy will decrease with time, that is the inertial stable situation. In the northern hemisphere $f>0$, the increasing or decreasing of δy is only depending on the sign of the term $f-\frac{\partial u_g}{\partial y}$, thus we can get the criteria for inertial instability:

$$f-\frac{\partial u_g}{\partial y} \begin{cases} >0 & \text{inertial stable} \\ =0 & \text{neutral} \\ <0 & \text{inertial unstable} \end{cases} \quad (6.4.9)$$

Since $\partial M/\partial y = \partial(u-fy)/\partial y = -(f-\partial u/\partial y)$, the criteria of (6.4.9) can also be written as follows:

$$\frac{\partial M}{\partial y} \begin{cases} <0 & \text{inertial stable} \\ =0 & \text{neutral} \\ >0 & \text{inertial unstable} \end{cases} \quad (6.4.10)$$

$(f-\partial u/\partial y)$ is the absolute vorticity of the zonal shear air flow. For the longitudinal shear air flow, the absolute vorticity is $f+\frac{\partial v}{\partial x}$, while for two dimensional situations

$$\zeta_a = f + \frac{\partial v}{\partial x} - \frac{\partial u}{\partial y} \quad (6.4.11)$$

Thus the criteria of the inertial instability may be written as follows:

$$\zeta_a \begin{cases} >0 & \text{inertial stable} \\ =0 & \text{neutral} \\ <0 & \text{inertial unstable} \end{cases} \quad (6.4.12)$$

§ 6.5 Conditional symmetric instability

6.5.1 The concept of symmetric instability

In a mean flow which posses vertical wind shear and is under the hydrostatic balance as well as the geostrophic balance, even it is initially static stable and inertial stable, when the buoyancy force is combined with the rotational effect, in other words, when the air suffers vertical disturbance and horizontal disturbance simultaneously, namely when the air moves along slantwise direction, it will lead to gravity (buoyancy)-inertial unstable. Since this kind of instability is a sort of axis-symmetric instability, it is called as "symmetric instability". It can be regarded as the instability of the slantwise motion.

For explaining the reason why the instability can occur when the buoyancy is combined with the rotational effect, let us look at an example as shown in Fig. 6.12 at first, which is a vertical section perpendicular to the mean westerly wind. In the diagram the imaginary isolines of the mean potential temperature and the isolines of the mean local westerly wind absolute momentum $\overline{M}$ (where $\overline{M} = \bar{u} - fy$) are plotted. Assuming that the above two physical quantities are both increasing with height and decreasing toward north, and there is a small air parcel at the location A. If the air parcel moves vertically, it will suffer a restoring force since it is static stable, namely $\partial \bar{\theta}/\partial z > 0$. If it moves horizontally, it will also suffer a restoring force since it is inertial stable, namely $\partial \overline{M}/\partial y < 0$. If the air parcel takes a slantwise displacement, for instance it moves from A to B, then $M < \overline{M}$, that means the absolute momentum of air parcel is less than that of the environment, while $\theta > \bar{\theta}$, that means the potential temperature of air parcel is greater than that of the environment. According to the equation (6.1.1), i.e., $\frac{dw}{dt} = \frac{g}{\theta}[\theta(z) - \bar{\theta}(z)]$, and the equation (6.4.4), i.e., $\frac{dv}{dt} = f(u_g - u) = f(\overline{M} - M)$, we know that when $\theta > \bar{\theta}$ and $M < \overline{M}$ it will lead to $\frac{dw}{dt} > 0$ and $\frac{dv}{dt} > 0$, that means the air parcel will be accelerated upward and northward, in other words, it will obtain a slantwise acceleration so that the slantwise convection will be developed. Obviously, when the air parcel moves slantwise in the dashed area in the diagram, this kind of instability will occur and the tilted convection will be formed. Therefore the symmetric instability is favorable for the slantwise convection development.

If we look at Fig. 6.12 carefully we can see clearly that the symmetric instability can only occur under the situation that the slope of the constant $\bar{\theta}$ surface is greater than that of the constant $\overline{M}$ surface. For explaining the reason, we can change the diagram in Fig. 6.12 into the diagrams in Fig. 6.13a and b respectively.

The situation presented in Fig. 6.13a shows the slope of the constant $\overline{M}$ surface is less than the slope of the constant $\bar{\theta}$ surface, therefore when the air parcel at the place

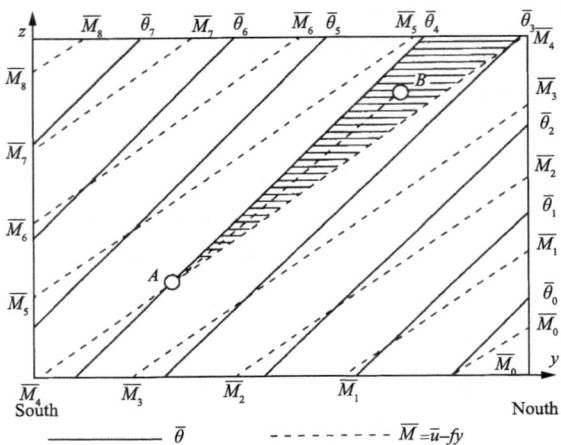

Fig. 6.12 Under the symmetric instability condition the vertical section of the potential temperature $\bar{\theta}$ and the absolute momentum $\bar{M}$.

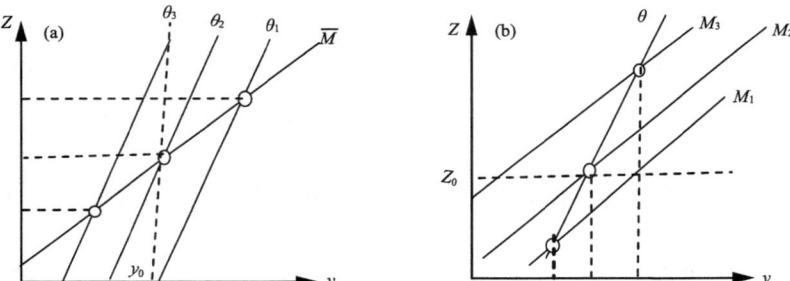

Fig. 6.13 (a) The gravity instability on the constant $\bar{M}$ surface and (b) the inertial instability on the constant $\bar{\theta}$ surface.

y_0 is lifted vertically it is static stable since $\left.\dfrac{\partial \bar{\theta}}{\partial z}\right|_{y_0} > 0$, however, when the air parcel moves along the constant $\bar{M}$ surface it is gravity instable since $\left.\dfrac{\partial \bar{\theta}}{\partial z}\right|_{\bar{M}} > 0$; Therefore the symmetric instable is the gravity instability on the constant $\bar{M}$ surface.

Similarly, the situation presented in Fig. 6.13b shows that the slope of the constant $\bar{M}$ surface is less than the slope of the constant $\bar{\theta}$ surface, when the air parcel at the height Z_0 moves horizontally it is inertial stable since $\left.\dfrac{\partial \bar{M}}{\partial y}\right|_{z_0} < 0$, however when the air parcel moves along the constant $\bar{\theta}$ surface it is inertial instable since $\dfrac{\partial \overline{M}}{\partial y}\Big|_{\bar{\theta}}>0$; Therefore the symmetric instable is also the inertial instability on the constant $\bar{\theta}$ surface.

6.5.2 The scale of symmetric instability

It can be seen from Fig. 6.12 that when the air parcel moves slantwise along the constant $\overline{M}$ line it is symmetric unstable. Meantime, since on the surface of constant $\bar{\theta}$, $\dfrac{\partial \overline{M}}{\partial y}\Big|_{\bar{\theta}}>0$, the symmetric instability is the inertial instability on the constant $\bar{\theta}$ surface, that is $f-\partial u/\partial y|_{\bar{\theta}}$ or $\delta y < \dfrac{\delta u \delta z}{\delta y\, f}$, hence we can obtain the horizontal scale of the symmetric instability as follows:

$$L < U_z D / f \tag{6.5.1}$$

where L, D are horizontal and vertical scale respectively, U_z is the vertical shear of the horizontal wind. The temporal scale of the symmetric instability will be written as

$$\tau = L/U \sim U_z D / (fU) \sim 1/f \tag{6.5.2}$$

With the above analysis, we know that symmetric instability has the same horizontal and temporal scale with the inertial instability. Furthermore, the scale of L and τ in equation (6.5.1) and (6.5.2) are same with the horizontal and temporal scales in the Table 1.3. Moreover, because $L = U_z = D/f \sim U/f$, and $R_0 = U/fL \sim 1$, so that the symmetric instability is a mesoscale instability. Over here we can further realize the physical meaning of the scale division given by Emanuel as mentioned in §1.1.

6.5.3 The criteria of symmetric instability

From the diagram in Fig. 6.9, it can be seen intuitively that

$$\begin{cases} \text{when } \dfrac{\partial z}{\partial y}\Big|_{\bar{\theta}} > \dfrac{\partial z}{\partial y}\Big|_{M}, \text{symmetric unstable} \\ \text{when } \dfrac{\partial z}{\partial y}\Big|_{\bar{\theta}} < \dfrac{\partial z}{\partial y}\Big|_{M}, \text{symmetric stable} \end{cases} \tag{6.5.3}$$

That is to say, when the slope of the constant potential temperature surface is greater than the slope of the constant $\overline{M}$ surface the atmosphere is symmetric unstable. Oppositely, when the slope of the constant potential temperature surface is less than the slope of the constant surface the atmosphere is symmetric stable. On account of

$$\dfrac{\partial z}{\partial y}\Big|_{\bar{\theta}} = -\dfrac{\partial \bar{\theta}/\partial y}{\partial \bar{\theta}/\partial z},\quad \dfrac{\partial z}{\partial y}\Big|_{M} = -\dfrac{\partial \overline{M}/\partial y}{\partial \overline{M}/\partial z} \tag{6.5.4}$$

where assuming $\partial\bar{\theta}/\partial y<0$, $\partial\bar{M}/\partial y<0$, and the thermal wind relationship is satisfied, namely

$$f\frac{\partial\bar{u}}{\partial z}=-\frac{g}{\theta_0}\frac{\partial\bar{\theta}}{\partial y} \quad (6.5.5)$$

Considering comprehensively the formulas (6.5.3)—(6.5.5), and substituting $M=u-fy$ into them, then we have

$$\frac{N^2}{(\partial\bar{u}/\partial z)^2}\left(1-\frac{\partial\bar{u}/\partial y}{f}\right)>1 \text{ (symmetric stable)} \quad (6.5.6)$$

where $N^2=\frac{g}{\theta_0}\frac{\partial\bar{\theta}}{\partial z}$. In the above formula, $N^2/(\partial\bar{u}/\partial z)^2=Ri$, in the northern hemisphere, generally $\partial\bar{u}/\partial y>0$, thus $\left(1-\frac{\partial\bar{u}/\partial y}{f}\right)\leqslant 1$, therefore if the formula (6.5.6) is satisfied, then we have $Ri>1$ (symmetric stable); Oppositely, we have $Ri<1$ (symmetric unstable).

The above criteria may also be derived by mathematics analysis and may obtain more other forms of the criteria. We will discuss the problem under two situations: non hydrostatic balance and static balance respectively.

(1) Non hydrostatic balance situation

Applying the following non hydrostatic, frictionless, Boussinesq equations:

$$\frac{du}{dt}-fv+\frac{1}{\rho}\frac{\partial p'}{\partial x}=0 \quad (6.5.7)$$

$$\frac{dv}{dt}+fu+\frac{1}{\rho}\frac{\partial p'}{\partial y}=0 \quad (6.5.8)$$

$$\frac{dw}{dt}-g\theta'/\theta_0+\frac{1}{\rho}\frac{\partial p'}{\partial z}=0 \quad (6.5.9)$$

$$\frac{\partial u}{\partial x}+\frac{\partial v}{\partial y}+\frac{\partial w}{\partial z}=0 \quad (6.5.10)$$

$$\frac{d\theta}{dt}=0 \quad (6.5.11)$$

In the above equations all signs are the common signs. Taking the Coriolis parameter f as a constant number and θ_0 is the typical value of the potential temperature θ.

Assuming the basic air flow $\bar{V}(x,z)$ is along y direction and it is independent of y. Also assuming that $\bar{V}(x,z)$ and the potential temperature $\Theta(x,z)$ are in thermal wind balance, i. e.

$$f\frac{\partial\bar{V}}{\partial z}=\left(\frac{g}{\theta_0}\right)\left(\frac{\partial\Theta}{\partial x}\right) \quad (6.5.12)$$

Considering a disturbance which is along the plane (x, z) and independent of y. The disturbance quantities are denoted by prime signs. Based on the continuity equation (6.5.10), the relations between the stream function ψ and the disturbance velocities perpendicular to the basic flow may be written as $u' = \partial\psi/\partial z$ and $w' = -\partial\psi/\partial x$. Seeking the partial derivatives of the equations (6.5.7) and (6.5.9) respect to z and x respectively and then after subtracting one from other we can get the following vorticity equation of the y component vorticity:

$$\frac{\partial}{\partial t}\left(\frac{\partial^2\psi}{\partial x^2}+\frac{\partial^2\psi}{\partial z^2}\right)=f\frac{\partial v'}{\partial z}-\left(\frac{g}{\theta_0}\right)\left(\frac{\partial \theta'}{\partial x}\right) \qquad (6.5.13)$$

The above equation denotes that the thermal wind balance is destroyed after the basic air flow disturbed, the circulation in the x-z plane traversed the basic flow will be increased. Now let us introduce the frequencies of the basic flow: F, S, N, where

$$F^2=f(f+\partial \bar{V}/\partial x), \quad S^2=f\partial\bar{V}/\partial z=\left(\frac{g}{\theta_0}\right)\left(\frac{\partial \Theta}{\partial x}\right), \quad N^2=\left(\frac{g}{\theta_0}\right)\left(\frac{\partial \Theta}{\partial z}\right)$$

The typical values of the frequencies for atmosphere are $F \sim 10^{-4}$ s^{-1}, $S \sim 0.5 \times 10^{-3}$ s^{-1} and $N \sim 10^{-2}$ s^{-1} respectively. It should be noted that, F^2/S^2 is the slope of the absolute vorticity vector, also is the slope of the constant $\bar{M}$ surface (i.e. the surface of $\bar{M}=fx+\bar{V}=$constant), while S^2/N^2 is the slope of the constant potential temperature surface in the basic air flow. The disturbance forms of the equations (6.5.8) and (6.5.11) may be written as follows:

$$\partial(fv')/\partial t=-F^2 u'-S^2 w' \qquad (6.5.14)$$

$$\partial(g\theta'/\theta_0)/\partial t=-S^2 u'-N^2 w' \qquad (6.5.15)$$

Assuming the scale of the basic flow frequency variation is much greater than that of the disturbance. Then substituting the equations (6.5.14) and (6.5.15) into the equation (6.5.13), we can obtain:

$$\frac{\partial^2}{\partial t^2}\left(\frac{\partial^2\psi}{\partial x^2}+\frac{\partial^2\psi}{\partial z^2}\right)=-N^2\frac{\partial^2\psi}{\partial x^2}+2S^2\frac{\partial^2\psi}{\partial x \partial z}-F^2\frac{\partial^2\psi}{\partial z^2} \qquad (6.5.16)$$

In a unbounded area, we can obtain a solution that is proportional to $\exp(i\sigma t)\exp\{ix(x\sin\varphi+z\cos\varphi)\}$. Here, φ is the angle between the disturbance displacement and the horizontal displacement, the frequency σ can be given by the following formula:

$$\sigma^2=N^2\sin^2\varphi-2S^2\sin\varphi\cos\varphi+F^2\cos^2\varphi \qquad (6.5.17)$$

Deriving the minimum value for σ^2, $\partial\sigma^2/\partial\varphi=0$, obtaining

$$\text{tg}2\varphi=2S^2/(N^2-F^2)$$

Eliminating φ, to get the minimum of σ^2, $\sigma^2_{\min}$, which can be denoted as follows:

$$2\sigma_{min}^2 = N^2 + F^2 - [(N^2+F^2)^2 - 4q]^{\frac{1}{2}} \qquad (6.5.18)$$

where
$$q = F^2 N^2 - S^4 \qquad (6.5.19)$$

q is in direct proportional to the Ertel potential vorticity of the basic air flow (Eliassen and Kleinchmidt, 1957). From formula (6.4.29), it can be seen that if and only if $N^2 + F^2 < 0$ or $q < 0$, the unstable ($\sigma_{min}^2 < 0$) situation might occur. This the another form of the criteria of the symmetric instability. From equation (6.5.18) it can get to know that the above criteria of the instability can be also written as

$$\frac{1}{Ri} > 1 \qquad (6.5.20)$$

where $Ri = F^2 N^2 / S^2$, thus the criteria of the symmetric instability is

$$q < 0 \quad \text{or} \quad Ri < 1 \qquad (6.5.21)$$

Because of $Ri = F^2 N^2 / S^4 = (F^2/S^2)/(S^2/N^2) < 1$, while (F^2/S^2) and (S^2/N^2) are the slopes of the constant $\overline{M}$ surface and the constant $\bar{\theta}$ surface respectively. Thus the criteria of the symmetric instability can also expressed as that when the slope of the constant $\overline{M}$ is less than that of the constant $\bar{\theta}$ surface, atmosphere is in symmetric unstable. Therefore when the slope of the constant $\bar{\theta}$ surface or the horizontal gradient is greater it is in favor of occurring of symmetric instability. The above analysis manifested there are good agreement between the different criteria for the symmetric instability.

(2) The situation under static balance

Omitting the term dw/dt in the equation (6.5.9), then the equation is changed into situation under static balance. Under this situation, the term $\partial^2 \psi/\partial x^2$ at the left side of the equation (6.5.16) will disappear and equation (6.5.17) will be changed as follows:

$$\sigma^2 = N^2 \text{tg}^2 \varphi - 2S^2 \text{tg} \varphi + F^2 \qquad (6.5.22)$$

Deriving the minimum value of the above formula, then we get the condition

$$\text{tg} \varphi = S^2 / N^2 \qquad (6.5.23)$$

This shows that under the situation in static balance, the direction of the minimum frequency of the symmetric disturbance is consistent with the slope of the isentropic surface, that is to say the disturbance is on the isentropic surface. Under this situation, both the disturbance temperature and disturbance pressure equals to zero. Hence the disturbance momentum equation may be written as follows:

$$\frac{\partial u'}{\partial t} = fv'$$

$$\frac{\partial v'}{\partial t} = -u'\left(f + \frac{\partial \bar{V}}{\partial x}\right) - w'\frac{\partial \bar{V}}{\partial z} \quad \text{or} \quad \frac{\partial v'}{\partial t} = -\zeta_\theta u'$$

where $\zeta_\theta = \frac{1}{f}\left(F^2 - \frac{S^4}{N^2}\right)$ is the absolute vorticity on constant Θ surface. So that the disturbance kinetic energy equation can be written as

$$\frac{1}{2}\frac{\partial}{\partial t}(u'^2 + v'^2) = -u'v'\frac{\partial \bar{V}}{\partial x} - v'w'\frac{\partial \bar{V}}{\partial z} \tag{6.5.24}$$

It follows that, the energy source of the symmetric disturbance is derived from the horizontal and vertical shear of the basic flow. When the vertical shear is in negative correlation with u' and w', the disturbance will increase.

6.5.4 The Concept of conditional symmetric instability

In recent years, many observational investigations noted that the frontal cloud and precipitation are concentrated in the area parallel to the front. The intervals of rain bands may be ranged from 80 km to 300 km, while their lengths are even more longer. The rain bands have a small angle intersected with the isothermal lines. In the abstract, the formation of the rain bands may include the instability of the Ekman layer in the front zone; the gravity waves formed on the front and the convection introduced by different advection etc., while Bennetts and Hoskins et al. (1979) proposed another notable possible cause. They think these rain bands are possibly caused by symmetric baroclinic instability. In the above section, we have discussed the basic concept and criteria of the symmetric instability. Roughly speaking, the criteria are that the horizontal temperature gradient is bigger; or the Richardson number is smaller; or the slope of the constant potential temperature surface is bigger than that of the constant M surface; or $q<0$ etc.

However, the dry air the symmetric instability conditions such as $Ri<1$ and $q<0$ etc. are difficult to be satisfied. Under this situation, if it takes no account of the latent heat releasing effect then the original symmetric stable atmosphere will be impossible to be changed into synoptic instable. For explaining this problem, now let us consider the original symmetric stable atmosphere of which the frequency N^2, F^2 and the potential vorticity q are positive everywhere and assume that in the further movement process, the frictional effect and the thermal source and sink are all ignored. Since under these restricts the potential vorticity q is conserved in the three dimensional motions, thus it will be positive everywhere. Suppose there is approximately thermal wind balance relationship ($S^2 = f\partial v/\partial z$), then according to the equation

(6.4.30), we have the relation $q < F^2 N^2$, therefore $F^2 N^2$ will be also kept as the positive value everywhere and no change. This means that under the assumptions of no friction, thermal source or sink and quasi-geostrophic balance, the original symmetric stable atmosphere will be impossible to be changed into symmetric instable. Bennetts and Hoskins (1979) therefore studied the possibility realized the symmetric instability in a moist atmosphere and introduced the concept of conditional symmetric instability (shorthand for CSI). In brief, we can say that the atmosphere is conditional symmetric instable if the original symmetric stable atmosphere is changed into symmetric instable due to the effect of releasing the latent heat.

6.5.5 The criteria of conditional symmetric instability

For considering the possibility of symmetric instability of the two dimensional air flow in the moist atmosphere saturated everywhere, we substitute the wet bulb potential temperature θ_W into the equations (6.5.7)—(6.5.11) to take place of the potential temperature θ and introduce two new frequency numbers:

$$N_W^2 = (g/\theta_0)(\partial \theta_W / \partial z) \qquad (6.5.25)$$

$$S_W^2 = (g/\theta_0)(\partial \theta_W / \partial x) \qquad (6.5.26)$$

By using the similar derivation method employed in the derivation of the equations (6.5.16)—(6.5.18), we can obtain the following equation of the stream function ψ:

$$\frac{\partial^2}{\partial t^2}\left(\frac{\partial^2 \psi}{\partial x^2} + \frac{\partial^2 \psi}{\partial z^2}\right) = -F^2 \frac{\partial^2 \psi}{\partial z^2} + (S_W^2 + S^2)\frac{\partial^2 \psi}{\partial x \partial z} - N_W^2 \frac{\partial^2 \psi}{\partial x^2} \qquad (6.5.27)$$

and the frequency equation:

$$\sigma^2 = N_W^2 \sin^2 \varphi + F^2 \cos^2 \varphi - 2S^2 \sin\varphi \cos\varphi \qquad (6.5.28)$$

As well as the equation of the minimum frequency:

$$2\sigma_{\min}^2 = N_W^2 + F^2 - [(N_W^2 + F^2)^2 - 4q_W]^{1/2} \qquad (6.5.29)$$

where q_W is the wet bulb potential vorticity

$$q_W = N_W^2 F^2 - S_W^2 S^2 \qquad (6.5.30)$$

According to the same discussing about the symmetric instability of the dry air, we know that for the moist atmosphere which is saturated everywhere the criteria of the symmetric instability are as follows:

$$q_W < 0, \text{ or } N_W^2 F^2 - S_W^2 S^2 < 0,$$

$$\text{or } Ri = (N_W^2 F^2)/(S_W^2 S^2) = (F^2/S^2)/(S_W^2/N_W^2) < 1$$

Because the ratios (F^2/S^2) and (S_W^2/N_W^2) are the slopes of the constant M and the constant surfaces respectively, when the slope of the constant θ_W surface is greater than that of the constant M surface, $Ri < 1$. Simultaneously, since the conditional symmetric instability is also the inertial instability on the constant θ_w surface, the above criteria may be also written as

$$\frac{\partial M}{\partial x}\Big|_{\theta_w} < 0, \text{ or } f\left(f + \frac{\partial v}{\partial x}\right)\Big|_{\theta_w} < 0, \text{ or } f\zeta_{\theta_w} < 0$$

The above discussion is based on the situation of the unlimited moist atmosphere without subsidence compensation motion. Under the situation, if the constant θ_w surface is more close to perpendicular than the absolute vorticity vector, then there is negative restoring force, so that the conditional symmetric instability will be generated.

In the limited atmosphere with subsidence compensation motion, the restoring force of the subsidence without latent heat releasing is positive. Thus the symmetric instability of the two dimensional air flow is depending on the restoring forces of both the ascending and descending flows simultaneously. The following is an indicative demonstration.

Fig. 6.14 is an illustration of the conditional symmetric instability in the limited moist atmosphere with subsidence compensation motion. In the diagram, $\bar{\theta}$, $\bar{\theta}_w$ and $\bar{M}$ denote the surfaces of the constant potential temperature, the constant wet bulb potential temperature and the constant momentum respectively. The constant $\bar{M}$ surface is parallel to the absolute vorticity vector ζ_a. The ascending branch BC of the circular flow tube is along the constant $\bar{\theta}_w$ surface and the descending branch AB is along the constant $\bar{\theta}$ surface (Bennetts and Hoskins, 1979).

Now let us consider the motion possibly occurred around the assumed flow tube in Fig. 6.14 and the conditions arising the symmetric instability. In the diagram the ascending motion along the constant $\bar{\theta}_w$ surface and the descending motion along the constant $\bar{\theta}$ surface are given. The slope of the constant $\bar{\theta}_w$ surface is greater than that of the constant $\bar{M}$ surface, while the slope of the constant $\bar{\theta}$ surface is less than that of the constant $\bar{M}$ surface, thus ascending branch of the circular flow tube is unstable. The ascending air parcel suffers a negative restoring force, namely it suffers the force in the same direction with the disturbance. According to the discussion similar to the equation (6.4.7), at this moment the motion equation of the air parcel in the ascending branch may be written as

$$\frac{du}{dt} = -f\zeta_{\theta_w}\delta x \quad (6.5.31)$$

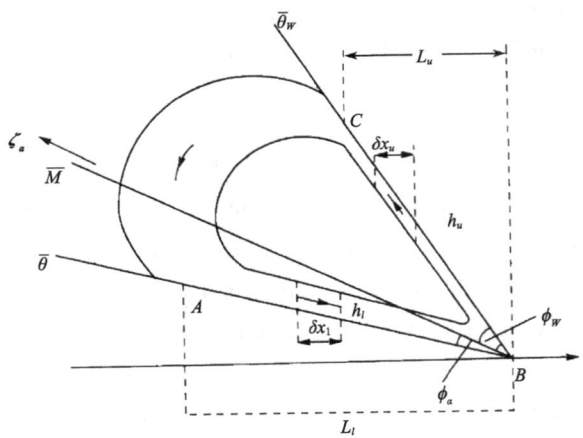

Fig. 6.14 Illustration of the conditional symmetric instability in the limited moist atmosphere with subsidence compensation motion.

where subscript u denotes the quantity on the ascending branch and δx_u denotes the horizontal displacement of the disturbed air parcel. Assume the disturbance pressure difference between BC on the ascending branch is $\Delta p'_u$ and the horizontal distance between BC is L_u as well as the disturbance pressure gradient force is $\Delta p'_u/L_u$. The pressure gradient force should be counteracted with the instable force, i. e.

$$-f\zeta_{\theta_W}\delta x_u = \Delta p'_u/L_u$$

or

$$\Delta p'_u = -f\zeta_{\theta_W}\delta x_u L_u = (-f\zeta_{\theta_W})\delta \tau L_u/h_u \qquad (6.5.32)$$

where $\delta\tau$ is the volume with unit cross section, h_u is the width of the flow tube, $\delta\tau = \delta x_u \cdot h_u$.

In contrast with the ascending branch, the descending branch is stable. Therefore the air parcel in the descending flow suffers positive restoring force in opposite direction with the disturbance. Similar to the above discussion, for maintaining the air parcel subsidence, it is necessary to keep a disturbance pressure difference $\Delta p'_l$ between A and B

$$\Delta p'_l = (f\zeta_\theta)\delta\tau L_l/h_l \qquad (6.5.33)$$

where the subscript l denotes the quantity of the subsidence branch; h_l is the width of the flow tube in the subsidence branch; and L_l is the horizontal distance of AB. Since the flow CA is wide, thus the difference between p'_c and p'_A could be ignored. Therefore for maintaining the instable circular circulation, it is required that the pressure difference used to balance with the instable force acting on BC has to be greater than the pressure difference used to balance with the stable force acting on AB, that is

$$(-f\zeta_{\theta_w})\delta\tau L_u/h_u > (f\zeta_\theta)\delta\tau L_l/h_l \qquad (6.5.34)$$

When h_u is more smaller, h_l is more bigger, the above inequality will be more easier to be satisfied. That is to say the more narrow the ascending branch and the wider the subsidence branch the easier to meet the condition for symmetric instability. The above inequation is the criteria of the symmetric instability under the situation with the subsidence competition motion. The criteria can be also written as follows:

$$\alpha f\zeta_\theta + f\zeta_{\theta_w} < 0 \qquad (6.5.35)$$

where $\alpha = (h_u L_l)/(h_l L_u) = O(1)$. It can be seen that the criteria of the symmetric instability under the situations of unlimited moist atmosphere without subsidence compensation and the dry atmosphere are the special situations of the criteria in equation (6.5.35), which can be also written as other three types:

$$\alpha(N_W^2/N^2)q + q_w < 0; \quad Ri^{-1} > (\alpha+1)/(\alpha+N^2/N_W^2); \quad \alpha\varphi_a < \varphi_w \qquad (6.5.36)$$

All of these are the criteria of the conditional symmetric instability for limited moist atmosphere with subsidence compensation motion. In the second type of equation (6.5.36) the factor S_W^2/S^2 is ignored.

6.5.6 The favorable situation for leading to CSI

Following is a discussion about the possibility to generate the conditional symmetric instability when the wet bulb potential vorticity q_w changes to a negative value ($q_w < 0$) from the initial state of $q_w > 0$.

For this purpose we will derive the generalized potential vorticity equation at first. Change the motion equation into vector form as follows:

$$\frac{\partial \mathbf{V}}{\partial t} + (\mathbf{V} \cdot \nabla)\mathbf{V} + 2\mathbf{\Omega} \times \mathbf{V} = -\frac{1}{\rho}\nabla p + \mathbf{k}\frac{g\theta}{\theta_0} + \mathbf{F} \qquad (6.5.37)$$

where $\mathbf{V}$ is three dimensional wind vector, $\mathbf{\Omega}$ is the Earth's rotational angle velocity, $\mathbf{F}$ is the frictional term, $\boldsymbol{\zeta} = \nabla \times \mathbf{V}$ denotes the vorticity of the three dimensional wind vector. Take the operation for the equation (6.5.37) by acting $\nabla \times$, then we can obtain the vorticity as the following form:

$$\frac{\partial \boldsymbol{\zeta}}{\partial t} - \nabla \times (\mathbf{V} \times \boldsymbol{\zeta}) = \nabla \times \left(\mathbf{k}g\frac{\theta}{\theta_0}\right) + \nabla \times \mathbf{F} \qquad (6.5.38)$$

The above formula may be rewritten as

$$\frac{\partial \boldsymbol{\zeta}}{\partial t} + (\mathbf{V} \cdot \nabla)\boldsymbol{\zeta} - (\boldsymbol{\zeta} \cdot \nabla)\mathbf{V} = -\left(\frac{g}{\theta_0}\right)\mathbf{k} \times \nabla\theta + \boldsymbol{\mathcal{F}} \qquad (6.5.39)$$

where $\mathfrak{J} = \nabla \times F$, the above formula can be also rewritten as

$$\frac{d\zeta}{dt} - (\zeta \cdot \nabla)V = -\left(\frac{g}{\theta_0}\right)k \times \nabla\theta + \mathfrak{J} \qquad (6.5.40)$$

Since it is difficult to satisfy the criteria of the symmetric instability in the dry atmosphere for the scale of 100 km, we will study the influence of the moist process now. In the moist atmospheric three dimensional motion, the variation of the wet bulb potential temperature (θ_w) is only caused by the non-adiabatic heating except the latent heat releasing, that is

$$\frac{d\theta_w}{dt} = Q \qquad (6.5.41)$$

The above equation may be rewritten into the following form:

$$\frac{\partial\theta_w}{\partial t} + V \cdot \nabla\theta_w = Q \qquad (6.5.42)$$

Take ∇ operation for the above formula, to get

$$\frac{\partial \nabla \theta_w}{\partial t} + (V \cdot \nabla)\nabla\theta_w + \frac{\partial\theta_w}{\partial x}\nabla u + \frac{\partial\theta_w}{\partial y}\nabla v + \frac{\partial\theta_w}{\partial z}\nabla w = \nabla Q \qquad (6.5.43)$$

Multiply the above formula by ζ, to get

$$\zeta \cdot \frac{d\nabla\theta_w}{dt} + \frac{\partial\theta_w}{\partial x}\nabla u \cdot \zeta + \frac{\partial\theta_w}{\partial y}\nabla v \cdot \zeta + \frac{\partial\theta_w}{\partial z}\nabla w \cdot \zeta = \nabla Q \cdot \zeta \qquad (6.5.44)$$

On the other hand, multiply the equation (6.5.40) by $\nabla\theta_w$ to obtain

$$\nabla\theta_w \cdot \frac{d\zeta}{dt} - \nabla\theta_w \cdot [(\zeta \cdot \nabla)V] = \left(\frac{g}{\theta_0}k\right) \cdot (\nabla\theta_w \times \nabla\theta) + \mathfrak{J} \cdot \nabla\theta_w \qquad (6.5.45)$$

Add the formula (6.5.44) and (6.5.45) together, and multiply the sum by $f(g/\theta_0)$, and then to introduce the wet bulb potential vorticity equation

$$q_w = f(g/\theta_0)\zeta \cdot \nabla\theta_w = f\zeta N_w^2 + (g/\theta_0)\{k \times (\partial V/\partial z)\} \cdot \nabla_2\theta_w \qquad (6.5.46)$$

where ζ is the 3D absolute vorticity vector, ζ is the vertical component of ζ, k is the unit vector in the vertical direction, ∇_2 is the sign of horizontal gradient, $V=(u, v, 0)$, then we get

$$\frac{dq_w}{dt} = f(g^2/\theta_0^2)k \cdot (\nabla\theta_w \times \nabla\theta) + f(g/\theta_0)\zeta \cdot \nabla Q + f(g/\theta_0)\mathfrak{J} \cdot \nabla\theta_w \qquad (6.5.47)$$

The equation (6.5.47) is the wet bulb potential vorticity equation. The second and third terms in the right side of the above equation are the variation of the wet bulb potential vorticity q_w induced by the effects of non-adiabatic heating and friction respectively. The first term in the right side of the equation denotes the variation of q_w

when there is an angle existed between the surfaces of θ and θ_w in the horizontal direction. When the atmosphere is adiabatic and frictionless the individual change of q_w, dq_w/dt, will be only depending on the first term in the right side of the equation. As shown in Fig. 6.15 that if the moisture increase in the thermal wind direction, then q_w will decrease (namely, $dq_w/dt<0$) according to the equation (6.5.47). Under this situation, even the atmosphere is symmetric stable ($q_w>0$) at the initial time, while after a time period it will be changed into symmetric instable ($q_w<0$). The diagnosis shows that, if there is even an small angle in an order of 0.1° between θ and θ_w surfaces, it will lead in a negative q_w in a duration of 1—2 days due to the effect of the first term in the right side of the equation (6.5.47). This shows that if the moisture increase in the thermal wind direction, then q_w will decrease. This is the favorable situation for the generating the conditional symmetric instability. It is also an useful criterion in operational application.

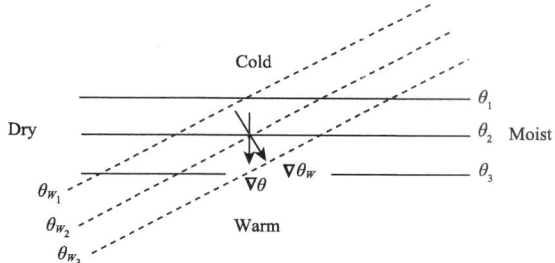

Fig. 6.15 The horizontal distribution of θ (solid lines) and θ_w (dashed lines) which is favorable for leading to $q_w<0$, the situation of conditional symmetric instability. (After Bennetts and Hoskins, 1979)

6.5.7 Applications

As state above, the conditional symmetric instability is one of the mechanisms for the developing of the slantwise convection. While the slantwise convections are also related with the strong weather such as the heavy rain and other severe convective weather. Thus the analysis of the conditional symmetric instability can be used for forecasting the convective weather.

Sometimes the atmosphere is convective stable, in other words in the sounding analysis there is no positive instability energy, so that it seems there is no possibility to occur convective weather, but actually if the atmosphere is conditional symmetric instable then the convective weather is still possible to form. This is because when the atmosphere is conditional symmetric instable, it may be the gravity unstable on the constant $\overline{M}$ surface, i. e. $\frac{\partial \overline{\theta}_e}{\partial z}|_{\overline{M}} < 0$, in other words when the air parcel rises slantwise along the constant $\overline{M}$ surface, it is conditional unstable. Thus if we make an analysis of the T-$\ln P$ diagram along the constant $\overline{M}$ surface, we can find the conditional instability energy clearly.

Emanuel (1983) analyzed a case of the convective precipitation process happened in south area of the United States on Dec. 3, 1982, The sounding analysis of Oklahoma City (OKC) showed the atmosphere is stable (Fig. 6.16a), while the vertical section across OKC showed that the atmosphere over OKC was symmetric instability since the slope of the constant θ_e (equivalent potential temperature) surface was greater than the slope of the constant $\overline{M}$ surface (Fig. 6.17). So that the T-$\ln P$ diagram made along the constant $\overline{M}$ surface showed the atmosphere was gravity unstable (Fig. 6.16b). This analysis revealed the somewhat secret mechanism of the precipitation process.

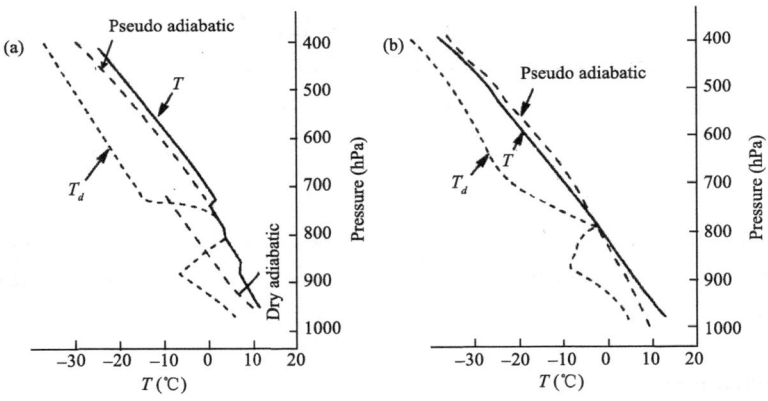

Fig. 6.16 (a) T-$\ln P$ diagram of OKC, at 0000 GMT Dec. 3, 1982; and (b) T-$\ln P$ diagram along the constant surface of M (M=50 m·s^{-1}) in Fig. 6.14 at the same time with (a). The solid line is temperature, dot line is dew point temperature, dashed line is the pseudo moist adiabatic line, the dashed line with small circles is the dry adiabatic line. T=Temperature, T_d=Dew-point temperature. (After Emanuel, 1983)

The similar precipitation processes are founded in China quite often as well. For example, a heavy rain process occurred on Jun. 13, 1983, in Fuyang station of Anhui Province, the local rainfall in 12 hours was 170mm. Another example was the heavy

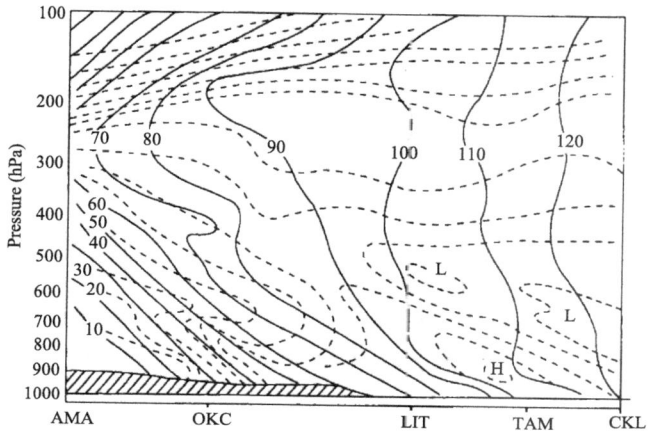

Fig. 6.17 The vertical section from AMA to CKL at 0000 GMT, Dec. 3, 1982. The dashed line is constant line of θ_e (K), solid line is constant line of M (m·s^{-1}). (After Emanuel, 1983)

rain process occurred on Oct. 11, 2003, in Chuangzhou, Hebei Province, the local 24 hours rainfall was 110.5 mm. The atmospheric stratifications of the both cases were convective stable but conditional symmetric instable.

§ 6.6 Kelvin-Helmholtz instability

If the vorticity is concentrated on a discontinuity line of the velocity, then the linear instability of the airflow, i. e. there is a maximum value of the shear vorticity on the line, will become most distinct. This kind of instability related with discontinuity is called Kelvin-Helmholtz instability (for short, the K-H instability). Following we will discuss the mechanism and criteria of the instability.

Firstly, let us consider a discontinuity surface, which separated two parts of the uniform, impressible and two dimensional moving fluids. Assume the line S is the discontinuity line. On the line we only need to consider the motion in a plane. The subscripts 1 and 2 denote the variables in two sides of the line S respectively. The following equations will be suitable for any points on line S.

$$\left.\begin{array}{l} \dfrac{\partial \mathbf{V}_1}{\partial t}+\mathbf{V}_1 \cdot \nabla \mathbf{V}_1 = -\nabla \dfrac{p_1}{\rho} \\ \dfrac{\partial \mathbf{V}_2}{\partial t}+\mathbf{V}_2 \cdot \nabla \mathbf{V}_2 = -\nabla \dfrac{p_2}{\rho} \end{array}\right\} \qquad (6.6.1)$$

Now introduce the signs

$$\frac{V_1+V_2}{2}=\overline{V}, \quad \frac{\partial}{\partial t}+\overline{V}\cdot\nabla=\frac{\delta}{\delta t} \tag{6.6.2}$$

And subtract the second formula in (6.6.1) from the first, to obtain

$$\frac{\delta}{dt}(V_1-V_2)=-(V_1-V_2)\cdot\nabla\overline{V}-\nabla\frac{p_1-p_2}{\rho} \tag{6.6.3}$$

The kinematic boundary condition is required that V_1 and V_2 have the same component in the direction perpendicular to line S, and the line S moves with velocity $\overline{V}$ in direction perpendicular to itself. Thus on the line S the points moved with velocity $\overline{V}$ will be kept on the line S, and the direction of V_1-V_2 will be in tangency of the line S.

Let r_a and r_b to denote the position vectors of the two points on the line S moved with same velocity V, then

$$\frac{\delta r_a}{dt}=\overline{V}_a, \quad \frac{\delta r_b}{dt}=\overline{V}_b, \quad \frac{\delta}{dt}(r_a-r_b)=\overline{V}_a-\overline{V}_b \tag{6.6.4}$$

Under the limit case of r_a-r_b, it can be written as

$$r_a-r_b=dr_s, \quad \overline{V}_a-\overline{V}_b=d\overline{V} \tag{6.6.5}$$

Thus, the final equation in (6.6.4) will be changed into the following form:

$$\frac{\delta}{dt}dr_s=d\overline{V}=dr_s\cdot\nabla\overline{V} \tag{6.6.6}$$

By means of the formulas (6.6.3) and (6.6.6), the rate of the mass variation dc can be calculated as follows

$$dc=(V_1-V_2)dr_s \tag{6.6.7}$$

and

$$\frac{\delta}{dt}dc=dr_s\cdot\frac{\delta}{dt}(V_1-V_2)+(V_1-V_2)\cdot\frac{\delta}{dt}pr_s$$

$$=-(V_1-V_2)\cdot\nabla\overline{V}dr_s\cdot\nabla\frac{p_1-p_2}{\rho}$$

$$+dr_s\cdot\nabla\overline{V}(V_1-V_2)=0 \tag{6.6.8}$$

Because dr_s and (V_1-V_2) have same direction, and the dynamic boundary condition that (p_1-p_2) has no gradient along the line S, the right side of the formula (6.6.8) equals zero. While this is just a special case of the Kelvin theory.

The rate of mass variation $dc=(V_1-V_2)dr_s$ represents the circulation around the line segment dr_s on the discontinuity line. From the formula (6.6.8), if the endpoints of the segment dr_s close to each other, then $|(V_1-V_2)|$ must increase, vise versa.

Hiland (1942) used to explain the mechanism of the K-H instability by means of the equation (6.6.8). Suppose the potential inner air flows are located in two sides of the line S. Then the only vorticity is the sliding vorticity along the line S. Every line element $d\mathbf{r}_s$ plays the effect of point vortex with the intensity of dc and generates a velocity field of $dc\mathbf{k}\times(\mathbf{r}-\mathbf{r}_s)_s/2\pi(\mathbf{r}-\mathbf{r}_s)^2$. Integrate all segment along the line S then we can obtain the total velocity. When $\mathbf{r}$ is on S, the integration is divergent, but exist a basic value which equals $\overline{V}$.

At beginning, suppose S is a straight line (a limitless stationary sliding vorticity line), separating two branches of the uniform and counterblast flows. Suppose there are unlimited special uniform points moving with velocity $\overline{V}$. These points separated S into many small segments with the same vorticity intensity dc.

As long as S is a straight line, $\overline{V}$ will be zero everywhere and nothing will happen. Now let us introduce disturbance and assume S take the shape of a sine curve with a small amplitude as shown in Fig. 6.18a. Now, at the inflection points B and D, $\overline{V}$ is still zero. While at the wave trough (point A) $\overline{V}$ points to right side; and at the ridge (point C) $\overline{V}$ points to left. Therefore the sliding vortex migrates along line S. It moves away from point D and concentrates to point B from two sides around the point B. This process is irreversible and will lead to the shear line curling around point B. While the point B likes the point D will keep stationary due to the symmetric reason. Hence at last as a result a set of the vortexes with same interval will be formed as shown in Fig. 6.18b. However, it is also an instable state. It may possibly be controlled by the similar development process of the waves with longer wavelength.

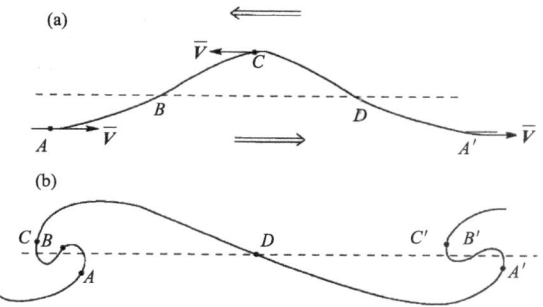

Fig. 6.18 K-H instability. (After Eliassen, 1983)

In the atmosphere, some small vortexes are usually founded on the line convection. This is possibly caused by K-H instability. Observations showed that sometimes the shear air flows may be maintained steadily on the line convection, so that it is very often to see the radar echo belt is curled. Sometimes by means of Doppler radar we can see the stream field of the small vortexes on the line elements as shown in Fig. 5.65 and Fig. 5.66. Browning et al. (1970) believed that the K-H instability may be the possible mechanism to cause the break of the horizontal shear and generate the small vortexes.

References

BENNETTS D A, HOSKINS B J, 1979. Conditional symmetric instability—a possible explanation for frontal rainbands[J]. Q J Roy Meteor Soc, 105:945-962.

BROWNING K A, 1971. Structure of the atmosphere in the vicinity of large amplitude Kelvin — Helmholtz billows[J]. Q J Roy Meteor Soc, 97: 283-299.

BROWNING K A, HARROLD T W, STARR J R, 1970. Richardson number limited shear zones in the free atmosphere[J]. Q J Roy Meteor Soc, 96: 40-49.

CHARNEY J G, ELIASSEN A, 1964. On the growth of the hurricane depression[J]. J Atmos Sci, 21:68-75.

ELIASSEN A, 1983. Hydrodynamic Instability. Mesoscale Meteorology[M]. Sweden: SMHI.

EMANUEL K, 1983. On assessing local conditional symmetric instability from atmosphere sounding [J]. Mon Wea Rev, 111:2016-2033.

EMANUEL K, 1984a. Conditional Symmetric Instability, Dynamics of Mesoscale Weather Systems-NCAR Summer Colloquium Lecture Notes[C]. 11 June—6 July Boulder Colorado, 159-183.

EMANUEL K, 1984b. Symmetric Instability, Dynamics of Mesoscale Weather System-NCAR Summer Colloquium Lecture Notes[C]. 11 June—6 July Boulder Colorado, 145-158.

HOWARD L N, 1961. Note on a paper of John W. Mile[J]. J Fluid Mech, 10:509-512.

KESSLER E, 1987. Thunderstorm Morphology and Dynamics[M]. University of Oklahoma Press.

KUO H L, 1961. Convective in conditionally unstable atmosphere[J]. Tellus, 13:441-459.

LILLY D K, 1986. Atmospheric Instability. Mesoscale Meteorology and Forecasting[M]. Am Meteor Soc.

LINDZEN R S, 1974. Wave-ClSK in the tropics[J]. J Atmos Sci, 31:156-179.

OGURA Y, LIOU M T, 1980. The structure of a mid-latitude squall line: A case study[J]. J Atmos Sci, 37:553-567.

RAYMOND D J, 1975. A model for predicting the movement of continuously propagating convective storms[J]. J Atmos Sci, 32:1308-1317.

RAYMOND D J, 1976. Wave-ClSK and convective mesosystems[J]. J Atmos Sci, 33:2392-2398.

SAITOH S, TANAKA H, 1988. Numerical experiments of Conditional Symmetric baroclinic instability as a possible cause for frontal rain-band formation. Part II :Effects of water vapor supply[J]. J Met Soc Japan, 66:39-53.

SHOU S W, LI Y H, 1999. Study on moist potential vorticity and symmetric instability during a heavy rain event occurred in the Jiang-Huai Valleys[J]. Adv Atmos Sci, 16(2):312-321.

XU Q, CLARK J H E, 1985. The nature of symmetric instability and its similarity to convective inertial instability[J]. J Atmos Sci, 42:2880-2883.

Chapter 7
Factors Effecting the Development of MCSs

In this chapter the factors effecting the development of MCSs, such as the potential instability, the vertical wind shear and vertical motions etc. will be discussed.

§ 7.1 The relationship between the atmospheric potential instability and convection

In the above chapter we have discussed the atmospheric potential instability. Following we will further discuss the relationship between the atmospheric potential instability and convection.

Suppose the air parcel in motion has no exchanges of heat, water vapor, mass and momentum with environment and has no friction effect as well as to assume the quasi stationary condition is satisfied, i. e. there is no difference of the pressure between the air parcel and its environment, then the vertical motion equation can be written as follows:

$$\frac{dw}{dt} \approx g\left(\frac{\theta - \bar{\theta}}{\bar{\theta}}\right) \tag{7.1.1}$$

or

$$\frac{dw}{dt} \approx g\left(\frac{T - \bar{T}}{\bar{T}}\right) = g\frac{\Delta T}{\bar{T}} \tag{7.1.2}$$

where θ and T are the potential temperature and temoerature of the air parcel and $\bar{\theta}$ and $\bar{T}$ are the potential temperature and temoerature of the environment respectively. Integrading right side of the above equation for the height of z, then we can get

$$E = \int_{z_0}^{z} g\frac{\Delta T}{\bar{T}} dz = -\int_{p_0}^{p} R \cdot \Delta T d\ln p \tag{7.1.3}$$

While interacting left side of the equation (7.1.2) for the height of z, then we can get the increment (ΔE_k) of the kinetic energy (E_k) of the vertical motion of the air parcel. Where ΔE_k may be expressed as:

$$\Delta E_k = \int_{z_0}^{z} \frac{dw}{dt} dz = \int_{t_0}^{t} \frac{dw}{dt} w\, dt = \int_{w_0}^{w} w\, dw = \frac{1}{2}(w^2 - w_0^2)$$

$$= \Delta\left(\frac{w^2}{2}\right) = E_k - E_{k0} \tag{7.1.4}$$

So that we get:
$$E=\Delta E_k \tag{7.1.5}$$
This means that under the situation without effect of friction the instability energy E equals the kinetic energy increment of the air parcel lifted from the initial height z_0 to the height z. Therefore we understand that the kinetic energy of the vertica accerelation motion of the air parcel is converted from the instability energy. The bigger the instability energy the stronger the vertical accerelation and the severer the convective weather. In Fig. 7.1, the area encircled by $FABCF$ is the representative of the positive instability energy marked by A_+, while the area encircled by $FLDF$ is the representative of the negative instability energy marked by A_-. When $A_+ > A_-$, the situation is favorable to form convective weather.

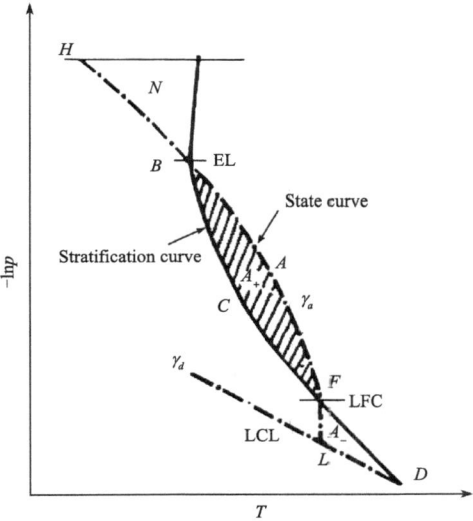

Fig. 7.1 T-ln$p$ diagram.
(LCL=Lifted condensation level, LFC=Level of free convection, EL=equilibrium level)

§ 7.2 The factors influencing on convective clouds

The development of the convective clouds will be influenced by many factors such as the instability, the loading of liquid water in the cloud on the entrainment effect, the merging of convective clouds and effect of the descending flows etc. The effects may be descussed theoretically as follows:

7.2.1 The effect of the loading of liquid water in cloud

The vertical motion equation of the air parcel under the consideration of the effect of the liquid water loading in the cloud may be written as follows:

$$\frac{dMw}{dt} = M\left(\frac{T'_v - T_v}{T_v}\right)g - Mlg - MKw \tag{7.2.1}$$

where M is mass, w is vertical velocity, T'_v and T_v are the virtual temperature of the air parcel and environment respectively, l is the mixing ratio of the liquid water in the air parcel, g is the gravity acceralation, K is the frictional coefficient. In the equation (7.2.1), the first term in right side is the Archimede's bouyancy force, the second term is the dragging force caused by the liquid water, the third term is the frictional force. Under the condition without water content, no frictional force and $T = T_v$, the equation (7.2.1) for unit mass air parcel may be simplified as

$$\frac{dw}{dt} = g\frac{\Delta T}{T} \tag{7.2.2}$$

To integrate the equation (7.2.2) for the height z, we can get

$$w_z = \left[w_0^2 - 2R\int_{p_0}^{p_z}\Delta T d\ln p\right]^{\frac{1}{2}} \tag{7.2.3}$$

where w_0 is the vertical velocity at the initial height above surface. When $w_0 = 0$, the value of w_z equals the area of the instability energy. It can be seen that when $w_0 = 0$, the air parcel can reach the height that the positive energy area P is balanced with the negative energy area N. Accorging to the observational data, the actual cloud top is corresponding to the height of the point B in Fig. 7.1. The height is therefore usually called as the upper limit of convertion.

In the equation (7.2.2) it is assumed that the air parcel does not contain the liquid water. This is one of the reasons the theoretical value of the cloud top is higher than the actual situation. Now let us consider the air parcel contained liquid water. Under this situation the second term in the equation (7.2.1) $Mlg \neq 0$. This model considered the loading water is called as "a loaded moist adiabatic (LMA)" (Chisholm, 1973).

From equation (7.2.1), when $M = 1$ we can get

$$\frac{dw}{dt} = g\left[\frac{T'_v - T_v}{T_v} - l\right] \tag{7.2.4}$$

From equation (7.2.4) we can get

$$w_z = \left[w_0^2 - 2R_d\int_{p_0}^{p_z}(T'_v - T_v - lT_v)d\ln p\right]^{\frac{1}{2}} \tag{7.2.5}$$

where p_0 is the pressure at the cloud bottom, which is indicated by the lifted condensation level (LCL); p_z is the pressure at the height z; l is the mixing ratio of the liquid water in the air parcel (in unit: g • kg^{-1}); T'_v and T_v are the virtual temperature the air parcel and environment respectively (in unit: K); w_0 is the vertical velocity at the cloud bottom (unit: m • s^{-1}). Chisholm et al. (1973) pointed out that based on the aircraft observation for the hailstorms in Alberta Canada, the values of w_0 are normally 4—6 m • s^{-1}, with the mean value about 5 m • s^{-1}; R_d is the gas constant of the dry air.

For examining the above LMA model, Chisholm et al. (1973) made a comparism between the cloud tops observed by radar and calculated by the LMA model respectively based on the data from the 29 Alberta hailstorms occurred during 1967—1968. The results showed that 75% of the differences between the observational value and the calculated value of the cloud top are less than 0.8 km. Since the calculated cloud top is normally higher than the radar observed cloud top and the later is normally lower than the actual cloud top, so that the calculated cloud top is more closer to the actual height of the cloud top.

The above discussion showed that it is necessary to consider the effect of the loaded water in the air parcel when we estimates the height of the cloud top, since there are normally large amount of the precipitation matters such as hailstones, snow crystals and rain droplets etc. in the severe convective storm clouds.

7.2.2 The effect of the entrainment of the air in the environment of the cloud

There are some other reasons to cause the height cloud top calculated based on the simple air parcel model as shown in equation (7.2.1). One of the reasons is the effect of the entrainment (Fig. 7.2). The entrainment of the air in the environment of the cloud will decrease the temperature difference between inner and outer cloud and thus to decrease the height of the cloud top.

Based on the cloud column model, in which the cumulus cloud is regarded as a cloud column, Malkus et al(1959). derived the relationship between the mass entrainment rate $\left(\frac{1}{M}\frac{dM}{dz}\right)$ of mass (M) and the diameter of the cloud diameter (D) as follows:

$$\frac{1}{M}\frac{dM}{dz} \approx \frac{1}{D} \qquad (7.2.6)$$

According to the above equation it is clear that the mass entrainment rate will be

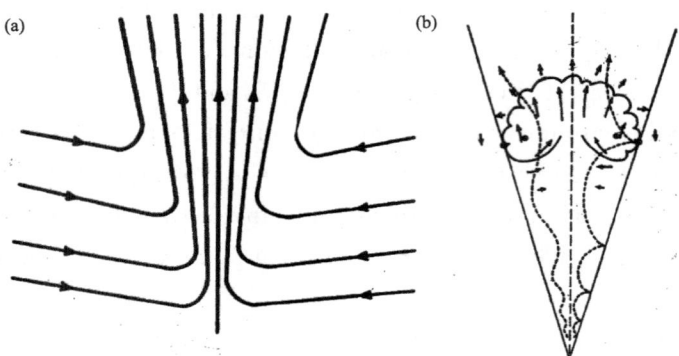

Fig. 7.2 The concepts of (a) entrainment into a draft (After Stommel, 1947); and (b) bubble convection (After Scorer and Ronne, 1956). In (b) partical trajectories are shown by dashed lines, instantaneous circulations by arrows.

decreased with the cloud diameter increasing. Since in the center part of the severe thunderstorms such as the supercell storm and MCC etc. with big diameter the entrainment is small, the vertical velocity in the center part of this kind of severe convective storms may be well described by the simple air parcel model without consideration of entrainment effect.

The subsidence and downdraft in the outer and inner of the cloud are also important effects on the development of the convective clouds. The subsidence in the sourrounding of the cloud will decrease the temperature difference between inner and outer of the cloud and therefore the bouyancy for developing vertical velocity will be decreased. The cold downdraft in the low level of the thunderstorm will cause the mesoscale high pressure and outflows as well as the gust front with the characterist of density current flow. In general speaking, if the propagation velocity of the density current is matched with the whole storm motion, then the storm circulation will tend to be stationary. While if the gust front moves forward far away from the main precipitation core, then the supply of the moist air in the boundary layer for the storm will be cutted and thus the storm will be weakened and tend to dissipate.

§ 7.3 The effect of the vertical wind shear on propagation of convective storm

The convective cloud can be regarded as an obstacle relative to the environmental wind. The pressure exceeding and the pressure deficit will occur in the upstream and downstream of the obstacle respectively. The distribution of the induced hydrodynamic pressure $b = P - P_h$, i. e., the difference between the actual pressure P and the undisturbed environmental hydrostatic pressure P_h, is shown in Fig. 7.3 by the positive and negative signs. The vertical acceralation induced by the hydrodynamic pressure gradient can be approximately expressed as follows:

$$\frac{dw}{dt} \cong g \left(\frac{\Delta T}{T_0} + \frac{\partial b}{\partial P_a} \right) \qquad (7.3.1)$$

where ΔT is the the difference between the virtual temperature of the air parcel and the environmental temperature T_0.

If there is a stronger vertical wind shear of the environment wind velocity (V_e) outside of the cloud, while the vertical shear of the mean velocity inner the cloud (V_c) is relatively weaker due to the vertical mixing up and down. If the hydrodynamic pressure decreases with height, then there is an upward action force, and if the following equation is satisfied

$$\frac{\partial b}{\partial P_h} > -\frac{\Delta T}{T_0} \qquad (7.3.2)$$

then an upward net acceleration will be caused. The distribution of the wind as shown in Fig. 7.3 is favorable to cause air ascending in the downstream side and descending in the upstream side of the cloud and favorable to cause the new convective clouds in downstream side. Fig. 7.4 is also an illustration to show the effect of the environmental vertical wind shear on the propagation of the storm. From Fig. 7.4 we can see when the environmental vertical wind shear is weak, the propagation of the storm has no priority direction, when the environmental vertical wind shear is strong, the propagation of the storm has priority direction.

Under the situation of organized convection in mid-latitude, accompanied with the wind speed variation the wind direction is also usually veering with height. Suppose the wind velocity in cloud is simply to be set as the the mean wind velocity of the whole cloud layer as shown in Fig. 7.5, then the relative motion in the low level will enter the cloud from the right side of the storm while oppositely from left side at high level. Accordingly, the maximum pressure exceeding area at low level will occur in the right side of the storm and below the upper air pressure deficit area. Thus the new cloud will easy to grow in the right side of the mean wind direction.

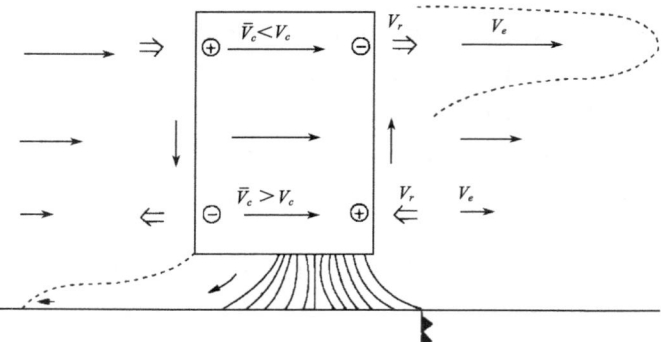

Fig. 7.3 Illustration of the effect of the vertical mixing of the horizontal momentum in the vertical section of the storm on the storm propagation. (the arrow V_r indicates the relative motion of the air in outside of the cloud related with the air inside of the cloud; The positive and negative signs indicate the hydrodynamic pressure occurred in the boundary of the cloud, V_c is the mean wind velocity in the cloud; V_e is the environmental wind velocity outside of the cloud) (Adapted from Newton, 1960)

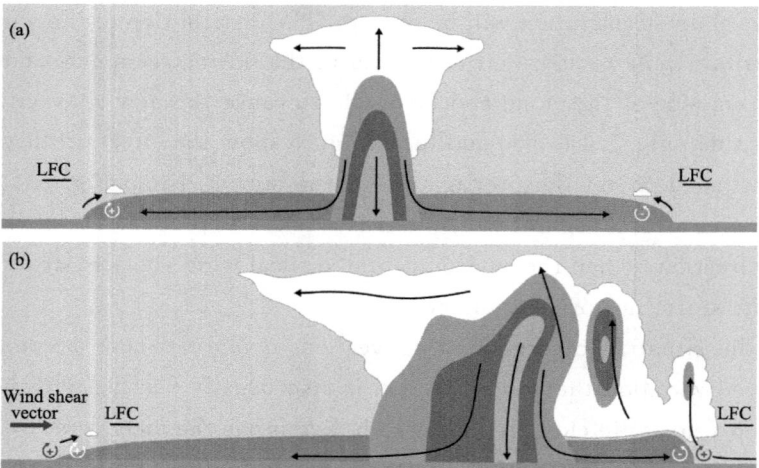

Fig. 7.4 An illustration shown the effect of the environmental vertical wind shear on the propagation of the storm. When the environmental vertical wind shear is weak the propagation of the storm has no priority direction (a); while when the environmental vertical wind shear is strong, the propagation of the storm has priority direction (b). (After Markowski, 2011) (see the color illustrations)

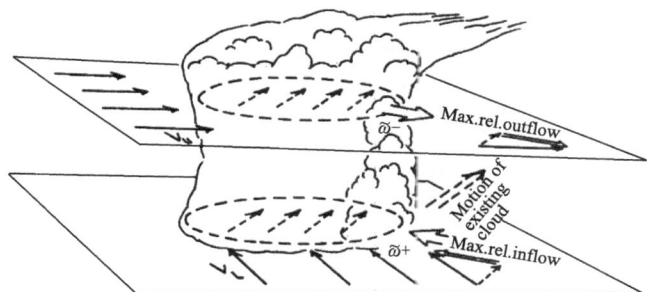

Fig. 7.5 Illustration of the distribution of the relative inflow and outflows in the storm when the wind direction veering with the height. (Adapted from Newton, 1960)

§ 7.4 The comprehensive effect of the environmental thermal and dynamic conditions on the intensity and types of the convective storms

The intensity and feature of the convective storm are closely related with the thermodynamic structure of environment.

One of the parameters to characterize the environmental thermodynamic structure is the atmospheric static stability. The convective available potential energy (CAPE), which is expressed by equation (7.4.1), may be used to show the atmospheric instability.

$$CAPE = \int_{LFC}^{EL} g \frac{(\theta_p - \theta_e)}{\theta_e} dz \qquad (7.4.1)$$

where θ_p is the potential temperature of air parcel, θ_e is potential temperature of environment. CAPE expressed the work done for the environment by the unit mass air parcel, which is lifted from the level of free convection (LFC) to the equilibrium level (EL). CAPE is equivalent to the positive energy area in the sounding analysis. If the influences of the pressure gradient, water loading and the mixing effect are neglected, then there is a relationship between CAPE and the maximum vertical velocity (W_{max}) as follows:

$$W_{max} \approx (2CAPE)^{\frac{1}{2}} \qquad (7.4.2)$$

Chisholm et al. (1973) calculated the energy in various types of convective storms by using LAM model and divided the storms according to the energy into three categories: the low energy, middle energy and high energy storms respectively. The cloud top temperature and the penetration depth of the cloud penetrated into the stratosphere for various types of the convective storms are compared each other as shown in Table 7.1. The supercell storms including the tornadic supercell storms are normally the high energy storms, while the multicell storms and ordinary cell storms are normally the middle energy storms and low energy storms respectively.

Except the temperature stratification, the moisture stratification is also an important characteristic of the environment thermodynamic structure. In general speaking, the abundant moisture in the boundary is favorable to increase the intensity of severe storms. Meanwhile the mid level dry air is also favorable to increase the instability and to cause the storm development. After the storm formed the dry inflow in midlevel will cause the falling rain droplets evaporation and form the cold outflow and severe damage wind. The bundling severe downdraft is called downburst. The downburst is normally the straight line wind. It is usually the divergent flow and different from tornado, which is usually convergent flow.

Table 7.1 The features of three types of convective storms. (After Chisholm, 1973)

Types of storms	Maxmum energy of the air parcel* ($J \cdot g^{-1}$)	Temperature at the top of storm cloud $T(°C)$	The depth penetrated into stratosphere (km)
Low energy	0.0—0.2	$T \geqslant -40$	No penetration
Mid energy	0.2—0.45	$-60 \leqslant T < -40$	< 0.75
High energy	> 0.45	$T < -60$	> 0.75

* i. e. the instability energy area in the T-$\ln P$ diagram.

Downburst is usually occurred under the situation with midlevel dry air. Generally speaking, all the severe convective storms with strong surface wind are formed under the environment with dry air at mid level and has relative higher cloud base. This feature is distict with the rain storms, for which the favorable situation is deeper moist layer and lower cloud base.

The vertical wind shear in the storm environment, especially in the layer from the surface to 6 km above surface, will influence on the storm types, organization, motion and many other features distinguishly (Fig. 7.6). But the observations showed that this kind of influences will be possibly corrected by the thermodynamic factors. Thus it is necessary to consider the effects of both the instability energy and the vertical wind shear comprehensively. This comprehensive effect may be expressed by the Bulk Richardson Number (BR).

$$BR = CAPE / \left(\frac{1}{2}\overline{U}_z^2\right) \qquad (7.4.3)$$

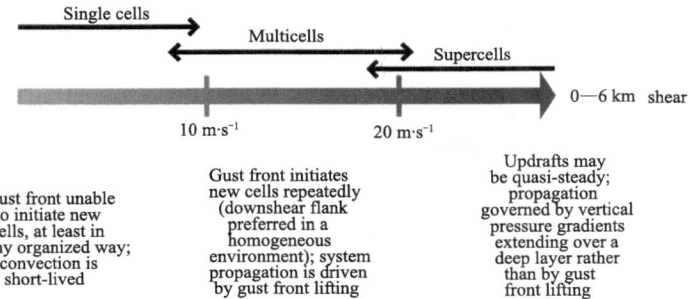

Fig. 7.6　The relationship between 0—6 km shear and the type of the storm.
(After Markowski, 2011)

where $\overline{U}_z^2$ is the difference between mean wind velocity weighted by density in the lowest 6 km atmosphere above surface ($\overline{U}_{6000}$) and the mean wind velocity in the lowest 500 m atmosphere above surface ($\overline{U}_{500}$), i. e.

$$\overline{U}_z = \overline{U}_{6000} - \overline{U}_{500} \qquad (7.4.4)$$

There is a close relationship between the value of BR and the types of the convective storms. Under the situation with certain $CAPE$ the weaker vertical wind shear is corresponding to the ordinary convective storms, while the mid and strong vertical wind shear will be corresponding to the multicell and supercell storms respectively. Therefore the ordinary storm, multicell storm and supercell storms will be corresponding to the bigger value, middle value and small value of BR respectively. According to the analysis given by Weisman and Klemp (1982) based on the observation for 10 supercell storms, 9 multicell storms and the other cases as well as the numerical simulation, the multicell storms are normally formed under the condition with $R > 30$, while the supercells are normally formed under the condition with $10 < R < 40$.

Normally there is a clear relationship between BR and the type of the storm, but the relationship between BR and the intensity of storm is not certainly. For example, a smaller $CAPE$ (<1000 m² · s⁻²) and a middle wind shear (4×10^{-3} s⁻¹) may make a smaller BR, the value of which is maybe in the extent favorable to form the supercell, but it is very often that no supercell storm occurs actually since the $CAPE$ is too small, in other words the energy is too small for causing a severe convective storm. Similarly, although a BR composed by a bigger $CAPE$ (>3500 m² · s⁻²) and a middle vertical wind shear maybe is not in the range for causing the supercell, but it is still possible to form severe convective storm.

§ 7.5 The effect of the vertical wind shear on the organization and splitting of the storms

7.5.1 The effect of the vertical wind shear on the organization of the thunderstorm

Chisholm et al. (1972) presented the hodographes for the ordinary storms, the multicell storms and the supercell storms respectively as shown in Fig. 7.7. The figure shows that the ordinary storm, the multicell storms and the supercell storms have the weaker, the medium and the strongest vertical wind shear background.

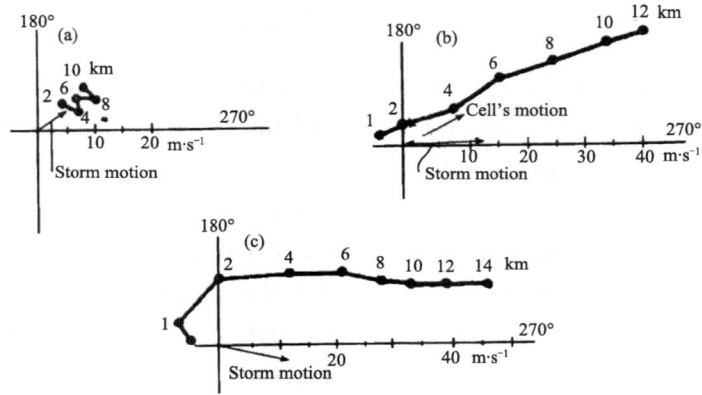

Fig. 7.7 The typical hodographes from the hailstorm research program of Canada, for (a) ordinary storms; (b) multicell storms and (c) supercell storms respectively. (After Chisholm and Renick, 1972)

Two mechanisms may possibly cause storm types influenced by the vertical wind shear. One of them is the ability of gust front to trigger new convection and another is the vertical dynamic forcing due to the difference of airflows between storm and environment.

First, since the storm types are depending on the ability of gust front to trigger new convection. While the later is related with vertical wind shear. In the environment without vertical wind shear, the cold pool caused by the downdraft from the convective cloud will spread on surface uniformly in all directions. In the convergence zone along the ambient of the cold pool the new convective cells will be formed. After the new cells formed they will be no motion since they are in a no vertical wind shear environment. But the gust front still moves outward continuously, so that the new cells themselves will be located in the cold and stable environment behind the gust front and therefore their further development will be ceased. Since the new storms are difficult to be formed in the ambient of the single cell thunderstorms, so that the isolated storm will be presented as the only structure form. Now let us discuss the situation with medium vertical wind shear. Under such situation, the distribution of the convergence zone in front of the horizontal outflow component caused by the surface cold pool below the storm cloud presented non-symmetric form. The strongest convergence zone occurs in the down-shear direction of the organized cell. Meanwhile the new cells growed in the convergence zone will move along the same direction. So that the time duration experiencing the low level convergence and to obtain supply of the warm and moist air in the front of the gust front for the convective cells will be increased. Under the situation with appropriate vertical wind direction shear, the velocities of the cell motion and the gust front motion could be the same, this will cause the updraft develope continuously. Thus the multicell structure may be caused since the new convective cells successively developed in the convergence zone along the outflow boundary. This is the represents of the basic physical mechanisms of the multicell storm's formation and maintaining.

Second, except the above two physical mechanisms for triggering the new convective by the gust front, another physical mechanism will become more important when the vertical wind shear is getting more stronger. Due to the difference of airflows between storm and environment, it will cause the inflow and outflow relative to the storm. According to the Bernoulli's law, the positive dynamic pressure will be caused at the relative inflow area and the negative dynamic pressure will be caused at the relative outflow area. If there is relative outflow at high level and inflow at low level in the flank of storm, then there is negative dynamic pressure at high level and positive dynamic pressure at low level, thus the surface air will be accelerated upward under the action of the vertical pressure gradient force. If the environmental vertical wind shear can stretch to the height of 4—6 km above the surface, then a distinct vertical dynamic pressure gradient will be caused and the surface air will be lifted acceleratory. This kind of dynamic forcing will be favorable to maintain the updraft and to cause the propagation toward the direction deviating the average wind direction. Meanwhile this action can also lead the combination of the vertical motion and the vertical vorticity in the storm flank, which is caused by the slantwise motion of the horizontal vorticity tube. This means that the updraft is rotatory flow, which is the essential characteristic.

7.5.2　The effect of the vertical wind shear on the splitting of the convective cell

The convective cell sometimes can be splited, which is related with the vertical wind shear of the environment. The spliting process of the convective cell can be illustrated by Fig. 7.8. First an environment wind field with vertical wind shear is shown in Fig. 7.8a. It can be seen that a vortex tube with horizontal axis is formed due to the vertical wind shear. Company with the convective cloud developing the horizontal vortex tube will be raised upward so that two vortex tubes with vertical axis and opposite rotation directions will be formed. After the convective cell developed into certain stage, the precipitation began. Due to the dragging effect of the precipitational materials and the middle level cold and dry airflow entering the cloud the downdraft will

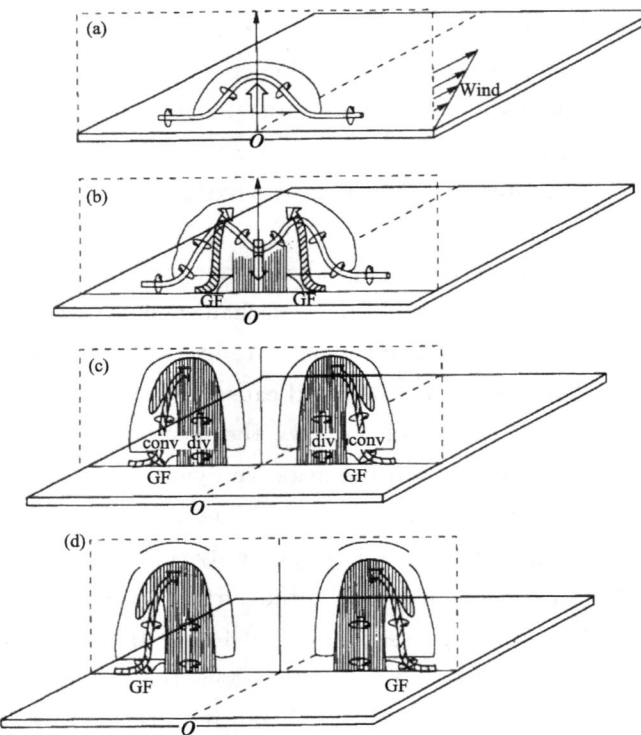

Fig. 7.8　Illustration of the thunderstorm splitting process. (The point O is located on the surface, at the center of the original storm; After split the storms move far away from the O point with time. GF indicates the gust front; Div and conv represent divergence and convergence respectively. The shadowed area is the precipitation area. (After Houze et al., 1982)

occur in the central part of the cloud. It will cause the up-convex shaped vortex tube to be changed into the concave shape, so that the downdraft also contains two opposite rotational vortex tubes as shown in Fig. 7.8b. Therefore two branches of the updraft will be formed in two sides of the downdraft, and each of the updrafts begins to develop and propagate respectively. As a result the thunderstorm cell is splited into two thunderstorm cells. One of them moves toward right, which is called right-moving storm. While another cell moves toward left called left-moving storm as shown in Fig. 7.8c, d. The above thunderstorm splitting process is also illustrated in Fig. 7.9.

For further demonstrate effect of the vertical wind shear on the organization structure and storm splitting, Weisman and Klemp et al. (1982, 1984) made the numerical experiences. They set the medium level unstable with the bouyancy energy about 2200 $m^2 \cdot s^{-2}$ as the thermodynamic condition of the environment and considered two kind situations of the vertical wind shear.

The first situation is the environment with the single wind direction, i.e. the wind directions are the same, while only the wind speeds are different in the upper and lower levels. Assuming the depth of the layer of wind shear is only 5 km and the wind speed is constant above 5 km. In such the wind shear environment as shown in Fig. 7.10a, if the wind shear is weak then only the short-life new convective cells can be produced and the gust front can only trigger the short-life new cells. The new cells are normally generated in the downshear direction of the initial cell. Since the stratification is stable they are difficult to be furtherly developed and will be soon located behind the gust front. When the intensity of the vertical wind shear increases, in the both left and right sides of the initial updraft column, relative to the wind shear vector, the low pressure begins to develop, which is strongest on the mid-level. Under the strong vertical wind shear condition, the low pressure may cause the convective cell to be splited and form two quasi-stationary cells. One of them moves toward right and presented cyclonic rotation, another one moves toward left and presented anticyclonic rotation.

The second situation to be considered is that in the lowest 5 km of atmospheric layer the wind vectors present rotation clockwisely as shown in Fig. 7.10b. Under this situation, In a weak shear condition, the newborn short-life cells will occur in the area along the gust front and in the right-ahead and left-ahead of the initial storm. In the strong wind shear condition, the mesoscale low pressure will occur in the right side of the initial updraft column due to the action of the dynamic pressure and will

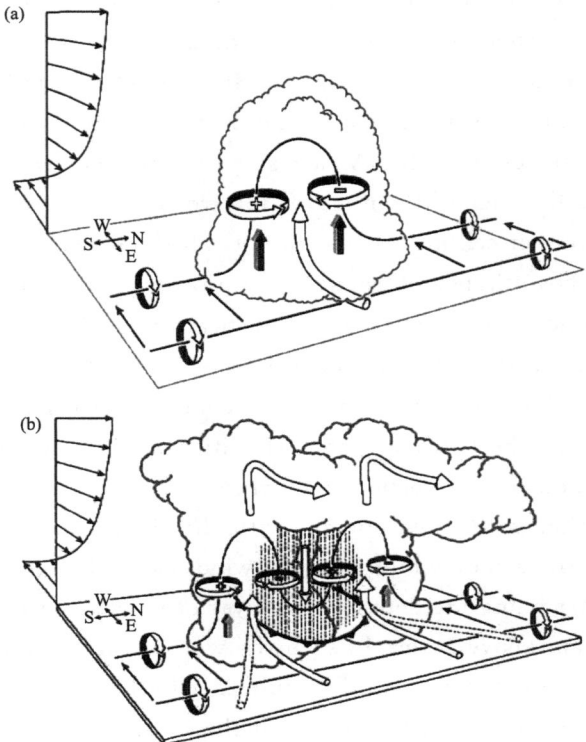

Fig. 7.9 Schematic of vertical vorticity generation through vortex tilting within finite convective lines and bow echoes. For (a) Conversion of crosswise horizontal vorticity of the environment to vertical vorticity in a cumulus cloud. Counter-rotating vorticity occur on either sides of the updraft. The positive and negative signs indicate the sense of the vertical vorticity. Mean environmental flow (shown by thin arrows) is unidirectional in the west-east direction and increasing with height. The vortex tube aligned north-south outside the cloud indicate the horizontal vorticity of the environment. Solid lines represent vortex line with the sense of rotation by circular arrows. Linear tilting of the environmental horizontal vorticity by cloud vertical air motion leads to the indicated vertical vorticity couplet. Shade arrows represent the vertical pressure gradient force associated with votices produced by tilting. Cylindrical arrow shows the direction of cloud relative airflow; (b) Continuetion of the cloud development shown in Fig. 7.9a. In an environment of unidirectional shear, tilting of the crosswise environmental horizontal vorticity of the environment by the initial updraft produces the two counter-rotating centers of vertical vorticity on the flanks of the updraft. The shaded arrows represent the forcing that promotes new updraft and downdraft growth on the flaks of the cumulonimbus. Vertical hatching shows rain. Cylindrical arrows show the direction of cloud relative airflow. The frontal symbol at the surface marks the boundary of the cold air spreading out beneath the storm, and the dashed cylindrical arrows indicate the shifted location of the storm inflow when updrafts become established on the storm flanks. (After Klemp, 1987)

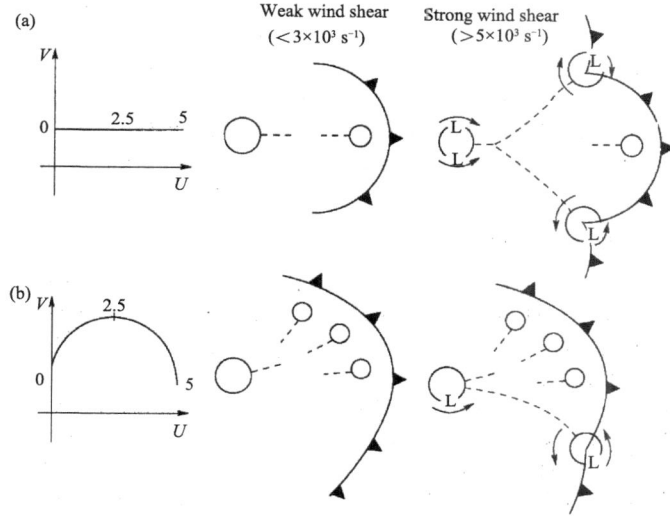

Fig. 7.10 An illustration about the variation of the convective cell under the situations with the pure wind direction shear and the wind direction veering with height clockwisely and under the conditions with weak and strong vertical wind shear respectively. The hodograph is in the left side of the figure; the big and small circles are representatives of the relative strong and weaker convective cells respectively; the dot line represents the path of the convective cell; the sawtooth lines indicates the gust front; L indicates the position of the mid-level low pressure; the arrows indicate the possible rotating direction. (Adapted from Klemp and Weisman, 1984)

form the quasi stationary cyclonic rotated updraft. However, at this moment the short life cells may still possible to be formed in the left side of the storm. If the wind changes with height anticlockwisely rather than clockwisely, then the development situation of the newborn in both left and right sides of the storm will be simply reversed. For instance, the left moving storm with anticyclonic rotated will occur. According to the climatic analysis, the most of low level environmental wind of the severe convective storms are rotating with height clockwisely. Hence, in the strong wind shear condition, the cyclonic rotating storms are much more than the anticyclonic rotating storms.

The close relationship between the development, motion and splitting of the storm and the vertical wind shear of the environment may be used for explaining the formation of the various types of convective precipitation observed in nature. Klemp and Weisman (1984) use the numerical simulation to study the variation of the convective precipitation types under the different types of wind shear profiles. The results demonstrated there is certainly the relationship between the vertical shear and the structure of the precipitation systems. For example, Fig. 7.11a and b showed the types of the hodographs and the types of the convective systems according to the model integration after 40 min, 80 min, and 120 min. In the diagram the isolines present the rainfall distribution at low level, similar to the radar reflectivity distribution, the shade area indicates the location that the updraft at the mid level exceed 5 m·s^{-1}, the number represents the central maximum vertical velocity, the saw-like line presents the position of the surface gust front. R indicates the Bulk Richardson number. Fig. 7.11a is the representative of the weak shear situation. It generates the short-life multicell thunderstorms. The new cells develop along the left flanking. The gust front moves gradually to the front of the storm. At 80 min, the initial updraft center has disappeared, the second center of updraft begins to form. At 120 min, the second

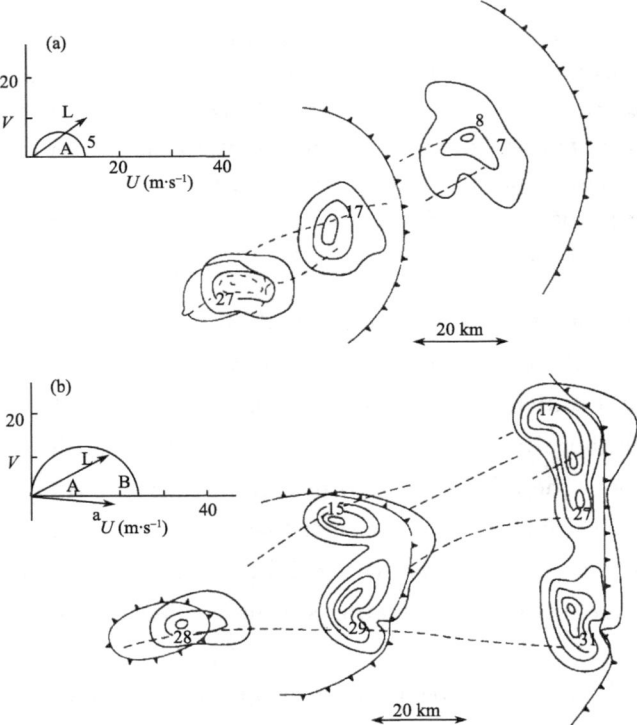

Fig. 7.11 The structures of the storms under the different vertical distribution of the aloft wind at the moments after the initial times of 40 min, 80 min and 120 min. (After Klemp and Weisman, 1984)

Chapter 7   Factors Effecting the Development of MCSs   • 259 •

updraft center has disappeared, replaced it with two weak updrafts, the gust front has moved out far away. During the whole life cycle, the rain area roughly moves with the mean wind. Fig. 7. 11b represents the situation that the supercell formed at the south end of the multicell storm.

§ 7. 6   The effect of the vertical wind shear on formation of tornado storms

7. 6. 1   The structure of tornadic storms

The severe local storm generating tornadoes is called tornadic storm, which is normally the supercell storm.

Fig. 7.12 illustrated a structure model of of the tornadic storm in different stage. In the formation stage there is a very strong rotationary updraft core in the center of the cloud body. In the updraft core the vertical motion is very strong so that the moisture condensates have no enough time to grow and in the radar PPI displayer will occur a no echo or weaker echo area (Y), which is called "eye" by Fujita. In the mature stage, the out part of the rotating updraft, i. e. the marked area by the letters ABCDE in Fig. 7.12, separated with the centre part (Y), due to the the "eye" moves toward the right to back side of the main cloud body. Thus a hock shaped projection may be found on the PPI echo picture. The hock arear is the location that the inflow and strong updraft located, on the upper level the hock echo disappeared and sometimes an echo hole may be presented.

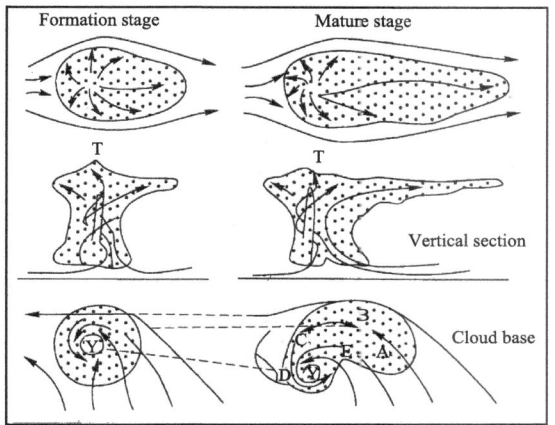

Fig. 7.12   A model of the rotational thunderstorm, Y is the eye area, ABCDE is the ambient echoes, T is the top of the cloud tower. (Adapted from Fujita, 1978)

The hock echo or the echo hole is one of the distinct symbols of the tornadic storm. Although not all the hock echo must accompanied with tornadoes, while there is a greater possibility to find the tornadoes near the path of the hock echo.

Why the eye can move to the edge of the main cloud body from the center of the main body of the cloud during the period from the formation stage to the mature stage? The problem may be explained as follows.

As we know that the eye area is actually the strong rotational vortex caused by the interaction between the rotating column and the environmental fluid field, in other words the interaction between the the convective cloud column and its ambient fluid field. The rotation of the vortex is normally cyclonic but sometimes may be anticyclonic. It is depending on the strength of the relative motion between the cloud column and environment.

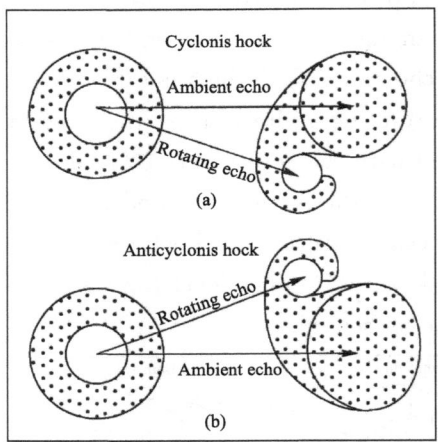

Fig. 7.13 An illustration of the location of the hock echo relative to the main echo moving from west to east. In the left side of the figure the small circle located in the center of cloud is the location of the eye circulation, after it moves out the hock echo formed. (Adapted from Fujita, 1978)

Assuming a rotating column is located in a homogenous motion fluid and its axis is perpendicular to the moving direction, then the circulation will be generated in its ambient. At the side along the direction same with the fluid motion the velocity will increase, while at the opposite side the velocity will decrease. According to the Bernoulli's law, the pressure will decrease at the side with larger velocity, while increase at the side with slower velocity. Thus a force will act on the column. Its direction will be perpendicular to the axis of the column and points to the side that the velocity increased. The force is called Magnus force. This phenomenon is called Magnus effect. When a rigid column body rotates in the motion fluid with speed u and density ρ, if the circulation intensity in the ambient of the rotating body is Γ, then the magnitude of the Magnus F will be as follows:

$$F = \rho u \Gamma \tag{7.6.1}$$

Due to the existing of the Magnus force effect, when the relative motion between the ambient wind field and the convective cloud is cyclonic circulation, the convective cloud will suffer a pressure gradient force pointed to the right of its moving direction, thus it will be pushed to the right side of the forward direction. Due to the effect of the fluid's Mixed viscous effect, the cyclonic circulation will be caused at the right-back side of the rotating convective cloud. The cyclonic circulation is called the tornadic cyclone (see the upper part of Fig. 7.13). Oppositely, when the relative motion between the ambient wind field and the convective cloud is the anti-cyclonic circulation, then the anti-cyclone will be caused, which is called tornadic anticyclone (see the low part of Fig. 7.13).

7.6.2 The generation of tornadoes

Same with cyclone the tornado is a vortex system rotating around a vertical axis. The difference is the scale. The former is a large scale system, while the latter is a small scale system. The magnitude order of vorticity $O(\zeta)$ in a cyclone is 10^{-5} s^{-1}, while the order in tornado is much bigger. If the radius of tornado is 100—1000 m, the tangential velocity is 50 m · s^{-1}, then its vorticity order may be estimated as 10^{-1}—10^{-2} s^{-1}. Furthermore, the difference between cyclone and tornado is also shown in the development speed. In general speaking, it will take about 10—20 h for a cyclone developed from the initial stage to the mature stage, in other words for the cyclone the order of the local change of vorticity $O(\frac{\partial \zeta}{\partial t})$ is about 10^{-10} s^{-2}. While tornado can be generated in ten minutes, its order of the local vorticity change is about 10^{-4} s^{-2}, this means that the local vorticity change for tornadoes is about one million times of the cyclone. For explaining the formation of tornado we have to understand what is the reasons to cause the vertical vorticity increasing rapidly.

From the frictionless horizontal motion equation

$$\frac{\partial u}{\partial t} = -u\frac{\partial u}{\partial x} - v\frac{\partial u}{\partial y} - w\frac{\partial u}{\partial z} - \frac{1}{\rho}\frac{\partial p}{\partial x} + fv \tag{7.6.2}$$

$$\frac{\partial v}{\partial t} = -u\frac{\partial v}{\partial x} - v\frac{\partial v}{\partial y} - w\frac{\partial v}{\partial z} - \frac{1}{\rho}\frac{\partial p}{\partial y} - fu \tag{7.6.3}$$

to differentiate (7.6.2) with respect to y, and to differentiate (7.6.3) with respect to x, and to subtract (7.6.2) from (7.6.3) then we can obtain the vorticity equation as follows:

$$\frac{\partial \zeta}{\partial t} = -\left(u\frac{\partial \zeta}{\partial x}+v\frac{\partial \zeta}{\partial y}+w\frac{\partial \zeta}{\partial z}\right)-f\left(\frac{\partial u}{\partial x}+\frac{\partial v}{\partial y}\right)-\left(u\frac{\partial f}{\partial x}+v\frac{\partial f}{\partial y}\right)$$
$$-\zeta\left(\frac{\partial u}{\partial x}+\frac{\partial v}{\partial y}\right)+\left(\frac{\partial u}{\partial z}\frac{\partial w}{\partial y}-\frac{\partial v}{\partial z}\frac{\partial w}{\partial x}\right)+\left[\frac{\partial p}{\partial x}\frac{\partial \left(\frac{1}{\rho}\right)}{\partial y}-\frac{\partial p}{\partial y}\frac{\partial \left(\frac{1}{\rho}\right)}{\partial x}\right] \quad (7.6.4)$$

where $\zeta=\frac{\partial v}{\partial x}-\frac{\partial u}{\partial y}$. By analyzing the terms in right side of the equation we know that for the small scale convective process we have

$$O\left(\frac{\partial u}{\partial x}+\frac{\partial v}{\partial y}\right)\sim 10^{-2}\,\mathrm{s}^{-1},\ O(f)\sim 10^{-4}\,\mathrm{s}^{-1}$$

Therefore the order of the second term in right side of the equation (7.6.4) is $10^{-6}\,\mathrm{s}^{-2}$, which is much smaller than the order of the local vorticity change in tordado parent cloud. This means the effect of the Earth rotation (the Coriolis force) is not important for the formation of tonadoes. By estimating the third term in right side of the equation (7.6.4), i.e. the term about the advection of the Coriolis parameter the conclusion should be the same. Meanwhile we can find that the fourth, the fifth and the sixth terms will play an important role respectively for the formation of tornadoes.

Here let us look at the fifth term at first. In the cumulonimbus we have

$$O\left(\frac{\partial u}{\partial z}\frac{\partial w}{\partial y}\right)\sim O\left(\frac{\partial v}{\partial z}\frac{\partial w}{\partial x}\right)\sim 10^{-4}\,\mathrm{s}^{-2}$$

Obviously, the order of the term has agreed basicly with the rate of vorticity change for the formation of tornado.

Then, let us look at the fourth term. In the thunderstorm the order of divergence $O\left(\frac{\partial u}{\partial x}+\frac{\partial v}{\partial y}\right)$ is nearly $10^{-2}\,\mathrm{s}^{-1}$. If the cloud cluster has already possed certain rotatory and the vorticity order $O(\zeta)$ is about $10^{-2}\,\mathrm{s}^{-1}$, then the order of the fourth term can be $10^{-4}\,\mathrm{s}^{-2}$, thus the order of this term has also agreed with the rate of vorticity change for the formation of tornado. Therefore this term, i.e. the wind divergence factor, can also play important role for the formation of the tornadoes. However, it should be pointed out that the divergence factor can affect only under the situation that the vorticity is already big enough. Hence this term can not be the major factor to form the tornado. The actual observation showed that the tornado is not firstly formed at the low level where the convergence is maximum, which showed precisely the divergence factor is not the major cause for forming the tornado.

On the contrary, the fifth term, i. e. the twisting term presented the interaction between two horizontal vorticity, may be regarded as major factor for the development of tornado. This just proves the action of the twisting term. For the tornadic storm, this term has important effect on the generating of the vertical component of the vorticity. If $\partial w/\partial x = 2\times 10^{-3} \text{s}^{-1}$, $\partial v/\partial z = 5\times 10^{-3} \text{s}^{-1}$, then the incresement of the vertical vorticity generated by the twisting term can be 10^{-5}s^{-2}, after 300 s, the vorticity will increase $3\times 10^{-3} \text{s}^{-1}$, which is the vorticity value of a mesoscale cyclone. After that the air flow convergence in the mesoscale cyclone will cause the vorticity furtherly be concentrated.

The sixth term in the right side of the equation (7.6.4) is the force tube term. The effect of this term is to generate the horizontal component of vorticity. Then by the action of the twisting term to change the horizontal component of the vorticity into the vertical component of vorticity, thus this term is also important for the formation of the tornado.

The above analyses showed that the fifth term in the right side of the equation (7.6.4) plays major role to cause the generation of the vertical component of the vorticity. Now we are going to furtherly explain why in the cumulonimbus in the air mass far away from the frontal zone the possibility to develop strong vertical vorticity is very small, while in the cumulonimbus in the area near the front zone is favorable to form the tornadoes. For this purpose, we change the fifth term in the right side of the equation (7.6.4) into the following form:

$$\frac{\partial u}{\partial z}\frac{\partial w}{\partial y} - \frac{\partial v}{\partial z}\frac{\partial w}{\partial x} = \left[\frac{\partial \mathbf{V}}{\partial z} \times \mathbf{grad} w\right]_z \qquad (7.6.5)$$

where the right side of the above equation is the the projection on the vertical axis Z of the product of the vector that the horizontal wind $\mathbf{V}$ changes with height ($\frac{\partial \mathbf{V}}{\partial z}$), and the horizontal gradient vector of the vertical velocity component ($\mathbf{grad} w$).

Within the air mass the wind variation with the height in free atmosphere is little. But within the cumulonimbus the air flow distribution may be presented as Fig. 7.14a and the distributions of the vectors $\frac{\partial \mathbf{V}}{\partial z}$ and $\mathbf{grad} w$ in horizontal plane may be presented as Fig. 7.14b, in which the the vectors $\frac{\partial \mathbf{V}}{\partial z}$ and $\mathbf{grad} w$ are parallel to each other, but their directions are opposite. Thus the vector production in the equation (7.6.5) and the tendency to form the vorticity in the cloud are very little. Therefore we can make conclusion that within the cumulonimbus in air mass the possibility to form a tornado is very little.

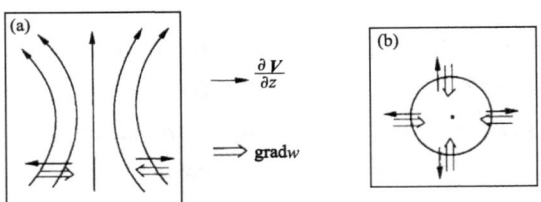

Fig. 7.14 (a) The air flow, $\frac{\partial V}{\partial z}$ and **grad**w in the vertical section near the center of the cumulonimbus inner the air mass; and (b) the distribution of $\frac{\partial V}{\partial z}$ and **grad**w in the horizontal plane.

It can be seen from the equation (7.6.5) that only the locations, where the directions of the vectors $\partial V/\partial z$ and **grad**w are different and with bigger intersection angle or near the right angle, the tornadoes may possibly to be formed. For the frontal clouds, the vertical change of the wind is depending on the ambient large scale parameters and the horizontal temperature gradient. In the cloud the vectors of the variation with height of the geostrophic wind have the same magnitude and the direction will be agreed with the isotherms (in frontal zone). The vectors of **grad**w will point to the center of the cloud accordancely as shown in Fig. 7.15.

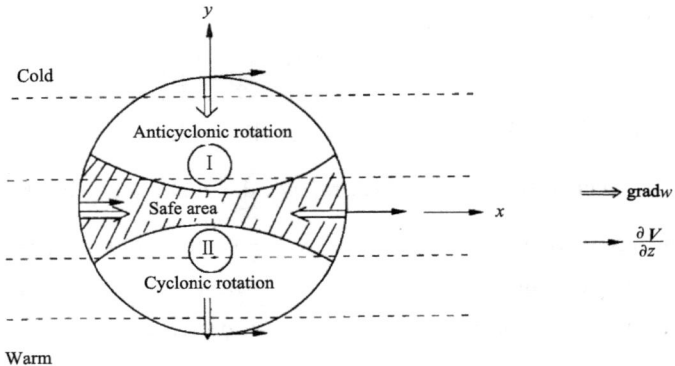

Fig. 7.15 Illustration of vortex development in the cumulonimbus on front when tornado forms. The circle represents the cloud area.

There are three zones, i. e. the Zone Ⅰ, Zcne Ⅱ and the safety zone in Fig. 7. 15. So called the safety zone implys the area where the tornadoes are normally difficult to be formed, since in the area the intersection angle between two vectors $\partial \mathbf{V}/\partial z$ and **grad**w is very small. Furthermore, since the vertical velocity gradient vector **grad**w in the center of cloud is very small or even equals zero, the tendency to form the vortex is very little. While in the areas Ⅰ and Ⅱ, the vectors $\partial \mathbf{V}/\partial z$ and **grad**w are nearly orthogonality, hence the tendency to form the tornadic vortex is much greater. And the projections of the product of the above vectors in the Zone Ⅰ and Zone Ⅱ are positive and negative respectively, thus the cyclonic tornado can be formed in Zone Ⅱ and the anticyclonic tornadoes can be formed in Zone Ⅰ.

References

ANTHES R A, 1977b. Hurricane model experiments with a new cumulus parameterization scheme [J]. Mon Wea Rev, 105:287-300.

ANTHES R A, ORVILL H D, RAYMOND D J, 1986. Mathematical Modeling of Convection. Thunderstorm Morphology and Dynamics[M]. Kessler E (ed.). Norman: University of Oklahoma Press:313-357.

AUSTIN P M, HOUZE R A Jr, 1972. Analysis of the structure of precipitation patterns in New England[J]. J Appl Meteor, 11:926-935.

BENNETTS D A, HOSKINS B J, 1979. Conditional symmetric instability—a possible explanation for frontal rainbands[J]. Q J Roy Meteor Soc, 105:945-962.

BETTS A K, 1974. Thermodynamic classification of tropical convective soundings[J]. Mon Wea Rev, 102:760-764.

BLUESTEIN H B, 1986. Fronts and Jet Streams: A Theoretical Perspective. Mesoscale Meteorology and Forecasting[M]. Am Meteor Soc.

BOSART L F, 1975. New England coastal frontogenesis[J]. Q J Roy Meteor Soc, 101:957-978.

BOSART L F, VAUDO C J, HELSDON J H Jr, 1972. Coastal frontogenesis[J]. J Appl Meteor, 11:1236-1258.

BROWNING K A, HARDMAN M E, HARROLD T W, et al, 1973. The structure of rainbands within a mid-latitude depression[J]. Q J Roy Meteor Soc, 99:125-231.

CARLSON T N, 1980. The role of the lid in severe storm formation: some synoptic examples from SESAME[C]. 12th Conf, On Severe Local Storms, 221-223.

CHARBA J, 1974. Application of gravity current model to analysis of squall line front[J]. Mon Wea Rev, 102:140-156.

CHARNEY J G, ELIASSEN A, 1964. On the growth of the hurricane depression[J]. J Atmos Sci,

21:68-75.

CHISHOLM A J, ENGLISH M, 1973. Alberta hailstorms[M]. AMS Met Monographs, 14:101.

CHISHOLM A J, RENICK J H, 1972. The kinematics of multicell and supercell Alberta hailstorms[R]. Alberta Hail Studies Research Council of Alberta Hail Studies, Rep 72—2, Edmonton Canada, 24-31.

COTTON W R, 1984. In"Proceeding Nowcasting-II symposium"[R]. Sweden:3-8.

ELIASSEN A, 1983. Hydrodynamic instability. Mesoscale Meteorology[M]. Sweden:SMHI.

ELLROD G P, MARWITZ J D, 1976. Structure and interaction in the subcloud region of thunderstorms[J]. J Appl Meteor, 15:1084-1091.

EMANUEL K A, 1982. Inertial instability and mesoscale convective systems. Part II: Symmetric CISK in a baroclinic flow[J]. J Atmos Sci, 39:1080-1097.

EMANUEL K A, 1983. On assessing local conditional symmetric instability from atmosphere sounding[J]. Mon Wea Rev, 111:2016-2033.

EMANUEL K A, FANTINI M, THORPE A J, 1987. Baroclinic instability in an environment of small stability to slantwise moist convection. Part I: Two dimensional models[J]. J Atmos Sci, 44:1559-1573.

EMANUEL K A, 1994. Atmospheric Convection[M]. Oxford: Oxford University Press.

EMANUEL K A, 1984. Symmetric Instability, Dynamics of Mesoscale Weather System-NCAR Summer Colloquium Lecture Notes[C]. 11 June—6 July Boulder Colorado, 145-158.

FUJITA T T, 1958. Structure and movement of a dry front[J]. Bull Am Meteor Soc, 39:574-582.

FUJITA T T, 1978. Manual of downburst identification for Project NIMROD[R]. SMRP Research Paper No 156, Dept of Geophysical Sci Chicago University.

GOFF R C, 1976. Vertical structure of thunderstorm outflows[J]. Mon Wea Rev, 104:1429-1440.

HOBBS P V, PERSSON P O G, 1982. The mesoscale and microscale structure and organization of clouds and precipitation in midlatitude cyclones. Part V:The substructure of narrow cold-frontal rainbands[J]. J Atmos Sci, 39: 280-295.

HOSKINS B J, 1975. The geostrophic momentum approximation and the semigeostrophic equations[J]. J Atmos Sci, 32:233-242.

HOUZE R A Jr, HOBBS P V, 1982. Organization and structure of precipitating cloud systems[J]. Advances in Geophysics, 24:225-316.

HOUZE R A Jr, HOBBS P V, DAVIS W M, 1976. Mesoscale rainbands in extratropical cyclones[J]. Mon Wea Rev, 104:868-878.

HOWARD L N, 1961. Note on a paper of John M Mile[J]. J Fluid Mech, 10:509-512.

KESSLER E, 1987. Thunderstorm Morphology and Dynamics[M]. University of Oklahoma Press.

KEYSER D, 1986. Fronts-Observations. Mesoscale Meteorology and Forecasting [M]. Am Meteor Soc.

KEYSER D, SHAPIRO M A, 1986. A review of the dynamics of upper-level frontal zones[J]. Mon

Wea Rev, 114:452-498.

KOSS W J, 1976. Linear stability of CISK-induced disterbances Fourier component eigenvalue analysis[J]. J Atmos Sci, 33:1195-1222.

KUO H L, 1961. Convective in conditionally unstable atmosphere[J]. Tellus, 13:441-459.

LILLY D K, 1986. Atmospheric Instability. Mesoscale Meteorology and Forecasting[M]. Am Meteor Soc.

LINDZEN R S, 1974. Wave-CISK in the tropics[J]. J Atmos Sci, 31:156-179.

LINDZEN R S, TUNG K K, 1976. Banded convective activity and ducted gravity waves[J]. Mon Wea Rev, 104:1602-1607.

MADDOX R A, 1980. Mesoscale convective complexes[J]. Bull Am Meteor Soc, 61:1374-1387.

MADDOX R A, 1984. Mesoscale convective complexes in the midlatitudes[R]. NOAA ERL, USA.

MARKS F D Jr, AUSTIN P M, 1979. Effects of the New England coastal front on the distribution of precipitation[J]. Mon Wea Rev, 107:53-67.

MALKUS J S, WITT G, 1959. The Evolution of a Convective Element: A Numerical Calculation, The Atmosphere and the Sea in Motion, Rossby Mem[M]. Bolin B(ed.). New York: Rockefeller Inst Press: 425-439.

MONCRIEFF M W, 1978. The dynamical structure of two-dimensional steady convection in constant vertical shear[J]. Q J Roy Meteor Soc, 104:543-568.

MONCRIEFF M W, MILLER M J, 1976. The dynamics and simulation of tropical cumulonimbus and squall lines[J]. Q J Roy Meteor Soc, 102:373-394.

NINOMIYA K, 1971. Mesoscale modification of synoptic situation from thunderstorm development as revealed by ATS III and aerological data[J]. J Appl Meteor, 10(12):103-112.

OGURA Y, 1975. On the interaction between cumulus clouds and the large scale environment[J]. Pure Appl Geoph, 113:869-889.

OGURA Y, CHO H R, 1973. Diagnostic determination of cumulus cloud populations form observed large scale variables[J]. J Atmos Sci, 30:1276-1286.

OGURA Y, LIOU M T, 1980. The structure of a mid-latitude squall line: A case study[J]. J Atmos Sci, 37:553-567.

RAYMOND D J, 1975. A model for predicting the movement of continuously propagating convective storms[J]. J Atmos Sci, 32:1308-1317.

RAYMOND D J, 1976. Wave-CISK and convective mesosystems[J]. J Atmos Sci, 33:2392-2398.

REED R J, 1955. A study of a characteristic type of upper-level frontogenesis[J]. J Meteor, 12:226-237.

RHEA J O, 1966. A study of thunderstorm formation along dry lines[J]. J Appl Meteor, 5:58-63.

SAITOH S, TANAKA H, 1988. Numerical experiments of Conditional Symmetric baroclinic instability as a possible cause for frontal rainland formation. part II: Effects of water vapor supply[J]. J Met Soc Japan, 66:39-53.

SCHAEFER J T, 1974. A simulative model of dryline motion[J]. J Atmos Sci, 31:956-964.

SCHAEFER J T, 1975. Nonliner biconstituent diffusion: A possible trigger of convection[J]. J Atmos Sci, 32:2278-2284.

SHAPIRO M A, 1981. Frontogenesis and geostrophically forced secondary circulations in the vicinity of jet stream frontal zone systems[J]. J Atmos Sci, 38:954-973.

SHOU S W, Li Y H, 1999. Study on moist potential vorticity and symmetric in stability during a heavy rain event occurred in the Jiang-Huai Valleys[J]. Adv Atmos Sci, 16(2): 312-321.

SIMPSON J E, 1964. Sea-breeze fronts in Hampshire[J]. Weather, 19:208-220.

SIMPSON J E, MANSFIELD D A, MILFORD J R, 1977. Inland penetration of sea-breeze fronts[J]. Q J Roy Meteor Soc, 103:47-76.

UCCELLINI L W, JOHNSON D R, 1979. The coupling of upper and lower tropospheric jet streams and implications for the development of severe convective storms[J]. Mon Wea Rev, 107: 682-703.

WAKIMOTO R M, 1982. The life cycle of thunderstorm gust fronts as viewed with Doppler radar and rawinsonde data[J]. Mon Wea Rev, 110:1060-1082.

WALLINGTON C E, 1959. The structure of the sea breeze fronts as revealed by gliding flights[J]. Weather, 14:263-270.

WALLINGTON C E, 1965. Gliding through a sea breeze front[J]. Weather, 20:140-144.

WEISMAN M L, KLEMP J B, 1984. Characteristics of isolated convection stroms[R]. NCAR, USA.

XU Q, CLARK J H E, 1985. The nature of symmetric instability and its similarity to convective inertial instability[J]. J Atmos Sci, 42:2880-2883.

YAMASAKI M, 1968. Numerical simulation of tropical cyclone development with the use of primitive equations[J]. J Met Soc Japan, 46:178-201.

YANAI M, ESBENSEN S, CHU J H, 1973. Determination of bulk properties of tropical cloud clusters from large scale heat and moisture budgets[J]. J Atmos Sci, 30:611-627.

YANG G X, SHU C X, 1985. Large-scale environmental conditions for thunderstorm development [J]. Adv Atmos Sci, 2(4):508-521.

YIH C S, 1969. Fluid Mechanics[M]. McGraw Hill Book Company: 446.

Chapter 8
Mesoscale Weather Diagnosis Analysis

For deepen understanding the mechanisms and rules of the weather system development and variation we have to quantitatively analyze various physical parameters based on the atmospheric dynamic and thermodynamic principles. In this chapter we will briefly introduce some diagnoses methods which are frequently applied in the mesoscale meteorological studies.

§ 8.1 ω equation

The atmospheric vertical velocity (ω) is the important parameter for the formation and development of the precipitation and convective weather. One of the important tool for diagnosing the vertical velocity is the ω equation.

8.1.1 The quasi-geostrophic ω equation

Under the quasi-geostrophic assumption, by using the static equation and continuity equation, the vorticity equation and thermodynamic equation can be changed into ω equation and the geopotential tendency equation. The quasi ω equation can be written as the following form:

$$\sigma \nabla^2 \omega + f_0^2 \frac{\partial^2 \omega}{\partial p^2} = F \qquad (8.1.1)$$

where $\quad F = f_0 \dfrac{\partial}{\partial p}[\mathbf{V}_g \cdot \nabla (\zeta_g + f)] + \nabla^2 \left[\mathbf{V}_g \cdot \nabla \left(-\dfrac{\partial \phi}{\partial p}\right)\right] = F_1 + F_2 ,$

$$\sigma = -\frac{1}{\rho \theta} \frac{\partial \theta}{\partial p} = -\frac{RT}{p\theta} \frac{\partial \theta}{\partial p} = -\pi \frac{\partial \theta}{\partial p},$$

$$\pi = -\frac{RT}{(\theta p)}$$

Additionally, ϕ is geopotential, ζ_g is geostrophic wind vorticity. The quasi ω equation is a diagnosis equation, which shows the relationship between the large scale horizontal circulation and the secondary vertical circulation. Therefore, it can be used for diagnosing the vertical velocity according to the large scale horizontal advection motions. Suppose the atmospheric vertical motion is in a wave form then according to the relationship that the Laplacian of any physical parameter is proportional to the negative value of the parameter itself there should be the following relationship between the vertical velocity ω and the quasi geostrophic forcing term F:

$$\omega \propto -F$$

F contains two terms, i. e. $F=F_1+F_2$. Where F_1 is the differential motion advection of the geostrophic wind vorticity. When the positive vorticity advection increases with height, it is corresponding to the ascending motion. While the positive vorticity advection decreases with height, it is corresponding to the descending motion. F_2 is the Laplace of the temperature advection by geostrophic wind. The warm advection area is corresponding to the ascending area and the cold advection area is corresponding to the descending area. While the computational result of each term could not represent the real value of the vertical velocity ω. Because F_1 and F_2 are not independent each other. Each term contains part of components in another term to counteract it. This point of view can be explained by the following analysis.

Change the terms F_1 and F_2 into the following forms:

$$\begin{cases} F_1 = A+B+C \\ F_2 = A-B-2D \\ F = 2A+C-2D \end{cases} \quad (8.1.2)$$

where $A = f_0 \dfrac{\partial \mathbf{V}_g}{\partial p} \cdot \nabla \zeta_g$ is the relative vorticity advection by the thermo wind;

$B = f_0 \mathbf{V}_g \cdot \nabla \dfrac{\partial \zeta_g}{\partial p}$ is the thermo wind vorticity advection by the geostrophic wind;

$C = f_0 \beta \dfrac{\partial V_g}{\partial p}$ is the earth rotational vorticity advection by the thermo wind.

$D = \left[J\left(u_g, \dfrac{\partial u_g}{\partial p}\right) + J\left(v_g, \dfrac{\partial v_g}{\partial p}\right) \right]$ is the geostrophic deformation term.

From the above analysis it can clearly be seen that both in the terms of F_1 and F_2 there is a counteract term B. So that both the term F_1 and F_2 can not be used solely for determining the real ω. When the effects of D and C are smaller, $F \cong 2A$, it means that there is ascending motion at the place with positive cyclonic vorticity advection induced by the thermo wind. Under such a simplified condition, the distribution of ω can be estimated by using a geopotential height chart and thickness chart only. But under the situation with greater baroclinicity near the frontal zone, the terms D and A have same order of magnitude and should not be omitted.

The above statements have explained the disadvantages of the quasi geostrophic ω equation. Hoskins et al. (1978) derived the ω equation in another way by setting out the following quasi geostrophic, quasi static, isentropic and frictionless dynamic equations in a pressure (p)-coordinate system:

$$\left(\frac{\partial}{\partial t}+\mathbf{V}_g \cdot \nabla\right)u_g - fv_a = 0 \tag{8.1.3}$$

$$\left(\frac{\partial}{\partial t}+\mathbf{V}_g \cdot \nabla\right)v_g + fu_a = 0 \tag{8.1.4}$$

$$\left(\frac{\partial}{\partial t}+\mathbf{V}_g \cdot \nabla\right)\left(-\frac{\partial \phi}{\partial p}\right) - \sigma\omega = 0 \tag{8.1.5}$$

$$\frac{\partial u_a}{\partial x}+\frac{\partial v_a}{\partial y}+\frac{\partial \omega}{\partial p}=0 \tag{8.1.6}$$

$$\frac{\partial \phi}{\partial p}=-\alpha \tag{8.1.7}$$

$$fu_g = -\frac{\partial \phi}{\partial y}, \quad fv_g = \frac{\partial \phi}{\partial x} \tag{8.1.8}$$

$$f\frac{\partial u_g}{\partial p}=\frac{\partial}{\partial y}\left(-\frac{\partial \phi}{\partial p}\right), \quad f\frac{\partial v_g}{\partial p}=-\frac{\partial}{\partial x}\left(-\frac{\partial \phi}{\partial p}\right) \tag{8.1.9}$$

where $v_a = v - v_g$, $u_a = u - u_g$ are the ageostrophic components, $\sigma = -\frac{\alpha}{\theta} \cdot \frac{\partial \theta}{\partial p}$ is the static stability parameter, α is specific volume. Set $f=$constant, deriving the equation (8.1.3) for p and multiplied by f, to get

$$\left(\frac{\partial}{\partial t}+\mathbf{V}_g \cdot \nabla\right)f\frac{\partial u_g}{\partial p}-f^2\frac{\partial v_a}{\partial p}=-f\frac{\partial v_g}{\partial p}\cdot \nabla u_g \tag{8.1.10}$$

deriving the equation (8.1.5) for y, to get

$$\left(\frac{\partial}{\partial t}+\mathbf{V}_g \cdot \nabla\right)\left[\frac{\partial}{\partial y}\left(-\frac{\partial \phi}{\partial p}\right)\right]-\frac{\partial}{\partial y}(\sigma\omega)=-\frac{\partial v_g}{\partial y}\cdot \nabla\left(-\frac{\partial \phi}{\partial p}\right) \tag{8.1.11}$$

deriving the equation (8.1.4) for p and multiplied by f, to get

$$\left(\frac{\partial}{\partial t}+\mathbf{V}_g \cdot \nabla\right)f\frac{\partial v_g}{\partial p}+f^2\frac{\partial u_a}{\partial p}=-f\frac{\partial v_g}{\partial p}\cdot \nabla v_g \tag{8.1.12}$$

deriving the equation (8.1.5) for x, to get

$$\left(\frac{\partial}{\partial t}+\mathbf{V}_g \cdot \nabla\right)\left[\frac{\partial}{\partial x}\left(-\frac{\partial \phi}{\partial p}\right)\right]-\frac{\partial}{\partial x}(\sigma\omega)=-\frac{\partial v_g}{\partial x}\cdot \nabla\left(-\frac{\partial \phi}{\partial p}\right) \tag{8.1.13}$$

By using the thermo wind relationship and the equation $\frac{\partial u_g}{\partial x}+\frac{\partial v_g}{\partial y}=0$ for omitting the time derivative terms in the equations and get

$$\frac{\partial}{\partial x}(\sigma\omega)-f^2\frac{\partial u_a}{\partial p}=-2Q_1 \tag{8.1.14}$$

$$\frac{\partial}{\partial y}(\sigma\omega)-f^2\frac{\partial v_a}{\partial p}=-2Q_2 \tag{8.1.15}$$

where,
$$Q=(Q_1,Q_2)=\left[-\frac{\partial V_g}{\partial x}\cdot\nabla\left(-\frac{\partial\phi}{\partial p}\right),-\frac{\partial V_g}{\partial y}\cdot\nabla\left(-\frac{\partial\phi}{\partial p}\right)\right] \quad (8.1.16)$$

Deriving the equation (8.1.14) for x and deriving the equation (8.1.15) for y, and then omitting the terms contained u_a, v_a by using the equation (8.1.6), and finally to get the new quasi geostrophic equation in which the forcing term is expressed by the Q vector as follows:

$$\nabla^2(\sigma\omega)+f^2\frac{\partial^2\omega}{\partial p^2}=-2\nabla\cdot Q \quad (8.1.17)$$

The above equation indicates that the quasi geostrophic vertical motion on the f surface can be determined only by the divergence of the vector Q, so that the disadvantage of the traditional quasi geostrophic equation is overcomed. According to the following relationship

$$\omega\propto\nabla\cdot Q$$

it can be deduced that if $\nabla\cdot Q<0$, then $\omega<0$ (ascending), oppositely if $\nabla\cdot Q>0$, then $\omega>0$ (descending).

8.1.2 Ageostrophic ω equation

Considering the non adiabatic and ageostrophic features of the mesoscale weather processes, it often need to use the completed ageostrophic balanced equation, in which more dynamic and thermodynamic factors are included and the various basic observational data may be more efficiently used. The ageostrophic ω equation may be derived from vorticity equation, divergence equation and the thermodynamic equation included the diabatic effect. Krishnamurti (1968) derived the ω equation under the condition with the Rossby number less than 1 in the following form:

$$\nabla^2(\sigma\omega)+f^2\frac{\partial^2\omega}{\partial p^2}=\sum_{n=1}^{10}\delta_n F_n+\sum_{n=11}^{14}\delta_n F_n \quad (8.1.18)$$

where δ_n is the tracing function with the value 0 or 1, F is forcing factor. There are 14 forcing factors in the equation (8.1.18), their meanings are respectively explained as follows:

$F_1=f\frac{\partial}{\partial p}J(\psi,\eta)$ is the vertical variation of the advection of the rotationary wind vorticity.

$F_2=f\frac{\partial}{\partial p}(\nabla\chi,\nabla\eta)$ is the vertical variation of the advection of the divergent wind vorticity.

$F_3=\pi\nabla^2 J(\psi,\theta)$ is the Lapalace of the rotationary wind temperature advection.

$F_4 = \pi \nabla^2 (\mathbf{V}_\chi, \nabla \theta)$ is the Lapalace of the divergent wind temperature advection.

$F_5 = f \dfrac{\partial}{\partial p}(\nabla^2 \psi, \nabla^2 \chi)$ is the vertical variation of the product of the vorticity and divergence.

$F_6 = -\nabla f \cdot \nabla \dfrac{\partial}{\partial p}\left(\dfrac{\partial \psi}{\partial t}\right)$ is the effect of the Coriolis parameter.

$F_7 = -2 \dfrac{\partial}{\partial t}\dfrac{\partial}{\partial p} J\left(\dfrac{\partial \psi}{\partial x}, \dfrac{\partial \psi}{\partial y}\right)$ is the effect of the deformation.

$F_8 = fg \dfrac{\partial^2}{\partial p^2}\left(\dfrac{\partial \tau_y}{\partial x} - \dfrac{\partial \tau_x}{\partial y}\right)$ is the effect of the frictional stress.

$F_9 = -\dfrac{R}{C_p p} \nabla^2 H_S$ is the effect of the sensible heat of the underlying surface.

where H_S is the exchanging rate of the sensible heat of the underlying surface.

$F_{10} = -\dfrac{R}{C_p p} \nabla^2 H_R$ is the effect of the radiational heating rate (where H_R is the heating rate of radiation).

$F_{11} = -\dfrac{R}{C_p p} \nabla^2 H_L$ is the effect of the large scale condensational heating, where H_L is the heating rate of the large scale condensational heating.

$F_{12} = -\dfrac{R}{C_p p} \nabla^2 H_C$ is the effect of the convective condensationa heating, where H_C is the heating rate of the convective condensation heating.

$F_{13} = f \dfrac{\partial}{\partial p}\left(\nabla \omega, \nabla \dfrac{\partial \psi}{\partial p}\right)$ is the vertical variation of the twisting term of the vorticity tube.

$F_{14} = f \dfrac{\partial}{\partial p}\left(\omega \dfrac{\partial \nabla \psi}{\partial p}\right)$ is the vertical variation of the vertical advection of the vorticity.

Among the above 14 terms, the terms 1—10 do not contain ω; the terms 11—14 contain ω, they are called the feedback terms. ψ is stream function, χ is the velocity potential. In the completed ω equation (8.1.18), the tracing function δ_n in all terms are taken as 1. While if $\delta_1 = 1$, $\delta_3 = 1$, and the δ_n of the rest of the terms are taken as 0 and $\psi = \dfrac{\phi}{f}$, then the equation (8.1.18) will be simplified as the quasi geostrophic ω equation in following form:

$$\nabla^2 (\delta \omega) + f^2 \dfrac{\partial^2 \omega}{\partial p^2} = f \dfrac{\partial}{\partial p}[\mathbf{V}_g \cdot \nabla (\xi_g + f)] + \nabla^2 \left[\mathbf{V}_g \cdot \nabla \left(-\dfrac{\partial \phi}{\partial p}\right)\right] \quad (8.1.19)$$

For solving the equation (8.1.18), the boundary conditions may be taken as follows:

$\omega_0 = \omega_H + \omega_F$ $\quad\quad\quad$ $p = 1000$ hPa

$\omega_\Gamma = 0$ $\quad\quad\quad\quad\quad\quad$ $p = 0$ hPa

$\omega_L = 0$ $\quad\quad\quad\quad\quad\quad$ the lateral boundary condition

where ω_H is the vertical velocity caused by the topographic lifting, ω_F is the vertical velocity caused by the frictional effect in the boundary layer.

§ 8.2　Analysis of Q vector

8.2.1　The definition of the quasi geostrophic Q vector

As pointed out previously that there are two forcing terms in the traditional quasi geostrophic ω equation. When their signs are oppositely it is difficult to decide qualitatively the direction of the vertical motion. For solving the problem, Hoskins et al. (1978) derived new form quasi geostrophic ω equation as follows:

$$\mathbf{V}^2(\sigma\omega) + f^2 \frac{\partial^2 \omega}{\partial p^2} = -2\mathbf{V} \cdot \mathbf{Q}$$

or $\quad\quad\quad\quad\quad\quad\quad\quad -A^2\omega = -2\mathbf{V} \cdot \mathbf{Q}$ $\quad\quad\quad\quad\quad\quad\quad\quad$ (8.2.1)

where

$$\omega = \omega_0 \sin\left(\frac{2\pi}{L_x}x\right)\sin\left(\frac{2\pi}{L_y}y\right)\sin\left(\frac{\pi}{p_0}p\right)$$

$$-A^2\omega = \left(\sigma\mathbf{V}^2 + f^2\frac{\partial^2}{\partial p^2}\right)\omega = -\left[\sigma\left(\frac{2\pi}{L_x}\right)^2 + \sigma\left(\frac{2\pi}{L_y}\right)^2 + f^2\left(\frac{\pi}{p_0}\right)^2\right]\omega$$

$$A^2 = \sigma\left(\frac{2\pi}{L_x}\right)^2 + \sigma\left(\frac{2\pi}{L_y}\right)^2 + f^2\left(\frac{\pi}{p_0}\right)^2$$

$$\mathbf{Q} = (Q_x, Q_y) = \left(-\frac{\partial \mathbf{V}_g}{\partial x} \cdot \mathbf{V}\left(-\frac{\partial \phi}{\partial p}\right), -\frac{\partial \mathbf{V}_g}{\partial y} \cdot \mathbf{V}\left(-\frac{\partial \phi}{\partial p}\right)\right) \quad\quad (8.2.2)$$

The above vector $\mathbf{Q}$ is called the quasi geostrophic Q vector. Where σ is the static stability, $\sigma = -\frac{\alpha}{\theta}\frac{\partial \theta}{\partial p}$, $\alpha = \frac{RT}{P}$ is specific volume. The quasi geostrophic Q vector defineded by the equation (8.2.2) can be also expressed into component form as follows:

$$Q_x = -\frac{R}{P}\frac{\partial \mathbf{V}_g}{\partial x} \cdot \mathbf{V}T = -\frac{R}{P}\left(\frac{\partial u_g}{\partial x}\frac{\partial T}{\partial x} + \frac{\partial v_g}{\partial x}\frac{\partial T}{\partial y}\right) \quad\quad (8.2.3)$$

$$Q_y = -\frac{R}{P}\frac{\partial \mathbf{V}_g}{\partial y} \cdot \mathbf{V}T = -\frac{R}{P}\left(\frac{\partial u_g}{\partial y}\frac{\partial T}{\partial x} + \frac{\partial v_g}{\partial y}\frac{\partial T}{\partial y}\right) \quad\quad (8.2.4)$$

The equations (8.2.3) and (8.2.4) show that the quasi geostrophic Q vector is depending on the product of the horizontal gradient of geostrophic wind and the hori-

8.2.2 The physical meaning of the quasi geostrophic Q vector

Now let us discuss the physical meaning of the quasi geostrophic Q vector. For simplifying the problem, we assume that the x coordinate axis is parallel to the isoline of temperature at the given point, then we have $-\frac{\partial T}{\partial x}=0$, and therefore

$$Q_x=-\frac{R}{P}\frac{\partial v_g}{\partial x}\frac{\partial T}{\partial y}, \quad Q_y=-\frac{R}{P}\frac{\partial v_g}{\partial y}\frac{\partial T}{\partial y}, \quad Q=Q_x i+Q_y j$$

This shows that Q_y denotes the variation of the magnitude of the temperature gradient $-\frac{\partial T}{\partial y}$. To set $\frac{\partial T}{\partial y}<0$, when $Q_y>0$, it denotes $\left|-\frac{\partial T}{\partial y}\right|$ decreases (frontolysis). When $Q_y<0$, it denotes $\left|-\frac{\partial T}{\partial y}\right|$ increases (frontogenesis) as shown in Fig. 8.1a; Q_x denotes the variation of the direction of the temperature gradient $-\frac{\partial T}{\partial y}$. When $Q_x>0$, it denotes the gradient vector $-\nabla T$ takes a cyclonic rotation; When $Q_x<0$, it denotes the gradient vector $-\nabla T$ takes a anticyclonic rotation as shown in Fig. 8.1b. Thus the Q vector represents the geostrophic disturbances including divergence and convergence as well as wind shear, which will cause the variation of the temperature field including the variations of the magnitude and direction of the temperature gradient.

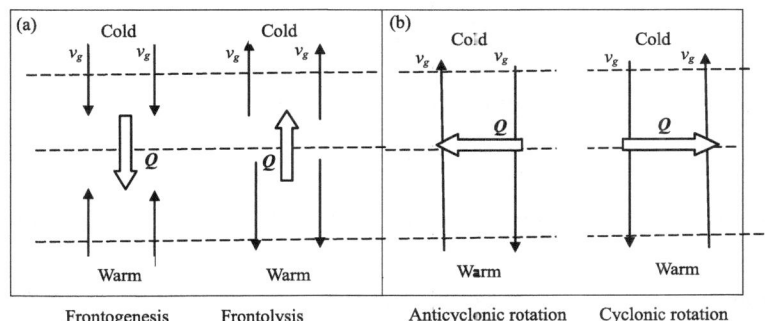

Fig. 8.1 The schematic diagram to denote the physical meaning of the Q vector components Q_x (b) and Q_y (a).

8.2.3 The relationship between the quasi geostrophic Q vector and the vertical velocity and frontogenesis or frontolysis

As it has been pointed previously that on the f plane the quasi geostrophic vertical motion is determined only by the Q vector divergence. When the ω field posses the wave feature, $\omega \propto \nabla \cdot Q$. Therefore it can be deduced that when $\nabla \cdot Q < 0$, $\omega < 0$ (ascending motion), and oppositely when $\nabla \cdot Q > 0$, $\omega > 0$ (descending motion). So that the vertical motion can be determined conveniently.

Moreover, because the Q vector determines the individual change of the fluid field and temperature field, i. e. the thermo wind field or the horizontal distribution of the temperature, so that the Q vector can be also used for forecasting the frontogenesis or frontolysis. Under the quasi geostrophic situation

$$\frac{D_g T}{Dt} = \left(\frac{\partial}{\partial t} + V_g \cdot \nabla \right) T = 0$$

Then

$$\frac{\partial}{\partial x}\frac{D_g T}{Dt} = \frac{\partial}{\partial x}\left(\frac{\partial}{\partial t} + V_g \cdot \nabla \right) T = \left(\frac{\partial}{\partial t} + V_g \cdot \nabla \right)\frac{\partial T}{\partial x} + \frac{\partial V_g}{\partial x} \cdot \nabla T = 0$$

$$\frac{\partial}{\partial y}\frac{D_g T}{Dt} = \frac{\partial}{\partial y}\left(\frac{\partial}{\partial t} + V_g \cdot \nabla \right) T = \left(\frac{\partial}{\partial t} + V_g \cdot \nabla \right)\frac{\partial T}{\partial y} + \frac{\partial V_g}{\partial y} \cdot \nabla T = 0$$

So that

$$\left(\frac{\partial}{\partial t} + V_g \cdot \nabla \right)\frac{\partial T}{\partial x} = \frac{P}{R}Q_x$$

$$\left(\frac{\partial}{\partial t} + V_g \cdot \nabla \right)\frac{\partial T}{\partial y} = \frac{P}{R}Q_y$$

$$\frac{D_g}{Dt}\left(\frac{R}{P}\nabla T\right) = Q \qquad (8.2.5)$$

The above equation is dot multiplied by ∇T, then we can get

$$\frac{D_g}{Dt}|\nabla T|^2 = \frac{2P}{R}Q \cdot \nabla T \qquad (8.2.6)$$

The left side of the formula (8.2.6) denotes frontogenesis function, it is thus clear that the Q vector can be used for deciding the frontogenesis or frontolysis. When the crossing angle between the Q vector and ∇T is less than 90°, $Q \cdot \nabla T > 0$, frontogenesis function is greater than 0, i. e. it is the frontogenesis situation. When the crossing angle is greater than 90°, $Q \cdot \nabla T < 0$, frontogenesis function is less than 0, i. e. it is the frontolysis situation. This conclusion has a good agreement with the discussion about Q_y previously. When the crossing angle between the Q vector and ∇T is less than 90°, $Q_y<0$, it is frontogenesis, when the crossing angle between the Q vector and ∇T is greater than 90°, $Q_y>0$, it is frontolysis (as shown in Fig. 8.2).

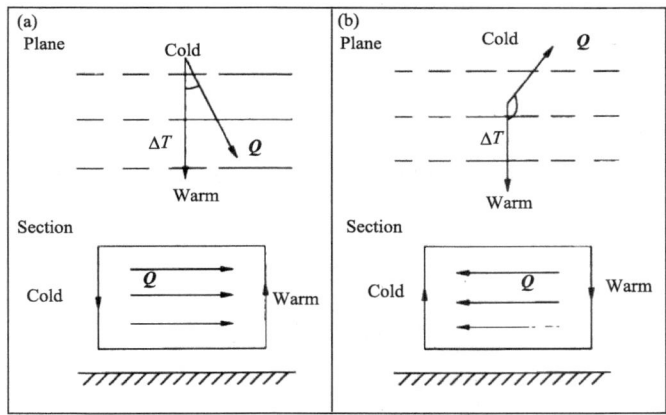

Fig. 8.2 The schematic diagram indicating the relationship between the Q vector and the frontogenesis(a) or frontolysis(b).

8.2.4 The relationship between the Q vector and the secondary circulation

As shown in the following equations that in the vertical section the zonal and meridional secondary circulation formatted by the vertical motion and the ageostrophic flow (u_a, v_a) are determined by Q_x and Q_y respectively,

$$\frac{\partial(\delta\omega)}{\partial x} - f^2 \frac{\partial u_a}{\partial p} = -2Q_x \quad (8.2.7)$$

$$\frac{\partial(\delta\omega)}{\partial y} - f^2 \frac{\partial v_a}{\partial p} = -2Q_y \quad (8.2.8)$$

where Q_x and Q_y are the zonal and meridional components of the Q vector respectively. It is clear that there is a close relationship between Q vector and the secondary circulation. Following let us discuss the relation between the direction of the secondary circulation and the direction of Q vector. When $\frac{\partial \omega}{\partial x}>0$ and $\frac{\partial u_a}{\partial p}<0$, (i. e., the zonal vertical circulation with low level westward flow and ascending motion to the west, descending motion to the east), then, $Q_x<0$, i. e. Q_x point to west. When $\frac{\partial \omega}{\partial y}<0$, $\frac{\partial v_a}{\partial p}>0$, (i. e. the meridional vertical circulation with low level northward flow, descending motion to the south and ascending motion to the north.), then $Q_y>0$, i. e. Q_y points to the north. Oppositely, we can also decide the directions of the zonal or meridional vertical circulation according to positive or negative signs of the components Q_x and Q_y. If the component of the Q vector is less than zero, then the direction of the vertical circulation is in clockwisely rotating; otherwise if the component of the Q vector is greater than zero, then the direction of the vertical circulation is in anticlockwisely rotating; In a word, the Q vector is always pointed to the ascending area and the back is toward to the descending area.

8.2.5 The method determining qualitatively the Q vector on synoptic maps

For the given point, taking the x axis along the direction of isotherm, and the y axis points to the side of cold air. Then according to the following expression

$$Q = \left(-\frac{R}{P}\frac{\partial v_g}{\partial x}\frac{\partial T}{\partial y} \right)i + \left(-\frac{R}{P}\frac{\partial v_g}{\partial y}\frac{\partial T}{\partial y} \right)j$$

Setting $f=$ constant, $\dfrac{\partial v_g}{\partial y} = -\dfrac{\partial u_g}{\partial x}$, we get:

$$Q = -\frac{R}{P}\left(\frac{\partial T}{\partial y}\right)\left(\frac{\partial v_g}{\partial x}i - \frac{\partial u_g}{\partial x}j\right)$$

or

$$Q = -\frac{R}{P}\left|\frac{\partial T}{\partial y}\right| k \times \frac{\partial V_g}{\partial x} \tag{8.2.9}$$

where k is the unit vector in vertical direction, $\dfrac{\partial V_g}{\partial x}$ is the variation rate of the geostrophic wind along the x direction. Thus we can estimate the Q vector according to the variation rate of the geostrophic wind on the synoptic map as shown in Fig. 8.3.

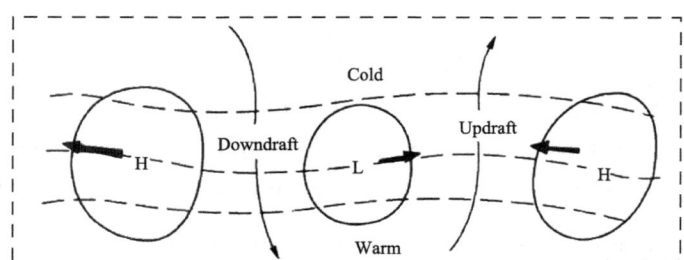

Fig. 8.3 The schematic diagram to estimate the Q vector on the synoptic map. L=low pressure, H=high pressure. (After Sanders and Hoskins, 1990)

8.2.6 The semi geostrophic Q vector ($Q^\wedge$), ageostrophic Q vector ($Q^\#$) and moist Q vector (Q^*)

In recent years, the Q vector theory has distinct progress. Some new concepts such as the semi geostrophic Q vector ($Q^\wedge$), ageostrophic Q vector ($Q^\#$) and moist Q vector (Q^*) etc. are proposed.

(1) The semi geostrophic Q vector ($Q^\wedge$)

The semi geostrophic approximation is also called the geostrophic momentum approximation. The dynamic equations under the semi geostrophic approximation condition contain much more information comparing with that under the quasi geostrophic approximation condition. It retained the geostrophic momentum advection caused by the ageostrophic wind and added the temperature advection caused by the ageostrophic wind and the β effect was considered. Based on the dynamic equation in the p coordinate system under the approximations of semi geostrophic, quasi static, isentropic and frictionless, the semi geostrophic Q vector (Q) can be expressed as follows:

$$Q^\wedge = (Q_x^\wedge, Q_y^\wedge) = \left\{ \frac{1}{2} \left[-\frac{\partial \mathbf{V}}{\partial x} \cdot \nabla \left(-\frac{\partial \phi}{\partial p} \right) - f \frac{\partial \mathbf{V}}{\partial p} \cdot \nabla v_g + v\beta \frac{\partial v_g}{\partial p} \right], \right.$$

$$\left. \frac{1}{2} \left[-\frac{\partial \mathbf{V}}{\partial x} \cdot \nabla \left(-\frac{\partial \phi}{\partial p} \right) - f \frac{\partial \mathbf{V}}{\partial p} \cdot \nabla u_g + v\beta \frac{\partial u_g}{\partial p} \right] \right\} \quad (8.2.10)$$

where $Q_x^\wedge$ and $Q_y^\wedge$ are the components of the semi geostrophic Q vector in x and y direction respectively. It's noteworthy that it is not only the geostrophic wind but also the real wind contained in expression of the semi geostrophic Q vector, which is the clear distinction with the quasi geostrophic Q vector. Similar to the quasi geostrophic, the following two expressions described the relationship between the directions of the semi geostrophic Q vector and the vertical circulation:

$$\frac{\partial(\delta\omega)}{\partial x} - f^2 \frac{\partial u_a}{\partial p} = -2Q_x^\wedge \quad (8.2.11)$$

$$\frac{\partial(\delta\omega)}{\partial y} - f^2 \frac{\partial v_a}{\partial p} = -2Q_y^\wedge \quad (8.2.12)$$

From the above two equations it is clear that the zonal and meridional vertical circulations are determined by $Q_x^\wedge$ and $Q_y^\wedge$ respectively. Thus the secondary circulations in the vertical section of any directions can be all determined by the component of the semi geostrophic Q vector in the direction. It is the same that the direction of the semi geostrophic Q vector is always point to the area of updraft and back to the downdraft area.

Similar to the quasi geostrophic Q vector, the ω equation denoted by the semi geostrophic Q vector can be written as follows:

$$\nabla^2(\sigma\omega) + f^2 \frac{\partial^2 \omega}{\partial p^2} = -2\nabla \cdot Q^\wedge \quad (8.2.13)$$

The physical meaning of the above equation is that under the condition of the semi geostrophic approximation, the vertical motion is determined only by the divergence of the semi geostrophic Q vector ($\nabla \cdot Q^\wedge$). When the ω field posses the wave feature, there will be the relationship between ω and the the divergence of the semi geostrophic Q vector ($\nabla \cdot Q^\wedge$) as follows: $\omega \propto \nabla \cdot Q^\wedge$, thus we can also conclude that if $\nabla \cdot Q^\wedge < 0$, then $\omega < 0$ (ascending); and if $\nabla \cdot Q^\wedge > 0$, then $\omega > 0$ (descending).

(2) Ageostrophic Q vector ($Q^\#$)

Based on the original equations in f plane p-coordinate system with the quasi static, isentropic, frictionless approximation, the ageostrophic Q vector ($Q^\#$) can be derived as the following expression:

$$Q^\# = (Q_x^\#, Q_y^\#) = \left\{ \frac{1}{2}\left[f\left(\frac{\partial v}{\partial p}\frac{\partial u}{\partial x} - \frac{\partial u}{\partial p}\frac{\partial v}{\partial x}\right) - h\frac{\partial \mathbf{V}}{\partial x} \cdot \nabla \theta \right], \right.$$
$$\left. \frac{1}{2}\left[f\left(\frac{\partial v}{\partial p}\frac{\partial u}{\partial y} - \frac{\partial u}{\partial p}\frac{\partial v}{\partial y}\right) - h\frac{\partial \mathbf{V}}{\partial y} \cdot \nabla \theta \right] \right\} \quad (8.2.14)$$

where $Q_x^\#$ and $Q_y^\#$ are the components of the ageostrophic Q vector in x and y directions respectively. It is noteworthy that all the terms in the ageostrophic Q vector expression include the real wind. This is the distict difference between the ageostrophic Q vector and the quasi geostrophic or semi geostrophic Q vectors.

Similar to the quasi geostrophic Q vector, we can also obtain the following relation between the ageostrophic Q vector and the secondary circulation:

$$\delta \frac{\partial \omega}{\partial x} - f^2 \frac{\partial u_a}{\partial p} = -2Q_x^\# \quad (8.2.15)$$

$$\delta \frac{\partial \omega}{\partial y} - f^2 \frac{\partial v_a}{\partial p} = -2Q_y^\# \quad (8.2.16)$$

It is clear from (8.2.15) and (8.2.16) that the zonal and meridional vertical circulations are decided by $Q_x^\#$ and $Q_y^\#$ respectively. Thus the secondary circulations in the vertical section in any direction are all deterned by the ageostrophic Q vector component in the direction. The direction of the ageostrophic Q vector is always toward to the ascending area and backward to the decending area.

As the same, the ageostrophic ω equation with the ageostrophic Q vector divergence as the only forcing term can be written as follows:

$$\nabla^2(\sigma\omega) + f^2 \frac{\partial^2 \omega}{\partial p^2} = -2\nabla \cdot Q^\# \quad (8.2.17)$$

It is clear that when the ω posses wave feature, we can get

$$\nabla \cdot Q^\# \propto \omega \quad (8.2.18)$$

Thus, if $\nabla \cdot Q^\# < 0$, then $\omega < 0$, while if $\nabla \cdot Q^\# > 0$, then $\omega > 0$. So that the ageostrophic Q vector divergence can be used for diagnosing the secondary circulation and vertical velocity. Because of the existing of the ageostrophic Q vector divergence, it will motivate the secondary circulation and let the large scale motion to be adjusted to counteract the effect of the thermo wind. Along with the increasing of the secondary circulation, the large scale motion will finally set the new thermo wind balance. The ageostrophic Q vector divergence plays important role during the processes that the thermo wind balance is destroyed and reconstructed constantly.

(3) Moist Q vector (Q^*)

Based on the original equation considering adiabatic heating caused by the condensation of the water vapor in the atmosphere, the ageostrophic moist Q vector (Q^*) may be derived and get the expression as follows:

$$Q^* = (Q_x^*, Q_y^*) = \left\{ \frac{1}{2} \left[f\left(\frac{\partial v}{\partial p}\frac{\partial u}{\partial x} - \frac{\partial u}{\partial p}\frac{\partial v}{\partial x}\right) - h\frac{\partial \mathbf{V}}{\partial x} \cdot \nabla \theta - \frac{\partial}{\partial x}\left(\frac{LR\omega \partial q_s}{C_p P \partial p}\right) \right], \right.$$
$$\left. \frac{1}{2}\left[f\left(\frac{\partial v}{\partial p}\frac{\partial u}{\partial y} - \frac{\partial u}{\partial p}\frac{\partial v}{\partial y}\right) - h\frac{\partial \mathbf{V}}{\partial y} \cdot \nabla \theta - \frac{\partial}{\partial y}\left(\frac{LR\omega \partial q_s}{C_p P \partial p}\right) \right] \right\} \quad (8.2.19)$$

where Q_x^*, Q_y^* are the components of the ageostrophic moist Q vector (Q^*) in x and y directions respectively. It is noteworthy that in the expression of the moist Q vector, not only the real wind but also the condensation latent heating terms are taken account, which is the distinct feature of the ageostrophic moist Q vector different from other Q vectors, since it is more close to the real atmospheric situation. The zonal and meridional vertical circulations can be determined by Q_x^* and Q_y^* respectively. Where Q_x^* and Q_y^* can be expressed as follows:

$$2Q_x^* = f^2 \frac{\partial u_a}{\partial p} - \sigma \frac{\partial \omega}{\partial x} \quad (8.2.20)$$

$$2Q_y^* = f^2 \frac{\partial v_a}{\partial p} - \sigma \frac{\partial \omega}{\partial y} \quad (8.2.21)$$

Similarly, the moist Q vectors point always toward the ascending motion area and backward to descending motion area.

The ageostrophic ω equation, in which the only forcing term is the moist Q vector divergence, can be written as follows:

$$\nabla^2(\sigma\omega) + f^2 \frac{\partial^2 \omega}{\partial p^2} = -2\nabla \cdot Q^* \quad (8.2.22)$$

As the same, when $\nabla \cdot Q^* < 0$, then $\omega < 0$; when $\nabla \cdot Q^* > 0$, then $\omega > 0$.

8.2.7 Resolution of the Q vector and its application

In recent years, some scientists proposed the theory and method of the resolution of the Q vector. Normally, the method of the Q vector partitioning is to separate the Q vector into two components, which are along or perpendicular to the isoline of temperature or the isoline of potential temperature or sometimes the isoline of the constant geopotential height respectively in a natural coordinate system.

For example, as shown in Fig. 8.4, the moist Q vector can be separated in a natural coordinate system along the isoline of the potential temperature. In the figure, n is the unit vector in direction of $\nabla\theta$, and $n=\dfrac{\nabla\theta}{|\nabla\theta|}$. Where n rotates 90° anti clockwisely, it can get the unit vector s, i.e., $s=k\times n$. The overall moist Q vector indicated by Q^*, the component of Q^* in n direction is indicated by Q_n^*. $Q_n^* = \left(\dfrac{Q^*\cdot\nabla\theta}{|\nabla\theta|}\right)\dfrac{\nabla\theta}{|\nabla\theta|}$ or $Q_n^* = \left(\dfrac{Q^*\cdot\nabla\theta}{|\nabla\theta|}\right)n$; the component of Q^* in s direction is indicated by Q_s^*. $Q_s^* = \dfrac{Q^*\cdot(k\times\nabla\theta)}{|\nabla\theta|}\left[\dfrac{(k\times\nabla\theta)}{|\nabla\theta|}\right]$. $Q^* = Q_n^* n + Q_s^* s$

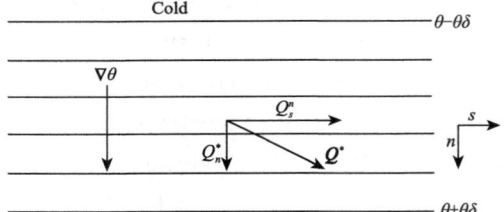

Fig. 8.4 The schematic diagram of the partitioning of the moist Q vector Q^*

The researches showed that the Q vector resolution method is helpful to reveal some meaningful features that are difficult to be revealed by the total Q vector itself. For example, when studying the physical mechanism of the frontogenesis and the vertical motions, Keyser et al. (1992) found out that when the Q vector is separated into two parts, which are parallel to and perpendicular to the isoline of temperature respectively, it will lead to a significant scale separation of the vertical motion distribution corresponding to the baroclinic distribution. The distribution of the component perpendicular to the isoline of the temperature of the Q vector posses the scale of the frontal zone and with belt structure. While the distribution of the component of the Q vector parallel to the isoline of the temperature posses the the synoptic scale and the circular structure. By using the idea of the partitioning analysis of the Q vector, Yue et al. (2002) analyzed the roles of the different scale vertical motions played in the heavy rain process caused by Meiyu front in China, and obtained good results helpful for deepen understanding the mechanism of the heavy rain process.

8.2.8 *C* vector

Xu (1992) induced the vertical ageostrophic vorticity equation based on the quasi geostrophic Q vector equation and combined it with the two component equations of the quasi geostrophic Q vector to get the completed three dimensional quasi geostrophic diagnostic equation, i.e. the *C* vector equation.

The quasi geostrophic (QG) equation can be written as the following form:

$$f^2 u = -(\partial_t + \mathbf{V}_g \cdot \nabla)\partial_x \phi_g \tag{8.2.23a}$$

$$f^2 v = -(\partial_t + \mathbf{V}_g \cdot \nabla)\partial_y \phi_g \tag{8.2.23b}$$

$$N^2 w = -(\partial_t + \mathbf{V}_g \cdot \nabla)\partial_z \phi_g \tag{8.2.23c}$$

$$\nabla \cdot \mathbf{V} = 0 \tag{8.2.23d}$$

where $\mathbf{V} \equiv (u, v, w)$ is the ageostrophic wind, $\mathbf{V}_g \equiv (u_g, v_g, 0) \equiv \mathbf{k} \times \nabla \phi_g / f$ is the geostrophic wind, ϕ_g is the geopotential, $\nabla \equiv (\partial_x, \partial_y, \partial_z)$. Where $Z \equiv [1-(p/p_0)^\kappa]C_p/\gamma$ is the pseudoheight coordinate, $\gamma \equiv g/\Theta_0$, Θ_0 is the reference temperature constant, $N^2 = N_d^2 \equiv \gamma \partial_z \Theta_g$ is the mean (dry) thermal stratification, Θ_g is the horizontal mean value of the potential temperature θ_g in a limited area, $\theta_g \equiv \partial_z \phi_g / \gamma$ satisfied the relationship of the geostrophic wind $\mathbf{V}_g$ and thermo wind, f is the Coriolis parameter, g is the gravity acceleration and is set as the constant.

From $\nabla \times [(8.2.23a), (8.2.23b), (8.2.23c)]$ we can get the ageostrophic pseudo potential vorticity equation as follows:

$$-\partial_z(f^2 v) + \partial_y(N^2 w) = 2C_1 \equiv 2Q_2 \tag{8.2.24a}$$

$$-\partial_z(f^2 u) + \partial_x(N^2 w) = 2C_2 \equiv -2Q_1 \tag{8.2.24b}$$

$$\partial_x(f^2 v) + \partial_y(f^2 u) = 2C_3 \tag{8.2.24c}$$

where $(Q_1, Q_2, 0) \equiv \mathbf{Q}$, i.e. the two dimensional geostrophic forcing vector defined by Hoskins et al. (1978). $\mathbf{C} \equiv (\mathbf{C}_H, C_3) \equiv (C_1, C_2, C_3)$, which is the three dimensional geostrophic forcing vectors, $\mathbf{C}_H = \mathbf{Q} \times \mathbf{k}$, and

$$C_1 \equiv -f\partial(u_g, v_g)/\partial(y, z) = -\gamma(\partial_y \mathbf{V}_g) \cdot \nabla \theta_g \tag{8.2.25a}$$

$$C_2 \equiv -f\partial(u_g, v_g)/\partial(z, x) = -\gamma(\partial_x \mathbf{V}_g) \cdot \nabla \theta_g \tag{8.2.25b}$$

$$C_3 \equiv -f\partial(u_g, v_g)/\partial(x, y) = [(\partial_x \partial_y \phi_g)^2 - (\partial_x^2 \phi_g)(\partial_y^2 \phi_g)]/f \tag{8.2.25c}$$

The non-dimentional forms of the equations (8.2.24a—c) and (8.2.23d) are as follows:

$$\nabla \times V = 2Ro\ C \qquad (8.2.26a)$$
$$\nabla \cdot V = 0 \qquad (8.2.26b)$$

Where $Ro \equiv U/(fL)$ is the Rossby number. The ageostrophic vorticity is propotional to the C vector, the stream lines of the C vector can be regarded as the ageostrophic vortex line. The ω equation in the C vector form can be obtained from $k\nabla \times \rightarrow \times (8.2.26a)$:

$$\nabla^2 \omega = -2Ro k \cdot \nabla \times C = 2Ro(\partial_x C_2 - \partial_y C_1) \qquad (8.2.27)$$

It can be seen from the above equation that the vertical velocity is induced by the vertical vorticity of the C vector as shown in Fig. 8.5. Because C_H is the rotational Q vector, i. e., $C_H = Q \times k$, therefore the two dimensional vector C_H analysis is basicly the same with the Q vector analysis. The three dimensional C vector contains more information compared with the Q vector, so that the 3D C vector analysis should be more useful than the Q vector analysis.

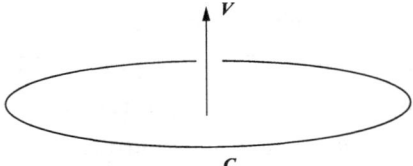

Fig. 8.5 The vertical velocity induced by the horizontal rotation of the C vector. (After Xu, 1992)

8.2.9 Application of the Q vector analysis

The Q vector analysis has been widely used in research and operational works. It is clear that the Q vector analysis can reveal the relations between the variation of the vertical motion and the precipitation systems. Furthermore, the Q vector analysis can be also used for the study on frontogenesis. The frontal circulation is related with the component of the Q vector magnitude, while the basic circulation is related with the component of the Q vector direction. The Q vector analysis can be used to decide the situation of frontogenesis more accurately, so that it is possibly to provide more important information for making a precipitation forecast. Many people found that it is possible to obtain more useful information about the generation or development or maintaining of the important weather systems such as the low pressure vortex, shear line etc. from the Q vector analysis. Some people analyzed the mechanism of the heavy rain enhancement related with typhoon and the relation between the heavy rain area and the secondary circulation on the Meiyu front based on the computation of the Q vector divergence and the temperature advection induced by the Q vector etc. Barnes et al. (1993) applied the Q vector analysis method on the case study of the weather and compared the results with the numerical forecasting and showed that the Q vector analysis can be very good applied on the short range weather forecasting operation service. Furthermore, the concepts of the semi-geostrophic, ageostrophic and moist Q vectors and the resolution method of Q vector etc. are applied gradually on the studys and operational services and obtained certain effects.

§ 8.3 Analysis of potential vorticity

8.3.1 The concept of potential vorticity

Rossby (1940) proposed that under the barotropic model, the proportion ratio of the absolute vorticity ζ_a and the thickness of the atmospheric column H is a constant. i. e.

$$\zeta_a/H = \text{constant} \tag{8.3.1}$$

where ζ_a/H is the simplest expression of the potential vorticity. Ertel (1942) proposed the concept of the generalized potential vorticity q, which can be expressed as follows:

$$q = \alpha \boldsymbol{\zeta}_a \cdot \nabla \theta \tag{8.3.2}$$

where θ is the potential temperature, α is specific volume, $\boldsymbol{\zeta}_a$ is the absolute vorticity vector. The generalized potential vorticity q is also called as the Ertel potential vorticity. It is the dot product of the absolute vorticity vector and the potential temperature gradient vector. Thus it is a physical parameter included both dynamic and thermodynamic factors. In the adiabatic, frictionless dry air the potential vorticity posses conservation strictly, i. e. $dq/dt = 0$. Under the static balance condition, the potential vorticity is the product of the absolute vorticity and the static stability:

$$q = (\zeta_\theta + f)\left(-g \frac{\partial \theta}{\partial p}\right) \tag{8.3.3}$$

where ζ_θ is the isentropic surface vorticity. In the constant pressure coordinate and the isentropic coordinate system the expressions of the potential vorticity are as follows respectively:

$$q = -g(f\boldsymbol{k} + \nabla_p \times \boldsymbol{V}) \cdot \nabla_p \theta \tag{8.3.4}$$

$$q = -g(f + \mathbf{k} \cdot \mathbf{V}_\theta \times \mathbf{V})/(\partial p/\partial \theta) \tag{8.3.5}$$

where $\mathbf{V}_p$ and $\mathbf{V}_\theta$ are the three dimensional gradient signs in the xyp and $xy\theta$ coordinate systems respectively. In these cases, the potential vorticities are called the constant pressure potential vorticity and the isentropic potential vorticity respectively.

For the typical midlatitude synoptic scale systems, $\zeta \ll f$, thus the equation (8.3.3) can be simplified as follows:

$$q \approx -gf \frac{\partial \theta}{\partial p} \tag{8.3.6}$$

Meanwhile, $\partial \theta/\partial p \approx -10$ K/100 hPa. In the northern hemisphere, $f > 0$, thus the value of q is normally positive, and its magnitude order can be estimated by using the following formula:

$$q = -(10 \text{ m} \cdot \text{s}^{-2})(10^{-4} \text{s}^{-1}) \left[-\frac{10 \text{ K}}{10 \text{ kPa}} \right] \frac{1 \text{ kPa}}{10^3 \text{ kg} \cdot \text{m} \cdot \text{s}^{-2} \cdot \text{m}^{-2}}$$

$$= 10^{-6} \text{ m}^2 \cdot \text{K} \cdot \text{s}^{-1} \cdot \text{kg}^{-1} = 1 \text{ PVU} \tag{8.3.7}$$

where PVU is the unit of potential vorticity. The distribution of the potential vorticity is normally increasing with latitude and height. In the troposphere, potential vorticity is normally less than 1.5 PVU, at the tropopause the potential vorticity may suddenly increase to 4 PVU, in the stratosphere the potential vorticity will quickly increase with height. The magnitude value of the potential vorticity in the low level of the troposphere near the equator area is nearly 0 PVU, while it is typically about 0.3 PVU in the midlatitude and it is about 1.0 PVU at the high troposphere in midlatitude area. The isoline of the potential vorticity (PV) equalling 2 PVU (PV=2PVU) is normally the representative of the boundary between the low potential vorticity air came from the low troposphere in low latitude and the high potential vorticity air from the high troposphere and stratosphere in high latitude area. In the area to the north of the subtropical jet stream, the constant PV surface with PV=2 PVU is close to the tropopause of the real atmosphere and is normally called the dynamic tropopause.

8.3.2 Analysis of the anomaly of PV

There are several PV analytical methods. One of the most frequent used methods is the analysis method of the isentropic potential vorticity (IPV), which is specifically by analyzing the isoline of the potential vorticity on the constant potential temperature surface, i. e. the isentropic surface. The isentropic surface is normally the constant potential temperature surface coincided with tropopause in the polar front area. In the north hemisphere the constant potential temperature surfaces with $\theta = 315$ K in winter season and $\theta=315$ K in summer season are taken. Because of the conservation of the potential vorticity, i. e., the potential vorticity of the moving atmosphere will keep constant under the adiabatic and frictionless conditions (Fig. 8.6), it is possible to trace the variation of the atmospheric disturbance by tracing the anomaly area of the potential vorticity, i. e., the areas with high or low values of the potential vorticity.

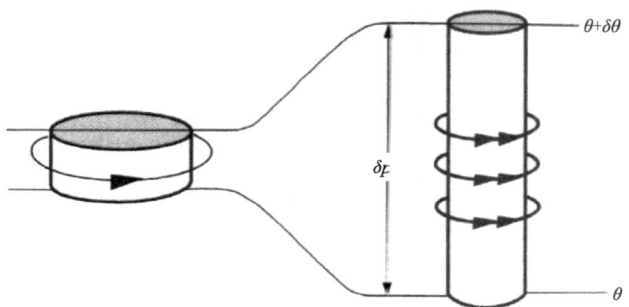

Fig. 8.6   A cylindrical column of air moving adiabatically, conserving potential vorticity. (After Holton, 2004)

An example is given in Fig. 8.7, in which the stretching from northwest to southeast and the fracturing processes of the high value area of the isentropic potential vorticity on the 300 K isentropic potential vorticity chart in the region from 40°N to north pole, 120°W—0°W, with the center of 60°W, and during the duration Sept. 20—25, 1982 was displayed.

8.3.3   The structure feature of the anomaly systems of the potential vorticity and potential temperature area at high and low levels

The potential vorticity posses two important features, i. e. the conservation and the invertibility. The invertibility means that under the given distribution of PV and boundary condition and assuming the atmospheric motion is balanced (for example geostrophic or gradient wind balances), the distributions of the wind, temperature and geopotential potential height etc. at the mean time may be retrieved. Based on the principle of the conservation and invertibility of the potential vorticity Hoskins et al. (1985) well explained the dynamic feature of the quasi geostrophic motion by using IPV analysis methodology and clearly revealed the structure feather and variation tendency of the upper and low level systems corresponding to the upper level potential vorticity anomaly and low level temperature anomaly. This kind of the analysis meth-

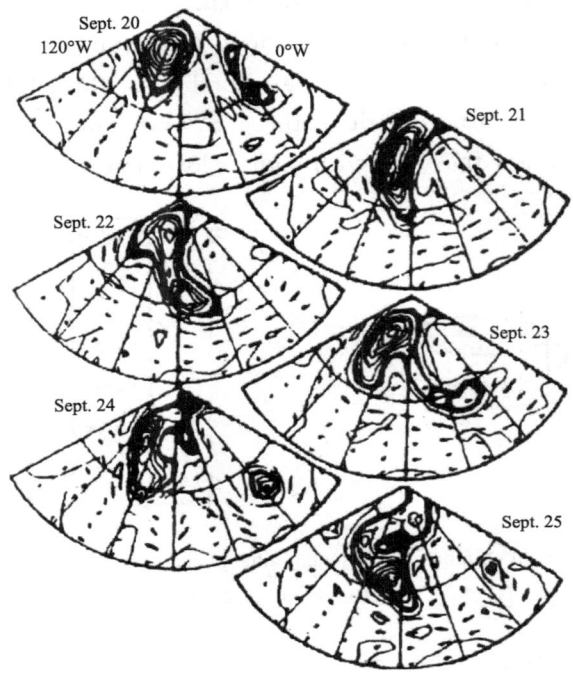

Fig. 8.7 The diagram showed the variation of the high value area of IPV on the 300 K isentropic surface chart, in the region 40°N—north pole, 120°W—0°W, with 60°W as the center and the duration Sept. 20—25, 1982. The interval of the isoline of IPV is 0.5 PVU, the blackened area denotes the value of IPV is 1.5—2.0 PVU, the arrows denote the wind vectors on the IPV surface. (After Hoskins et al., 1985)

od and theory is called PV thinking.

Fig. 8.8 is a schematic diagram of the structure feature of the upper and low level systems corresponding to the ideal upper level positive and negative PV anomaly and the low level positive and negative temperature anomaly.

Fig. 8.8a and Fig. 8.9 show that under the situation with upper level positive PV anomaly area, the PV magnitude value in the positive anomaly area of the potential vorticity is higher than that in the around area. In other words it is a high vorticity and high static stability area. Thus in the positive PV anomaly area the constant potential temperature surfaces will close up toward the PV anomaly center and increase the distances between the adjacent isentropic surfaces above and below the PV anomaly center. So that the static stability in the area will decrease. Meanwhile due to the effect of the PV conservation, it will cause the cyclonic vorticity increasing and lead to the occurence of the cyclonic circulation around the positive PV anomaly area.

Fig. 8.8b shows that under the situation with upper level negative PV anomaly area, the situation is just opposite with the above. The PV magnitude value in the negative anomaly area of the potential vorticity is lower than that in the around area. i. e., it is a low vorticity and low static stability area. Thus in the negative PV anomaly area the constant potential temperature surfaces will separate the negative PV anomaly center and decrease the distances between the adjacent isentropic surfaces above and below the PV anomaly center. So that the static stability in the area will increase. Meanwhile due to the effect of the PV conservation, it will cause the anticyclonic vorticity increase and result in the occurence of the anticyclonic circulation around the negative PV anomaly area.

Fig. 8.8c shows that under the situation with the positive temperature anomaly area at low level, where the potential vorticity distribution is equally, the intervals between the isentropic surfaces will increase, so that the static stability will decrease, meanwhile the cyclonic vorticity will increase and cause the cyclonic circulation around the positive temperature anomaly area. Similarly, under the situation with low level negative temperature anomaly, the intervals between the isentropic surfaces will decrease, so that the static stability will increase, meanwhile the anticyclonic vorticity will increase and cause the anticyclonic circulation around the negative temperature anomaly area as shown in Fig. 8.8d.

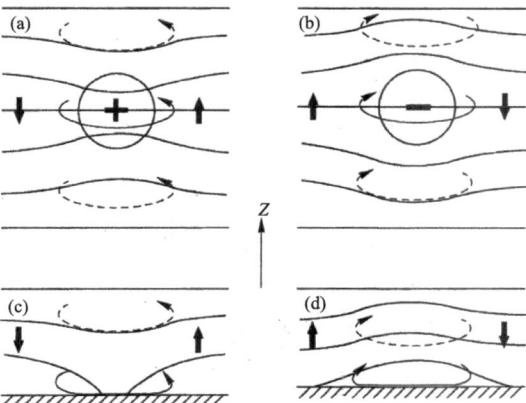

Fig. 8.8 A schematic diagram showing the isentropic surface and the circulation anomaly corresponding to the upper level positive and negative PV anomaly and the surface temperature anomaly. (After Hoskins, 1997)

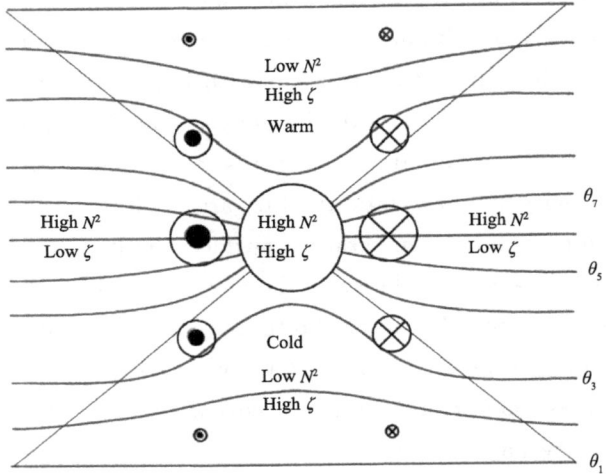

Fig. 8.9　Schematic diagram of the isentropic potential vorticity thinking.
⊙ denotes outflow, ⊕ denotes inflow. (After Hoskins et al., 1985)

The above *IPV* thinking includes the following key points: ① in the view of vorticity, the atmospheric structure typically consists the moving surface cyclones and anticyclones which are overlayed by the upper air troughs and ridges; ② The cyclonic and anticyclonic circulations will occur around the upper level positive and negative potential vorticity anomaly areas; the cyclonic and anticyclonic circulations will also occur corresponding to the surface positive and negative temperature anomaly. The overall wind field is the sum of the wind fields caused by the potential vorticity anomaly and temperature anomaly at the upper level and low level respectively; ③ Under the adiabatic and frictionless assumptions the local change of the potential vorticity will be caused by the potential vorticity advection on the isentropic surface; ④ The wind fields caused by the anomalys of the potential vorticity and temperature will change the distribution of the isentropic potential vorticity (*IPV*); ⑤ The distribution of *IPV* will be again related with the new induced wind fields. The successive interaction between the potential vorticity and temperature anomalys causes the development process of the weather systems. This process will be continued until the axis lines of the anomaly areas of upper and low levels are in the same vertical line.

　　The *IPV* thinking theory can be used for well explaining the process of the surface cyclone development. As shown in Fig. 8.10 it can be seen that when a upper level positive potential vorticity anomaly area, which is corresponding to the tropopause declining, moves eastward and overlays over the existing surface frontal zone at low level, the cyclonic circulation can be induced in the positive PV anomaly area and stretch downward. Its vertical stretching scale H_R is called the Rossby penetrating height, $H_R = f \dfrac{L}{N}$, where f, L and N are the geostrophic parameter, horizontal scale and bouyancy frequency respectively. The cyclonic circulation acts on the low level frontal zone and causes cold and warm advection. The warm advection will cause the low level positive temperature anomaly, thereby an additional cyclonic circulation will be caused in front of the existing low level cyclonic circulation, which will intensify the upper level cyclonic circulation, while which will prompt the intensifying of the low level cyclonic circulation and temperature advection, so that the surface cyclone will develop continuesly. This positive feedback process will be extended until the axis lines of the two anomalys at upper and low levels are located in the same vertical line.

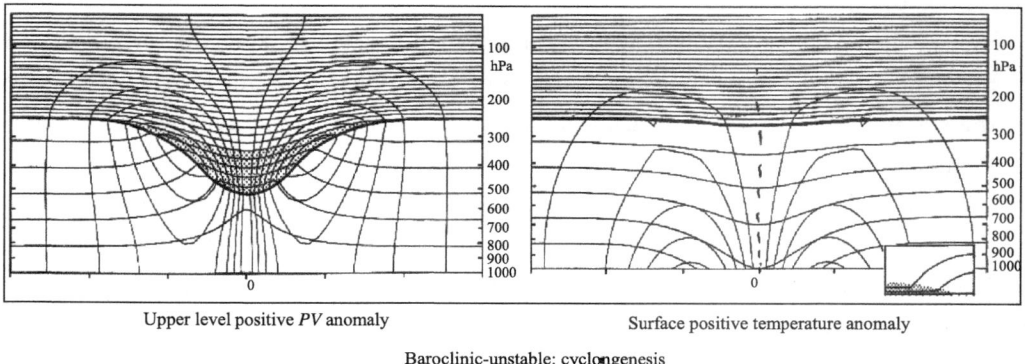

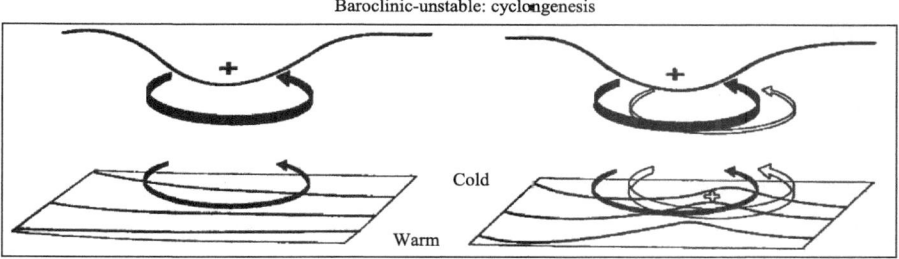

Fig. 8.10 The schematic diagram of the cyclone development process caused by the low level frontal zone overlayed by the upper level potential vorticty anomaly. (The upper level positive vorticity anomaly area is indicated sign + and the declining tropopause; on the surface map the arrows denote circulation and the isolines are the constant lines of potential temperature.) (After Hoskins et al., 1985)

8.3.4 Analysis of the moist potential vorticity

When studying the mechanism of precipitation it is necessary to consider the effect of the water vapor, thus the concept of the moist potential vorticity is induced. Instead of the potential temperature θ by the equivalent potential temperature θ_e, we can get the expreesion of the moist potential vorticity q_m as follows:

$$q_m = a\overline{\zeta}_a \cdot \nabla \theta_e \qquad (8.3.8)$$

Similarly, we can get the expression of the moist potential vorticity in the constant pressure and isentropic coordinate systems respectively as follows:

$$q_m = -g(f\mathbf{k} + \nabla_p \times \mathbf{V}) \cdot \nabla_p \theta_e = -g\zeta_p \frac{\partial \theta_e}{\partial p} - g\mathbf{k} \times \frac{\partial \mathbf{V}}{\partial p} \cdot \nabla_p \theta_e \qquad (8.3.9)$$

$$q_m = -g\zeta_\theta \frac{\partial \theta_e}{\partial p} \qquad (8.3.10)$$

where ζ_p and ζ_θ are the projections of ζ_a on the vertical direction respectively. If the effects of the diabatic heating and friction are not taken account, the moist potential vorticity will be conservation as well. The physical parameter q_m characterizes not only the dynamic and thermodynamic features of the atmosphere but also the the effect of the water vapor, thus the concept of the moist potential vorticity is widely applided especially in precipitation studies.

The potential vorticity, isentropic potential vorticity and moist potential vorticity are also denoted by the signs *PV*, *IPV* and *MPV* respectively. *MPV* can be usually divided into two parts, i. e. *MPV1*, which is usually called as the moist barotropic term and *MPV2*, which is usually called as the moist baroclinic term. Where $MPV1 = -g(\zeta+f)\frac{\partial \theta_e}{\partial p}$, which denotes effects of the inertial instability $(\zeta+f)$ and the convective instability $(-g\frac{\partial \theta_e}{\partial p})$;

$MPV2 = g\left(\frac{\partial v}{\partial p}\frac{\partial \theta_e}{\partial x} - \frac{\partial u}{\partial p}\frac{\partial \theta_e}{\partial y}\right)$, which contains the contributions of the moist baroclinic term $(\nabla_p \theta_e)$ and the vertical shear of the horizontal wind.

Similar to the concepts of the relative vorticity and the transport vorticity, we can also put forward the concepts of relative moist potential vorticity and the transport moist potential vorticity, the expressions of which may be written respectively as follows:

$$(MPV2)_{re} = -g\zeta \frac{\partial \theta_e}{\partial p} + g\left(\frac{\partial v}{\partial p}\frac{\partial \theta_e}{\partial x} - \frac{\partial u}{\partial p}\frac{\partial \theta_e}{\partial y}\right) \qquad (8.3.11)$$

$$(MPV)_{am} = -gf\frac{\partial \theta_e}{\partial p} \tag{8.3.12}$$

where $(MPV2)_{re}$ denotes the relative moist potential vorticity, and $(MPV)_{am}$ denotes the transport moist potential vorticity, which means the moist potential vorticity when the atmosphere is stationary (i. e. $u=0$, $v=0$), so that it implys also the background moist potential vorticity of the atmosphere. It is also clear that the relative moist potential vorticity equals to the potential verticity moist (MPV) subtracted by the background potential vorticity of the atmosphere. Thus the relative moist potential vorticity is called disturbantion of the atmospheric moist potential vorticity. The relative potential vorticity clearly reflects the dynamic effect on the development of the severe convective storms.

On the isentropic surface, when the crossing angle between the θ_e surface and the constant pressure surface is small, taking θ_e as the vertical coordinate, then the gradient of θ_e along the moist isentropic surface will be nearly zero, under this situation the isentropic dry and moist potential vorticity can be written in the simple form as shown by the formula (8.3.10).

Based on the original equations Wu et al. (1995) derived the moist potential vorticity equations as follows:

$$\frac{dp_m}{dt} = \alpha(\nabla p \times \nabla a) \cdot \nabla \theta_e + \alpha \nabla \theta_e \cdot F_\zeta + \alpha \zeta_a \cdot \nabla Q \tag{8.3.13}$$

where $p_m = \alpha \zeta_a \cdot \nabla \theta_e$.

For the dry air the specific humidity equals zero, so that the formula (8.3.13) becomes the dry air potential vorticity equation. According to the potential vorticity equation we can study the variation and development process of the potential vorticity and the factors influencing the variation.

8.3.5 Research progress of the potential vorticity theory

Afterwards Rossby and Ertel, the concept and theory of the potential vorticity is widely applied in the research areas. Following is a simple review: Kleinschmidt (1957) explained the cyclone genesis by using the PV anomaly of upper troposphere. Reed and Sanders (1953) studied the frontogenesis of the upper level troposphere by using the potential vorticity theory. Bennets and Hoskins (1979) proposed the concept of moist potential vorticity. Hoskins (1985) proposed that if the effects of diabatic heating and friction are neglected, the constant pressure surface potential vorticity and the moist potential vorticity have conservativeness, and when the moist potential vorticity is negative the atmosphere may present symmetric instable. Hoskins provided the criterion of the conditional symmetric instability by using the moist potential vorticity and derived the moist buble potential vorticity equation and used it for analyzing the situation favorable to develop the conditional symmetric instability. Hoskins also pointed out that the potential vorticity and temperature disturbances at high and low levels respectively will cause the development of the cyclonic circulations according to the PV thinking theory. Xu(1989) discussed the relations between the moist symmetric instability and the frontogenesis. Xu (1992) also studied the relations between geostrophic potential vorticity (GPV) and the precipitation belts. Cho and Chan (1991) analyzed the relationship between the meso-β scale PV anomaly and the rain bands, and pointed out that the meso-β scale PV anomaly is an effective mechanism for the formation of the rain bands. The works of Xu, Chan and Cho closely connected the PV concept and the precipitation and let the PV theory more valuable for the practice applications. Davis and Rossa (1998) proposed the concept of PV frontogenesis, which provided a new point that the upper troposphere frontogenesis can be regarded as the intensified process of the PV gradient on the isentropic surface. Davis and Emanuel (1991) raised an important point of view that the upper level PV development may be strongly influenced by the low level anomalys. Molinari et al. (1998) analyzed the interaction between the tropical storm and upper troposphere positive PV anomaly and pointed that the smaller scale upper level positive PV anomaly overlayed on the low level tropical cyclone center may be favorable to intensify the tropical cyclone. But they also pointed out that the large scale PV anomaly seems not favorable to the development of the tropical cyclone.

In China, since the 1980s many researches about the potential vorticity theory are majorly related with precipitation. For example, Yang et al. (1981) analyzed the onset of Indian monsoon by using the potential vorticity theory. Liu et al. (1996) analyzed the relationship between the moist potential vorticity and heavy rain caused by front and pointed out that it is possible to determine the position of the heavy rain area by using MPV of the low troposphere. Hou (1991) proposed a conceptual model for describing the activity of the cyclone in Yangtze River Basin in summer season based on the analysis of the IPV chart and the potential vorticity. Lu and Ding (1991) pointed out that the potential vorticity can be used for tracing the the cold air. Wu et al. (1995, 1997) proposed the slantwise vorticity development (SVD) theory, they suggested that the moist potential vorticity conservation may be expressed as follows:

$$a\zeta_\theta |\nabla \theta_e| = \text{constant}$$

This means that all the factors including the convective instability and the vertical wind shear as well as the moist baroclinicity can bring up the growth of the cyclonic vorticity. The growing of the cyclonic vorticity will be much stronger when the slope of the isentropic surface is bigger. Liu and Shou (2002) analyzed the effect of the dry intrusion on the development of the cyclone. The dry intrusion implys the dry downdraft, which is characterized by low relative humicity and high PV, from the mid-high troposphere. The dry intrusion will cause the low level vorticity increase and therefore will bring up the cyclone development and favorable to cause the formation of heavy precipitation and severe convective weather. Gao et al. (2002) studied the moist potential vorticity anomaly with heat and mass forcings in torrential rain systems and developed the MPV substance impenetrability theory for providing a useful tool diagnosis analysis.

To sum up, due to the greater progress of the concept and theory of the potential vorticity, it has been an important research tool in meteorological field nowadays. The conservation and retrievable properties are the major principles of the potential vorticity, which may be used for describing the atmospheric dynamic processes. Just like the effect of the vorticity in the barotropic atmosphere, the potential vorticity is a very useful tool in the research of the weather phenomena in the baroclinic atmosphere.

§ 8.4 Analysis of the helicity

8.4.1 The concept of helicity

Helicity is a parameter which can be used for estimating the strength of the storm inflow and the magnitude of the vorticity component along the inflow direction. Moffert (1978) defined the helicity as the volume integral of the dot product of the wind vector and the vorticity vector. Its expression may be written as following:

$$H_T = \iiint_\tau \mathbf{V} \cdot \nabla \times \mathbf{V} \, d\tau \qquad (8.4.1)$$

where $\mathbf{V}$ is the three dimensional wind vector, $\nabla \times \mathbf{V}$ is the vorticity vector, and the dot product of the above two vectors is called the local helicity (H_D) or the density of helicity, which can be expressed as

$$H_D = \mathbf{V} \cdot \nabla \times \mathbf{V} \qquad (8.4.2)$$

Afterwards, Brandes(1989) proposed the concept of the storm relative helicity, it is defined as

$$H_{s-r-T} = \int_0^h (V - C) \cdot \omega dz \tag{8.4.3}$$

The corresponding local storm relative helicity is

$$H_{s-r-D} = (V - C) \cdot \omega \tag{8.4.4}$$

where C is the moving velocity of the storm, $(V-C)$ is the wind velocity relative to the storm, $\omega = \nabla \times V$ is the three dimensional vorticity vector, h is the thickness of the inflow layer of the storm. Another relevante concept of the helicity is the mean storm relative helicity, which is defined as

$$H_{s-r-M} = \frac{1}{h}\int_0^h (V - C) \cdot \omega dz \tag{8.4.5}$$

It is clear from the above definitions that the storm relative helicity(H_{s-r-T}, hereinafter reffered to as the total helicity simply) reflected the strength of the rotation of the environmental wind field in a certain thickness atmospheric layer and the maganitude of the vorticity transported into the convective storm from its environment. The unit of it is $m^2 \cdot s^{-2}$; while the mean storm relative helicity(H_{s-r-M}, hereinafter referred to as the mean helicity simply) is just an height average of the total helicity, its unit is $m \cdot s^{-2}$; the local storm relative helicity(H_{s-r-D}, hereinafter referred to as the local helicity simply) is refered to as maganitude of the total helicity in the atmospheric layer with unit thickness at certain height, its unit is also $m \cdot s^{-2}$. Obviously the above three kinds of the helicitity are related each other but they are also different.

8.4.2 Simplifying of the helicity

For computing the helicity, it is normally taken the simplified form of the helicity. In a local rectangular coordinate system, $\nabla \times V = \xi i + \eta j + \zeta k$, where

$$\xi = \frac{\partial w}{\partial y} - \frac{\partial v}{\partial z}, \quad \eta = \frac{\partial u}{\partial z} - \frac{\partial w}{\partial x}, \quad \zeta = \frac{\partial v}{\partial x} - \frac{\partial u}{\partial y}$$

In general speaking, the vertical component of the vorticity is smaller than the vertical wind shear at least one order, thus the vertical component of vorticity may be neglected comparing with the horizontal component of the vorticity. Meanwhile the variation of the vertical velocity in horizontal direction is not big before the convection coming up, thus the horizontal vorticity may be simplified as $\xi \approx -\frac{\partial v}{\partial z}$ and $\eta \approx \frac{\partial u}{\partial z}$, then we have $V \approx V_H = ui + vj$ and $\omega_H = \nabla \times V \approx k \times \frac{\partial V_H}{\partial z}$ (where V_H is the horizontal wind vector, ω_H is the horizontal vorticity vector), thus the formulas (8.4.3—8.4.5) may be changed as follows:

$$H_{s-r-T} = \int_0^h (V_H - C) \cdot k \times \frac{\partial V_H}{\partial z} dz \qquad (8.4.6)$$

$$H_{s-r-M} = \frac{1}{h} \int_0^h (V_H - C) \cdot k \times \frac{\partial V_H}{\partial z} dz \qquad (8.4.7)$$

$$H_{s-r-D} = (V_H - C) \cdot k \times \frac{\partial V_H}{\partial z} dz \qquad (8.4.8)$$

where $(V_H - C) \cdot k \times \frac{\partial V_H}{\partial z} = (V_H - C) \cdot \omega = |(V_H - C)| \cdot |\omega_H| \cdot \cos\alpha$, α reflects the intersection angle between the vectors $(V_H - C)$ and ω_H. Thus it can be seen that the local storm relative helicity reflects the strength cf the storm inflow and the magnitude of the horizontal component vorticity along the inflow direction as mentioned above. In other words, it reflects the strength of the horizontal axis rotation and the motion along the rotational axis of the inflow as shown in Fig. 8.11. The storm relative helicity may be regarded as sum of the products of the relative velocity and the horizontal vorticity along the relative velocity direction in the atmospheric layer from the height 0—h. After the horizontal vorticity transported into the strorm it will be changed into the vertical vorticity (Fig. 8.12), so that it will be favorable to intensify the rotation of the storm and increase the weather intensity.

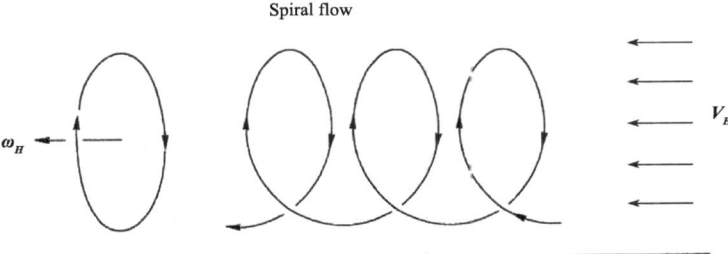

Fig. 8.11 The schematic diagram of the spiral motion when the horizontal vorticity vector and the horizontal air flow in the same direction. (After Doswell, 1991)

8.4.3 Application of helicity

In recent years a number of the researches about helicity are carried out in the world. For example, Lilly (1986) discussed the effect of the helicity on the formation of the supercells and proposed that the bigger helicity is favorable to increase the sta-

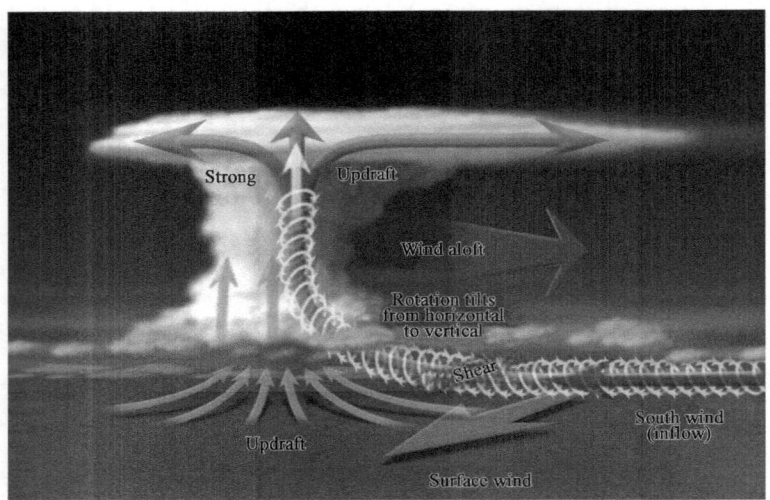

Fig. 8.12　Illustration of the storm relative helicity. (see the color illustrations)

bilization of the storm and to extend the life cycle of the storm. Rasmussen et al. (1998) gave a baseline climatology of Sounding-Derived Supercell and Tornado Forecast Parameters. As shown in Fig. 8.13, Woodall (1990) applied the helicity for forecasting tornadoes. Davies-Jones et al. (1990) proposed that the helicity can be taking as a parameter for forecasting the severe convective storms. Leftwich Jr. (1990) studied the indicating meaning of the helicity magnitude corresponding to different types of the convective weather. The National Severe Storm Forecasting Center (NSSFC) developed operational program of the helicity analysis, so that the helicity analysis can be conveniently used for the operational severe convective weather forecasting. Furthermore, the helicity is very often combined with the convective available energy to form a new parameter "energy-helicity" (see the next section), which is a significant indicator of severe weather.

§ 8.5　Analysis of the atmospheric instability

8.5.1　Atmospheric thermal instability

In previous chapters, we have discussed the concepts of the atmospheric static instability, the conditional instability and the convective instability etc. They are usually characterised by the vertical distribution of the thermodynamic parameters such as the potential temperature θ, the equivalent potential temperature θ_e, the saturated

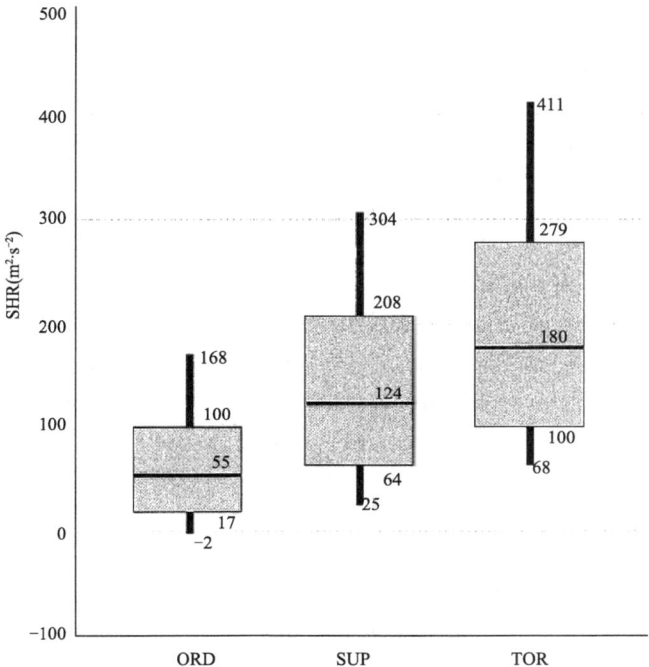

Fig. 8.13 A baseline climatology of sounding-derived supercell and tornado forecast parameters. ORD = ordinary thunderstorm, SUP = supercell storm, TOR = tornadic storm, SRH = storm-relative helicity. (After Rasmussen, et al., 1998)

equivalent potential temperature θ_e^*, the dry static energy E_d, the moist static energy E'_θ and the satuarated moist energy E_θ^* etc.

8.5.2 Atmospheric instability energy and its relation with convection

If the air parcel has no exchanges of heat, water vapor, mass, momentum with its environment, which is frictionless and quasi static, then the vertical motion equation can be written as following:

$$\frac{dw}{dt} = g\frac{T-\overline{T}}{T} = g\frac{\Delta T}{T} \tag{8.5.1}$$

where T and $\overline{T}$ represents the temperature of the air parcel and the environment respectively. The integration of the right side of the equation (8.5.1) for the height (Z) is the instability energy (E) as following:

$$E = \int_{Z_0}^{Z} g\frac{\Delta T}{T}dz = -\int_{p_0}^{p} R\Delta T d\ln p \tag{8.5.2}$$

While the integration of the left side of the equation (8.5.1) is the kinetic energy increment of the air parcel (ΔE_k):

$$\Delta E_k = \int_{z_0}^{z} \frac{dw}{dt} dz = \int_{t_0}^{t} \frac{dw}{dt} w\, dt = \int_{w_0}^{w} w\, dw = \frac{1}{2}(w^2 - w_0^2) \quad (8.5.3)$$

This shows that the kinetic energy of the convective motion comes from the the releasing of the instability energy.

8.5.3 Convective available energy (CAPE)

When the gravity force and buoyancy force are not equal to each other, part of the potential energy can be changed into the kinetic energy of the vertical motion, this portion of the potential energy is called the convective available energy (CAPE). It is the mechanical work done by the buoyancy force for the rising air parcel. It is also the measurement of the magnitude of the atmospheric instability. The positive CAPE is necessary for generating convection. CAPE may be defined as:

$$CAPE = g \int_{Z_{LFC}}^{Z_{EL}} \left(\frac{T_{vp} - T_{ve}}{T_{ve}} \right) dz \quad (8.5.4)$$

where T_{vp} and T_{ve} are the virtual temperature of the air parcel and the environment respectively. Z_{LFC} and Z_{EL} are the level of free convection and the equilibrium level respectively. If the drag force caused by the gravity force of the water droplets floating in the cloud is taken account, then the modificatory convective available potential energy MCAPE may be written as

$$MCAPE = g \int_{Z_{LFC}}^{Z_{EL}} \left(\frac{T_{vp} - T_{ve}}{T_{ve}} - g \right) dz$$

or

$$MCAPE = g \int_{Z_{LFC}}^{Z_{EL}} \frac{1}{T_{ve}} [(T_{vp} - T_{ve}) - T_{ve}(r_l - r_i)] dz \quad (8.5.5)$$

where r_l and r_i are mixing ratios of the liquid water and ice respectively. After considering the mean value theorem, we can get the following expression from (8.5.4):

$$CAPE = \overline{\left(\frac{T_{vp} - T_{ve}}{T_{ve}} \right)} \times (Z_{EL} - Z_{LFC}) \quad (8.5.6)$$

where $Z_{EL} - Z_{LFC} = \Delta H_{FCL}$ is called the thickness of the free convective layer. $\overline{(\ )}$ denotes the arithmetic average for ΔH_{FCL}. CAPE is propotional to the positive area of the instability energy in the $T\text{-}\ln p$ diagram (see Fig. 8.14). The magnitude of CAPE is depending on ΔH_{FCL} and the magnitude of the mean buoyancy in the thickness. Under the certain conditions, if the thickness of the free convective layer ΔH_{FCL} increases (or decreases) then the mean buoyancy in the free convective layer will certainly decrease (or increase). Thus it is necessary to consider the aspect ratio, i. e. the ratio between height and width, of the positive energy area on the T-$\ln P$ diagram when analysising $CAPE$, except considering the magnitude of $CAPE$. When the aspect ratio is different, the instability, for example the lifted index LI, may be significantly different although the magnitude of $CAPE$ is the same. For ignoring the influence of ΔH_{FCL} and highlighting the effect of the mean buoyancy force, Blanchard (1998) introduced the concept of the normalized convective available potential energy ($NCAPE$), which can be expressed as follows:

$$NCAPE = CAPE / \Delta H_{FCL} \qquad (8.5.7)$$

$NCAPE$ denotes the magnitude of mean acceleration or the mean buoyancy acting on the air parcel with unit mass in thickness of the free convective layer ΔH_{FCL}. Since the unit of $CAPE$ is $J \cdot kg^{-1}$ or $m^2 \cdot s^{-2}$, the unit of $NCAPE$ is $J \cdot kg^{-1} \cdot m^{-1}$ or $m \cdot s^{-1}$. If the free convection level and the equilibrium height are expressed by pressure, then $NCAPE$ can be written as:

$$NCAPE = CAPE / (P_{LFC} - P_{EL}) \qquad (8.5.8)$$

Its unit is $J \cdot kg^{-1} \cdot hPa^{-1}$.

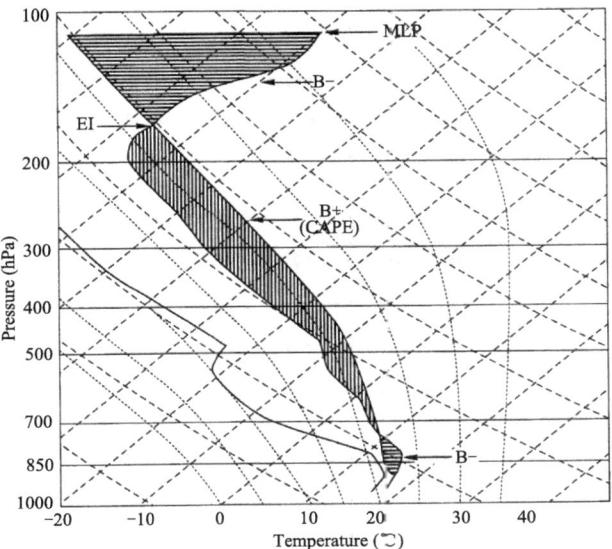

Fig. 8.14 The schematic diagram of the positive or negative energy areas and $CAPE$ in a skew T-$\ln p$ diagram.

8.5.4 Convection inhibition energy(CIN)

Colby (1984) proposed the concept of the convection inhibition energy (CIN), which is expressed as

$$CIN = g\int_{Z_i}^{Z_{LFC}} \frac{T_e - T_p}{T_B} dz \qquad (8.5.9)$$

where T_e and T_p denotes the temperature of environment and air parcel respectively. T_B is the mean temperature of the stable layer. Z_i and Z_{LFC} are the initial height and height of the free convection level respectively. CIN denotes the magnitude of the negative work done by the air parcel from the mean atmospheric boundary layer when it arrived the level of free convection (LFC) through the stable layer. CIN denotes also the criterion that is necessary to be exceeded for the air parcel to obtain the convective potentiality. In the T-$\ln P$ diagram, CIN is the negative energy area NA below the LFC. Thus (8.5.9) can be also written as follows:

$$NA = CIN = -\int_{P_{LFC}}^{P_i} (T_{vp} - T_{ve}) R_d \frac{dp}{p} \qquad (8.5.10)$$

where T_{vp} and T_{ve} are the virtual temperature of air parcel and environment respectively. P_i and P_{LFC} are the pressure at initial height and the pressure at LFC respectively.

Colby (1984) pointed out that in order to raise the air parcel from the atmospheric boundary through the stable layer, the air parcel has to poss enough kinetic energy of unit mass or a strong enough vertical upward velocity W_{CIN}, which has the following relationship with CIN:

$$W_{CIN} = \sqrt{2CIN} \qquad (8.5.11)$$

8.5.5 Descending convective available energy (DCAPE)

Since the evaporation of the liquid water in unsaturated air and the melting of the solid water below the melting level will absorb heat, so that it will cause the air cooling and generates the downdraft. The possible intensity of the downdraft is propotional to the descending convective available energy (DCAPE), of which the expression was given by Emanuel (1994) as following:

$$DCAPE = \int_{P_i}^{P_n} R_d (T_{\rho e} - T_{\rho p}) d\ln P \qquad (8.5.12)$$

where P_i and P_n are the pressure at the initial height of the descending air parcel and the pressure when the air parcel arrives the neutral layer or the surface ground respectively. $T_{\rho e}$ and $T_{\rho p}$ are the environment temperature and the density temperature of the air parcel respectively. The density temperature T_ρ implies the due air temperature when the density of the dry air equals the density of the moist air, which may contain the liquid and solid water, under the same pressure. T_ρ may be expressed as follows:

$$T_\rho \equiv T \frac{1+\gamma/\varepsilon}{1+\gamma_T} \qquad (8.5.13)$$

where T is temperature, γ_T is the mixing ratio of the water substances, $\gamma_T = \gamma + \gamma_l + \gamma_i$, where γ, γ_l, γ_i are the mixing ratios of the water vapor, liquid water, and ice respectively. ε is the rate between the gas constant of the dry air R_d and the gas constant of the water vapor. If there is no hydrolic substantials, $\gamma_T = \gamma$, then

$$T_\rho \equiv T \frac{1+\gamma/\varepsilon}{1+\gamma_T} = T \frac{1+\gamma/\varepsilon}{1+\gamma} \cong T(1+0.608\gamma) = T_v \qquad (8.5.14)$$

where T_v is the virtual temperature. Thus T_ρ is the generalization of T_v, in other words, T_v is the special case of T_ρ, which is widely used in the theory about the moist atmospheric convection. By using T_ρ, we can get

$$CAPE = g \int_{Z_{LFC}}^{Z_{EL}} \frac{1}{T_{\rho e}} (T_{\rho p} - T_{\rho e}) dz \qquad (8.5.15)$$

where $T_{\rho p}$ and $T_{\rho e}$ are the densities of the air parcel and the environment respectively.

If the vertical velocity at the initial point of the descending motion is zero without regard to other factors, then in terms of theory when the air parcel arrives to the neutral buoyancy layer or the surface ground, the vertical descending velocity caused by the negative buoyancy energy will be expressed as the following:

$$-W_{max} = \sqrt{2DCAPE} \qquad (8.5.16)$$

Since the value of the $DCAPE$ will increase with the decreasing of the relative humidity at the initial state, if the mid level inflow is colder and drier, the convection will be more stronger.

The convective circulation includes updraft and downdraft branches. If the influence of the water vapor on the density is neglected then the $CAPE$ of the updraft and the $DCAPE$ of the downdraft can be respectively expressed as follows:

$$CAPE = \int_{up} g \frac{T_u - T_e}{T_e} dz \qquad (8.5.17)$$

$$DCAPE = -\int_{down} g \frac{T_a - T_e}{T_e} dz \qquad (8.5.18)$$

where T_u and T_d are the temperature of the ascending air parcel and descending air parcel respectively. The subscripts up and down denote the updraft and downdraft branches respectively. Thus the total work done by the moving air parcel around the convective circulation will be as follows:

$$W = CAPE + DCAPE = \int_{Z_b}^{Z_t} g\left(\frac{T_u - T_d}{T_e}\right) dz \approx -\int_{up} RT_u d\ln P + \int_{down} RT_d d\ln P$$

$$\approx -RT d\ln P \approx p d\alpha \approx TCAPE \qquad (8.5.19)$$

where Z_b and Z_t are called the heights of the bottom and the top of the convection layer respectively. $TCAPE$ is the total convective available energy around a round of the reversible heat engine cycle. The energy can be converted into the kinetic energy of the updraft and downdraft. Larger $TCAPE$ is necessary for maintaining a strong circulation.

8.5.6 Some frequently used thermal stability indexes

The following thermal instability indexes are used frequently in the practical works.

(1) Lifted index(LI)

The lifted index (LI) is defined as

$$LI = T_{500} - T'_{500} \qquad (8.5.20)$$

where T_{500} and T'_{500} are the environmental temperature at 500 hPa and the temperature of the air parcel, when it is lifted to the height of 500 hPa with the moist adiabatic lapse rate starting from the level of free convection (LFC). Bigger positive value of LI implys bigger instability.

(2) Showalter Index(SI)

Showalter index (SI) is defined as

$$SI = T_{e500} - T_{p500} \qquad (8.5.21)$$

where T_{e500} is the environmental temperature at 500 hPa constant pressure surface, T_{p500} is the temperature of the air parcel which is lifted to the lifted condensation level (LCL) dry adiabatically at first, and then to the 500 hPa constant pressure surface with moist adabatical lapse rate. $SI < 0$ implys the atmosphere is unstable. The bigger the negative value of the SI, the stronger the atmospheric instability.

(3) The simplified Showalter Index(SSI)

$$SSI = T_{e500} - T'_{p500} \qquad (8.5.22)$$

where T'_{p500} is defined as the temperature of the air parcel originally at 850 hPa when it is lifted to 500 hPa dry adiabatically. SSI is always positive, the smaller value of SSI implys the bigger instability.

(4) Total Total Index(TT)

$$TT = T_{850} + T_{d850} - 2T_{500} \qquad (8.5.23)$$

where T_{850}, T_{d850} and T_{500} are the temperature and dew point temperature at 850 hPa constant pressure surface and the temperature at 500 hPa constant pressure surface respectively. The bigger TT, the stronger the instability. Sometimes, the surface temperature T_0 and dew point temperature T_{d0} are taken account in the TT index, this index is called the modified total total index and indicated by TT_{mod},

$$TT_{mod} = \frac{1}{2}[(T_0+T_{d0})+(T_{850}-T_{d850})]-2T_{500} \tag{8.5.24}$$

(5) A index

$$A=(T_{850}-T_{500})-[(T-T_d)_{850}+(T-T_d)_{700}+(T-T_d)_{500}] \tag{8.5.25}$$

The greater value of A is corresponding to the stronger instability.

(6) K index (the air mass index)

$$K=(T_{850}-T_{500})+T_{d850}-(T-T_d)_{700} \tag{8.5.26}$$

The greater value of K is corresponding to the stronger instability.

8.5.7 Assembly instability parameters combining thermal and dynamic factors

Very often the atmospheric thermal instability parameters and dynamic parameters such as the vertical wind shear etc. are combined to form some new assembly instability parameters with significant synoptic-dynamic meteorological meanings. Some of the new parameters are simply introduced as follows.

(1) Richardson Number(Ri)

Richardson number (Ri) is a well known index indicated the strength of turbulence and can be expressed as follows:

$$Ri = \frac{g}{T}\left(\frac{\Delta\theta}{\Delta z}\right) / \left(\frac{\Delta U}{\Delta z}\right)^2 \tag{8.5.27}$$

where Ri number indicates the relationship between static stability and vertical wind shear. It actually also indicates the relationship between $CAPE$ and kinetic energy. When Ri is small (for example $Ri < 0.25$), turbulence may be formed. Fritschi analyzed the relationship between Ri number and weather phenomena and gave the following critertia:

$0.25 \geqslant Ri \geqslant -1$ possible to form mid-latitude systematic convections
$Ri < -1$ possible to form air mass thunderstorms
$Ri < -2$ possible to form tropical cumulonimbus

(2) Bulk Richardson number (BRN)

Weisman and Klemp (1982) introduced the concept of Bulk Richardson Number and indicated it by BR or BRN:

$$BRN = CAPE / \left(\frac{1}{2}\overline{U}_z^2\right) \tag{8.5.28}$$

where $\overline{U}_z$ is the difference between the mean wind velocity of the lowest 6 km $\overline{U}_{6000}$ and the mean wind velocity of 500 hPa $\overline{U}_{500}$.

The ratio between $CAPE$ and BRN is called BRN-Shear($BRNSHR$).

$$BRNSHR = \frac{1}{2}(u^2 + v^2)^{\frac{1}{2}} \tag{8.5.29}$$

where u, v are $[(u,v)_{0-6} - (u,v)_{0-0.5}]$; where $(u,v)_{0-6}$ and $(u,v)_{0-0.5}$ are the u, v components of the 0 to 6 km and 0 to 0.5 km density weigted mean wind velocitty respectively.

(3) Simplified BRN($SBRN$)

Equation (8.5.28) can be written in the following form:

$$BRN = CAPE/0.5S^2 \tag{8.5.30}$$

where S is the vertical wind shear between the layer from boundary to the height 6 km over ground surface. Colquhoun and Riley (1996) designed a similar parameter called simplified BRN ($SBRN$):

$$SBRN = -883LI/(S6)^2 - 1.32(S6) + 41 \tag{8.5.31}$$

where LI is the lift index, $S6$ is the wind shear between ground surface and 600 hPa. Because there is a high correlation coefficient (about -0.86 — -0.90) between LI and $CAPE$, so that LI may be used to instead of $CAPE$ and make the calculation of BRN more convenient for operational application.

(4) Convective Richardson number (CRN)

Convective Richardson number (CRN) may be defined as the ratio between $CAPE$ and ΔU^2, where ΔU represented the wind velocity difference between the ground surface and the height of the storm cloud top

$$CRN = CAPE/\Delta U^2 \tag{8.5.32}$$

If $CAPE$ in (8.5.32) is insteaded by the modified $CAPE$ ($MCAPE$), then we get a new parameter called modified CRN ($MCRN$) expressed as following:

$$MCRN = MCAPE/\Delta U^2 \tag{8.5.33}$$

Hart and Korotky (1991) and Davis (1993) defined an index called energy-helicity index (EHI):

$$EHI = \frac{CAPE \times SRH}{1.6 \times 10^5} \tag{8.5.34}$$

When EHI is bigger, the potential to occur supercells and tornadoes will be higher.

(5) Vortex-genesis parameter (VGP)

Rasmussen and Wilhelmson (1983) derived the conversion rate that the horizontal vorticity converted into vertical vorticity due to the twist effect:

$$\left(\frac{\partial \zeta}{\partial t}\right)_{Twist} = \eta \cdot \nabla w \qquad (8.5.35)$$

where ζ is the vertical component of vorticity, η is the horizontal vorticity vector, w is vertical velocity. Based on this principle a new parameter called vortex genesis parameter (VGP) is induced, which may be expressed as following:

$$VGP = S(CAPE)^{\frac{1}{2}} \qquad (8.5.36)$$

where S is the mean wind shear of the atmospheric layer between the ground surface and height of h:

$$S = \int_0^h \frac{\partial v}{\partial z} dz \bigg/ \int_0^h dz \qquad (8.5.37)$$

8.5.8 Analysis of conditional symmetric instability (CSI)

Some of the CSI calculating methods which may be used in operational work are introduced as following.

(1) CSI analysis based on moist potential vorticity (MPV)

The CSI may be analyzed by using MPV. If $MPV<0$, then it is conditional symmetric unstable, while if $MPV>0$ then it is conditional symmetric stable. On the constant pressure surface MPV may be defined as

$$MPV = -\eta \cdot \nabla \theta_e \qquad (8.5.38)$$

where η is 3D vector of the absolute vorticity in a (x, y, p) coordinate system, θ_e is equivalent potential temperature. To expand equation (8.5.39) and assume the air flow is in geostrophic balance and neglect the terms related with vertical velocity and the variation along y direction, then we can get:

$$MPV = g\left(\frac{\partial M_g}{\partial p}\frac{\partial \theta_e}{\partial x} - \frac{\partial M_g}{\partial x}\frac{\partial \theta_e}{\partial p}\right) \qquad (8.5.39)$$

$$MPV = MPV_1 + MPV_2 \qquad (8.5.40)$$

where MPV_1 and MPV_2 equal to the first term and second term in equation (8.5.39) respectively and indicate the horizontal and vertical components of the moist vorticity respectively. $M_g = V_g + fx$ is absolute geostrophic momentum. θ_e is equivalent potential temperature. When $MPV<0$ and the atmosphere is convective stable then the atmosphere is conditional symmetric unstable.

CSI can also be determined based on the analysis of absolute vorticity on isentropic surface. It can be seen from equation (8.5.39) that the absolute vorticity on isentropic surface may be expressed as following:

$$(MPV)_{\theta_e} = \left(\frac{\partial M_g}{\partial x}\right)_{\theta_e} \left(-\frac{\partial \theta_e}{\partial p}\right) \tag{8.5.41}$$

The above equation shows that under the convective stable condition, when $\left(\frac{\partial M_g}{\partial x}\right)_{\theta_e} < 0$, may satisfy $(MPV)_{\theta_e} < 0$. This means the area with negative value of isentropic suface absolute vorticity should be a conditional symmetric unstable area.

About the influence of the M_g and θ_e fields on MPV we can also make a further analysis according to the following procedure. Firstly, to analyze the terms $\frac{\partial M_g}{\partial p}$ and $\frac{\partial \theta_e}{\partial x}$ in MPV_1, which indicate the vertical wind shear and the moist baroclinity of the atmosphere. Suppose x points to the warm air side, then $\frac{\partial \theta_e}{\partial x} > 0$. Therefore when the vertical wind shear increases the negative value of MPV_1 will be increased so that it is favorable to cause symmetric instability. Therefore if the horizontal gradient of θ_e is bigger, i. e. the slope of of the constant surface of θ_e, then the situation will be more favorable to cause greater symmetric instability. When the slope of the constant surface of θ_e is much greater than the slope of constant surface of M_g, the symmetric instability of atmosphere will be much greater.

Secondly, to consider the terms $\frac{\partial M_g}{\partial x}$ and $\frac{\partial \theta_e}{\partial p}$ in the term MPV_2, which indicate the horizontal variation of the absolute momentum and the convective instability of atmosphere respectively. In northern hemisphere the absolute vorticity is normally positive, hence $\frac{\partial M_g}{\partial x}$ is normally positive. When $\frac{\partial \theta_e}{\partial p} < 0$, i. e., the atmosphere is convective stable, MPV_2 will be positive and is not favorable to cause symmetric instability. When $\frac{\partial \theta_e}{\partial p} = 0$ or $\frac{\partial \theta_e}{\partial p} > 0$, i. e. the atmosphere is neutral or convective unstable, then the situation will be favorable to cause symmetric unstable. When $\frac{\partial \theta_e}{\partial p} = 0$, $MPV_2 = 0$, if $MPV_1 < 0$, then $MPV < 0$, hence the situation does not obstruct to cause symmetric instability. When $\frac{\partial \theta_e}{\partial p} > 0$, $MPV_2 < 0$, if $MPV_1 < 0$, then $MPV < 0$; if $MPV_1 > 0$, while $|MPV_2| > MPV_1$, then it is still $MPV < 0$, hence under such situation the atmosphere has both convective instability and symmetric instability, so that both vertical convection and slatwise convection will be possible. However, since the vertical convection has relative quickly growth rate, the atmospheric motion will present the feature of smaller temporal and spatial scale motions. Based on the above analysis it can be concluded that in the neutral or weaker convective instability condition it will be easier to form symmetric instability. This conclusion has agreement with that the symmetric instability formation is normally under the condition with smaller Richardson number ($Ri<1$).

(2) CSI eastimation based on the the absolute vorticity on isentropic surface

Since symmetric instability is also the inertial instability on isentropic surface as discussed previously, CSI may be also determined by means by analyzing the absolute vorticity on isentropic surface. If somewhere the absolute vorticity is negative then the atmosphere over the area should be conditional symmetric unstable.

(3) Estimating CSI by using slantwise CAPE (SCAPE)

Under the situation that the slope of the isentropic surface is greater than the slope of the constant surface of the absolute momentum (M), the air parcel rised along the constant M surface will be convective unstable. Based on this principle, we may estimate the CSI by using the slantwise CAPE (SCAPE), i. e. the convective available potential energy of a rising air parcel along the constant M surface. The parameter SCAPE may be represented by the total positive area between the state curve line of the air parcel and the environmental virtual temperature profile on the T-$\ln p$ diagram and may be expressed as following:

$$SCAPE = \int_{LFS}^{EL} \frac{g}{\theta_{v0}} (\theta_{vp} - \theta_{ve}) dz \qquad (8.5.42)$$

where LFC is the free slantwise convection level. EL is the convectional equilibrium level, θ_v is virtual potential temperature. θ_{v0} is the typical value of θ_v in the atmosphere. θ_{vp} and θ_{ve} are the θ_v of air parcel and environment respectively. The integration is made along the environmental constant absolute momentum surface. When SCAPE >0, the atmosphere is slantwise unstable.

SCAPE may be estimated by using single station sounding data. To assume the wind is in geostrophic balance, the horizontal and vertical gradients of the absolute momentum (M) as well as the vertical component of the absolute vorticity are all constant numbers, then the parameter SCAPE may be calculated by using the following equation:

$$SCAPE = \frac{1}{2} \frac{f}{\eta} (V_1 - V_0)^2 + \int_0^1 \frac{g}{\theta_{v0}} (\theta_{vp} - \theta_{ve}) dz \qquad (8.5.43)$$

where V_0 and V_1 are the wind velocity at the initial level and terminal level respectively. The first and second terms in the right side of the above equation represent the contributions of the vertical shear of the wind velocity and the buoyancy force for SCAPE respectively.

§ 8.6 Some case studies of the severe weather

8.6.1 Diagnostic analysis of a severe convective storm process with tornado

A tornado process occurred at Nantong of Jiangsu Province, China. Around 17:30, 12 Jul., 2004. Zhizhong and Henan villages, Baochang town, Haimen City in Nantong region suffered raid of tornado and hail. The hailstones were as large as broad beans and the hailstorm lasted for 10 minutes or so. The storm moved from northwest to southeast and large area of farmlands suffered serious loss.

Before and after the occurrence of Nantong's storm, it can be inferred from the CAPE value given by NCEP data that: at 14:00 of 12 Jul., the early phase of strong convective weather occurred at Nantong, it was located in the center of CAPE magnitude, which was roughly 2800 J • kg^{-1} and contained extremely large instability energy; At 20:00, the intensity of center CAPE magnitude diminished and moved slightly eastward to the region of Shanghai(Fig. 8.15). This phenomena is also coincident with the strong convective weather occurred in Shanghai around 19:00.

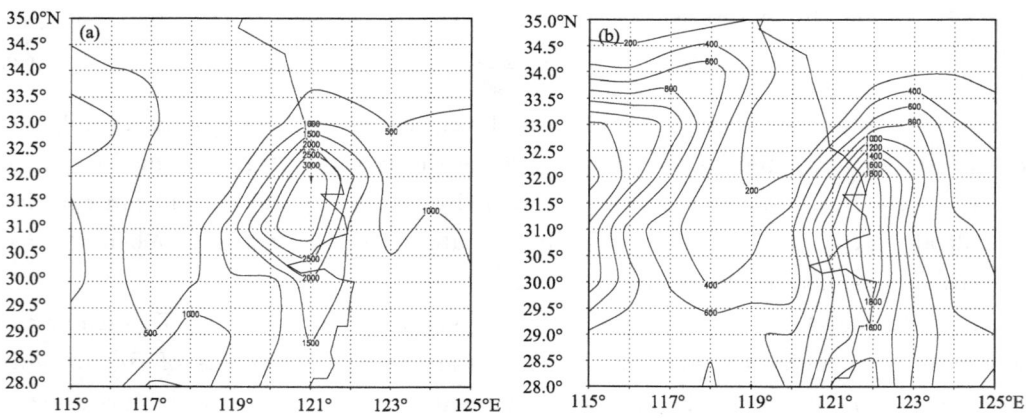

Fig. 8.15 Observed CAPE distributions on 1400 BST (a) and on 2000BST (b) 12 Jul., 2004 (unit: J • kg^{-1}). (After Zhang et al., 2004)

The following atmospheric convective energy parameters were diagnosed : ① Best Convective Available Potential Energy(BCAPE); ② Convection Inhibiting Energy(CIN); ③ Normalized Convective Available Potential Energy(NCAPE); ④ the Convective Available Potential Energy with Density Temperature; ⑤ Energy-Helicity Index (EHI) ($EHI = CAPE \times SRH/1.6 \times 10^5$); ⑥ Severe Weather Threatening Index (SWEAT).

From numerical simulation results, it can be found that, if BCAPE is calculated with density temperature, the result will be better than that calculated with virtual temperature. The central position and intensity correspondence of SWEAT and EHI indexes are both good, while we have considered more factors with them. They also have a certain degree of resolution and period of validity upon the types of severe weathers. Meanwhile, we found that, on the upper air of mono-stations, the time-varying rates of intensity for the above three parameters were the largest. The increasing rate of these energy parameters can effectively indicate the variation of weather. Besides, the spatial distribution and positions of CIN and NCAPE also have fairly good indicating significance for severe weather(Fig. 8.16).

It can be concluded from the observation analysis and simulation results that: the invasion of dry and cold air in upper layer and the intense convergence of warm and humid air at lower level brought about instable stratification. This provided energy condition for the generation of the tornado of Nantong on Jul. 12, 2004.

All above parameters have good correspondence upon the severe weather process. The effect will be better if the virtual temperature is replaced by density temperature. Concerning the vertical distribution of BCAPE, especially the lower level distribution, it is more reasonable to compute NCAPE than referring to the total value of BCAPE alone. EHI has synthetically considered SRH and BCAPE, which is favorable to distinguish the types of severe weathers. The time derivative of BCAPE, density temperature BCAPE, EHI and SWEAT indexes have distinct indicating significances upon severe weathers. Through utilizing numerical simulation and combining various convection parameters from every aspect, it is possible to capture the process of development of severe convective weather.

Calculating SWEAT index with simulation. This case has reconfirmed the statistical results of severe weathers of Jiangsu Province in the recent decade. The SWEAT index also has distinct indicating significance upon strong convective weathers of China. Here, although the wind direction shearing term of SWEAT index does not ac-

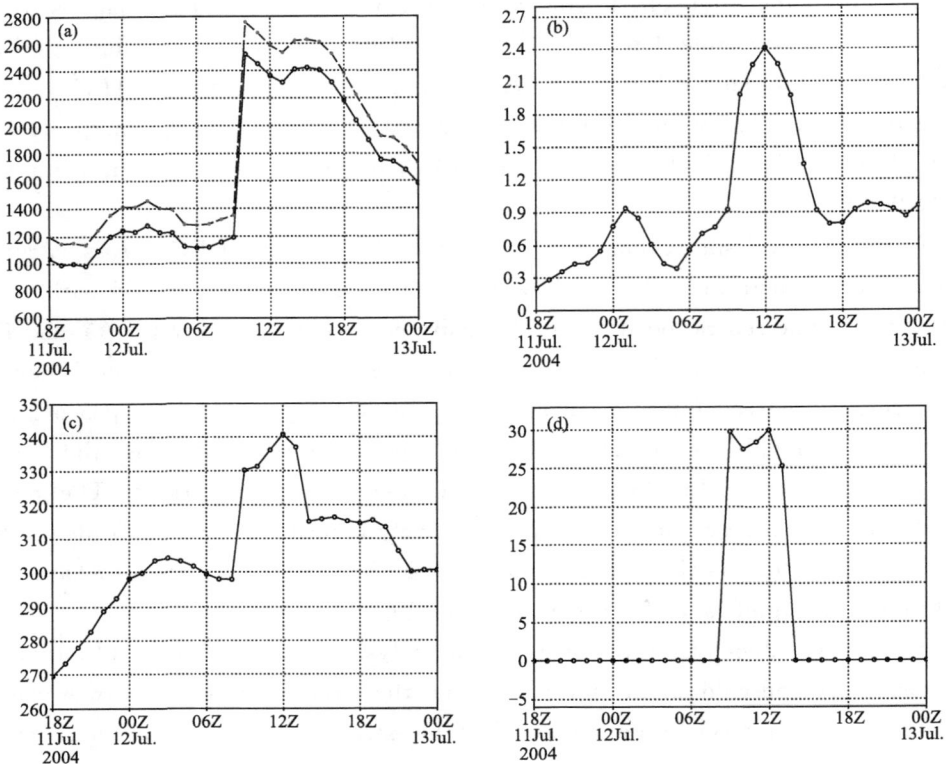

Fig. 8.16 The simulated temporal variation of energy parameters over Nantong during July 12, 2004. (a) BCAPE (dashed line, unit: $J \cdot kg^{-1}$) and BCAPE calculated by density temperature (solid line, unit: $J \cdot kg^{-1}$); (b) EHI (unit: $J \cdot kg^{-1} \cdot m^2 \cdot s^{-2}$); (c) SWEAT index; (d) Wind direction shearing tern of SWEAT index. (After Zhang et al., 2004)

count for a large proportion on the total magnitude, it has non-negligible indicating significance during its time-variation process. This indicates that the shearing of wind direction and speed have distinct effect on forming tornadoes.

8.6.2 Numerical simulation and diagnosis of a heavy rain process occurred in Nanjing on Jul. 5, 2003

On Jul. 5, 2003, a heavy rain process occurred near Nanjing, Jiangsu Province in eastern China. The precipitation area was very concentrated as shown in Fig. 8.17. It was mainly caused by a convective cloud cluster developed in this area.

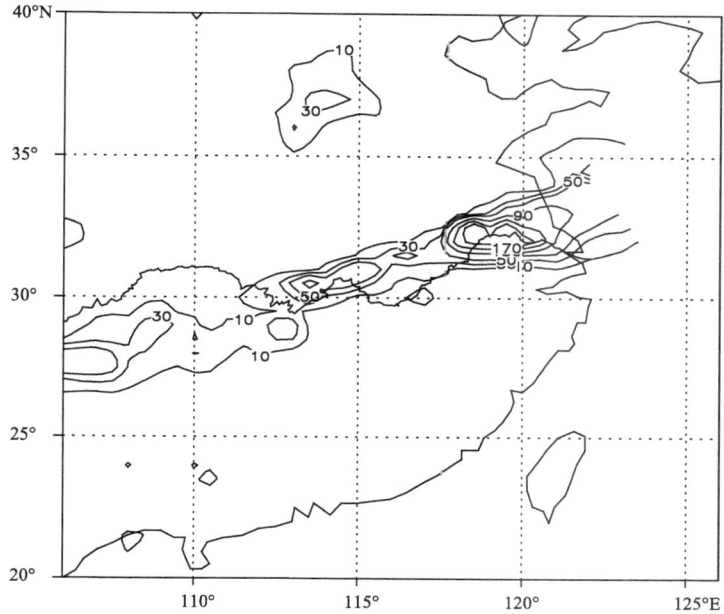

Fig. 8.17 Distribution of the 24 h rainfall, Jul. 5—6, 2003. (After Liao et al., 2005)

Based on the mesoscale numerical simulation, the heavy rain storm was closely related with a mesoscale low pressure with the diameter about 200 km. occurred near Nanjing on the surface map as shown in Fig. 8.18a and also a closed mesoscale cyclone over the surface mesoscale low on the 850 hPa constant pressure surface as shown in Fig. 8.18b. Fig. 8.19a—d showed the time-height(pressure) section of the vorticity, divergence, vertical velocity over the surface mesoscale low pressure companied with the mesoscale heavy rain-mass center and the precipitation intensity variation with time on Jun. 5, 2003. It is clearly that when the low level vorticity increases and the low level convergence as well as the upper level divergence increase and the vertical velocity increases the precipitation intensity will increase. While oppositely, when the low level vorticity decreases and the low level convergence as well as the upper level divergence decrease and the vertical velocity decreases the precipitation intensity will decrease. These results showed that the heavy rain process was caused by the mesoscale low pressure and cyclone. The rainfall area moved along the moving path of the mesoscale systems and the precipitation intensity variation was directly influenced by the intensity of the mesoscale systems.

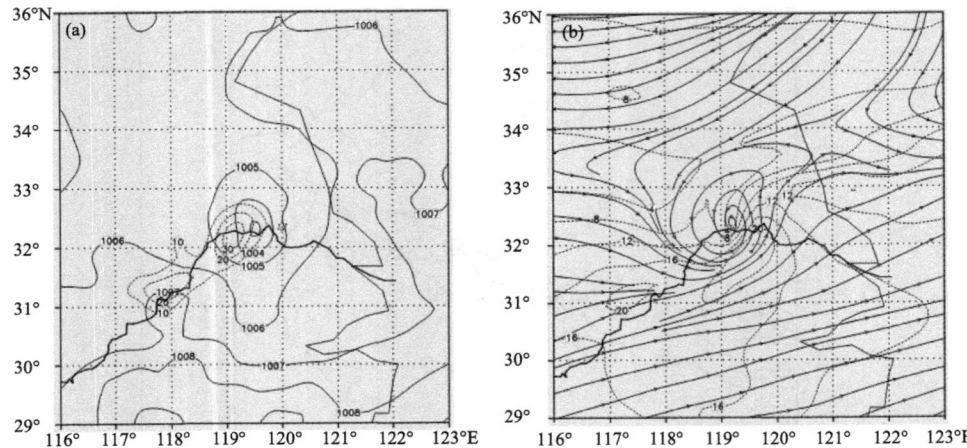

Fig. 8.18 (a) Surface map (the solid lines are isobars, the dashed lines are the contours of hourly rainfall); (b) the stream lines on 850 hPa constant pressure surface on 0800 BST June 5, 2003. (After Liao et al., 2005)

8.6.3 Diagnosis analysis of a severe storm process occurred in Shangqiu City of Henan Province in China on Jun. 3, 2009

On Jun. 3, 2009 a severe squall line with strong hail storm surprise attacked on Shangqiu City of Henan Province in China and brought up serious damage. According to the automatic meteorological station network data analyses on Jun. 3, 2009, the dew point temperature, pseudo potential temperature and specific humidity were distinctly higher than other places and there was a mesoscale shear line in the wind field near Shangqiu area (Fig. 8.20). The time-height (pressure) section of pseudo potential temperature (θ_{se}) during 00Z Jun. 3—00Z Jun. 4, 2009 over the area near Shangqiu showed that the instability ($\Delta\theta/\Delta z$) was very high at the time when the storm occurred near Shangqiu area (at 12Z, Jun. 3) (Fig. 8.21). The sounding data analysis of Xuzhou located near Shangqiu (Fig. 8.22) showed the CAPE was very high in this area. The vertical section of potential vorticity (PV) anomaly along the latitude of 34.5°N (Fig. 8.23) showed the column of high PV was stretching downward to the low level of the troposphere over the Shangqiu area. All the above factors should be the most favorable to form the severe storm near Shangqiu City.

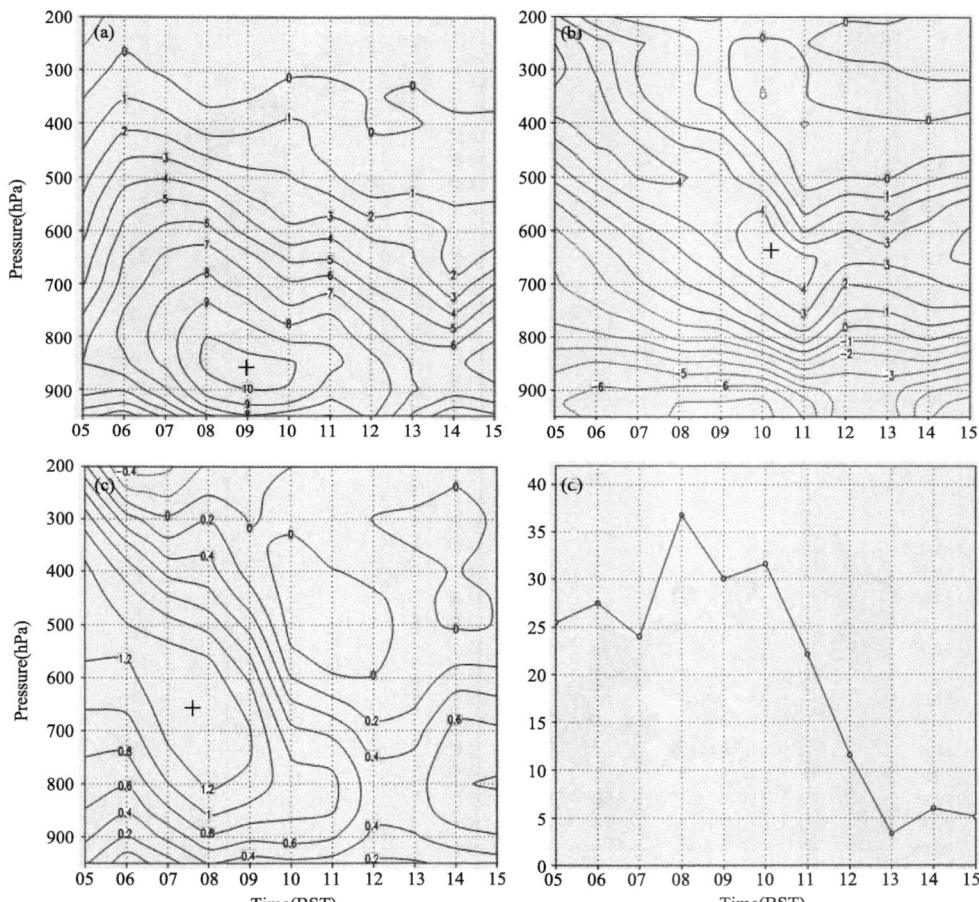

Fig. 8.19 (a) The time-height (pressure) section of the vorticity (in unit: 10^{-4} s^{-1}) over the surface mesoscale rain-mass center on Jun. 5, 2003; (b) same with (a) but for divergence (in unit: 10^{-4} s^{-1}); (c) same with (a) but for vertical velocity (in unit: m · s^{-1}); and (d) the precipitation intensity (in unit: mm · h^{-1}). (After Liao et al., 2005)

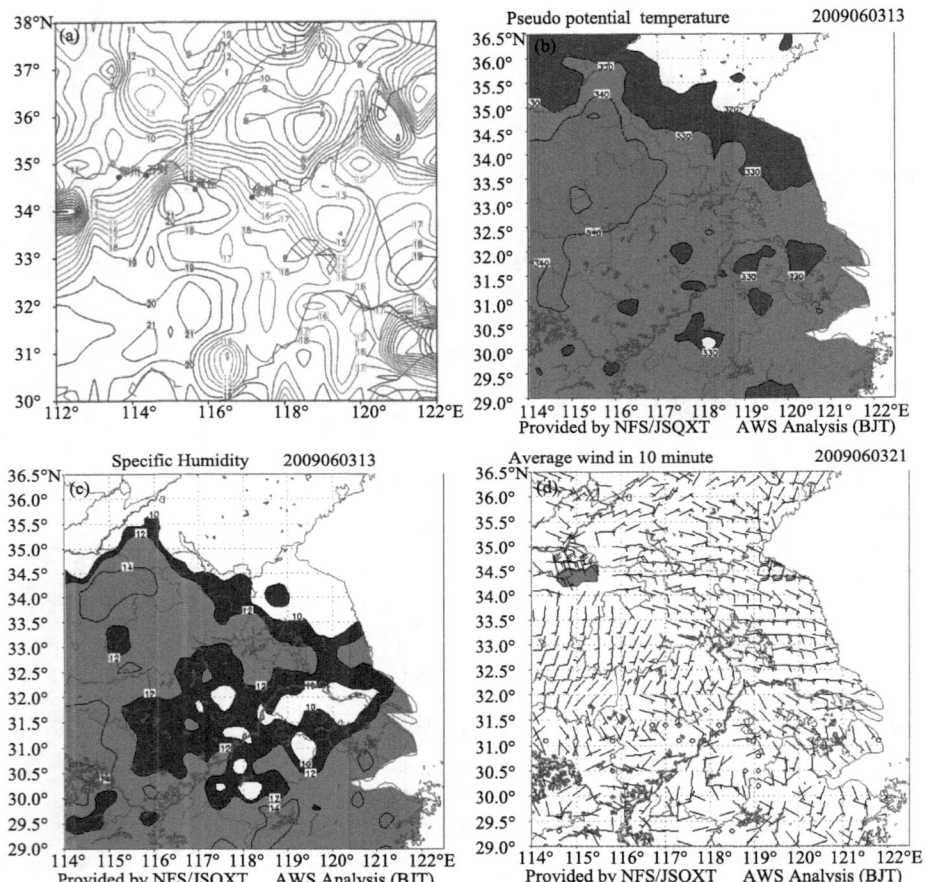

Fig. 8.20 Automatic meteorological station network data analyses on Jun. 3, 2009. (a) dew point temperature. The red point indicates the position of Shangqiu City; (b) pseudo potential temperature; (c) specific humidity; (d) wind field(see the color illustrations).

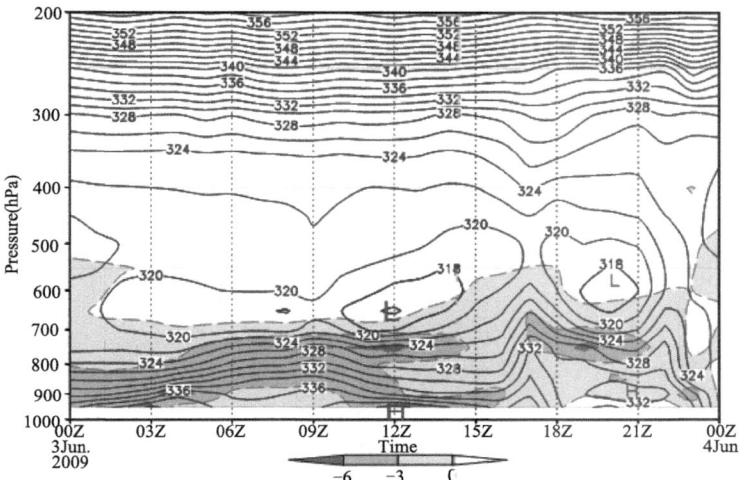

Fig. 8.21 The time-height (pressure) section of pseudo potential temperature Θ_{se} (solid lines, in unit: K) during 00Z Jun. 3—00Z Jun. 4, 2009 over the area near Shangqiu. Letters L and H indicate the low and high value of Θ_{se}, the shadow indicates the instability ($\Delta\theta/\Delta z$). (see the color illustrations)

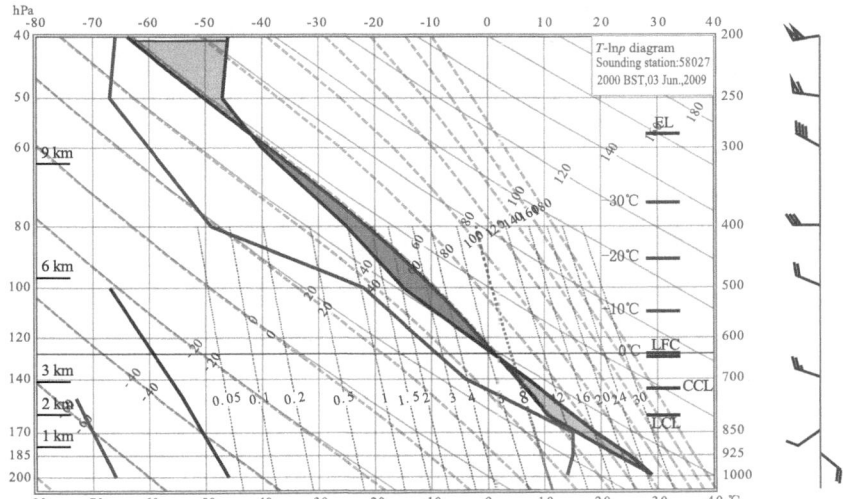

Fig. 8.22 The sounding analysis of Xuzhou near Shangqiu at 2000BST Jun. 3, 2009

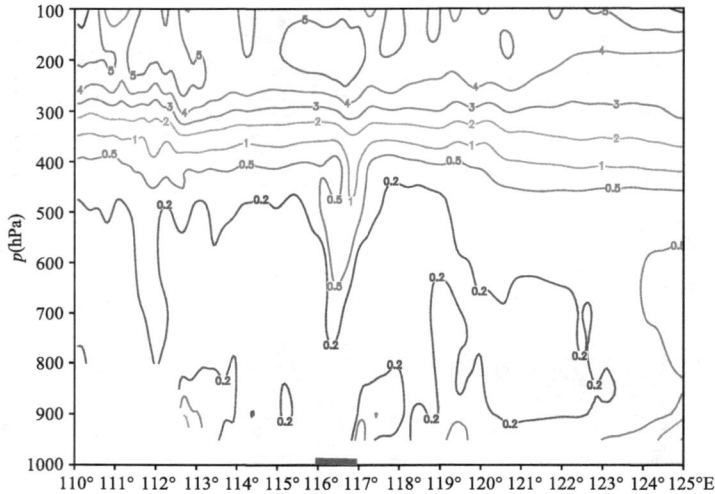

Fig. 8.23 The vertical section of potential vorticity anomaly along the latitude of 34.5°N. at 19BST Jun. 9, 2009. The thick line on the horizontal axis indicates the location of Shangqiu City.

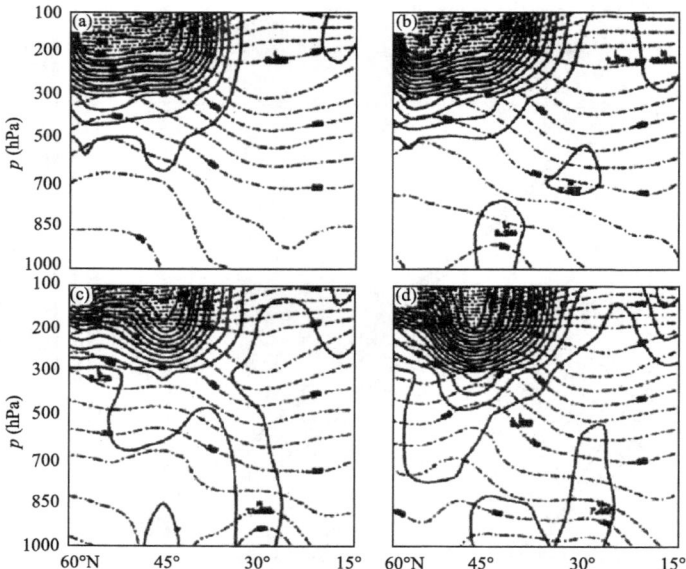

Fig. 8.24 The vertical section of potential vorticity anomaly along the longitude where the cyclone center located at 0500 Jul. 5, 1991 (a); 2000 Jul. 5, 1991 (b); 0500 Jul. 6, 1991 (c); 2000 Jul. 6, 1991 (d), respectively.

References

ARAKAWA A, 1993. Closure assumptions in the cumulus parameterization problem[J]. Meteorological Monographs, 24(46):1-16.

ARAKAWA A, CHENG M D, 1993. The Arakawa Schubert cumulus parameterization[J]. Meteorological Monographs, 24(46):123-136.

BARNES S L, 1972. Mesoscale objective analysis using weighted time series observations[R]. NOAA Technical Memorandum, ERL, NSSL-62.

BARNES S L, COLMAN B R, 1993. Quasigeostrophic diagnosis of cyclogenesis associated with a cutoff extratropical cyclone—The Christmas 1987 storm[J]. Mon Wea Rev, 121(6):1613-1634.

BENNETT D A, HOSKINS B J, 1979. Conditional Symmetric instability——a possible explanation for frontal rainbands[J]. Q J Roy Meteor Soc, 105:945-962.

CHAN D S T, CHO H-R, 1989. Meso-β scale potential vorticity anomalies and rainbands. Part I: Adiabatic dynamics of potential vorticity anomalies[J]. J Atmos Sci, 46(12): 1713-1723.

CHO H-R, CHAN D, 1991. Meso-β-scale potential vorticity anomalies and rainbands. Part II: Moist model simulations[J]. J Atmos Sci, 48(2):331-341.

CHO H-R, CAO Z H, 1998. Generation of moist potential vorticity in extratropical cyclones. Part II: Sensitivity to moisture distribution[J]. J Atmos Sci, 55:595-610.

CRESSMAN G P, 1959. An operational objective analysis system[J]. Mon Wea Rev, 87:367-374.

DAVIES H C, ROSSA A M, 1998. PV frontogenesis and upper tropospheric fronts[J]. Mon Wea Rev, 126:1528-1539.

DAVIES-JONES R, 1991. The frontogenetical forcing of secondary circulations. Part I: The duality and generalization of the Q vector[J]. J Atmos Sci, 48(4): 497-509.

DAVIS C A, 1992. Piecewise potential vorticity inversion[J]. J Atmos Sci, 49(16): 1397-1411.

DAVIS C A, EMANUEL K A, 1991. Potential vorticity diagnostics of cyclogenesis[J]. Mon Wea Rev, 119:1929-1953.

DU J, CHO H-R, 1996. Potential vorticity anomaly and mesoscale convective systems on the Baiu (Mei-Yu) front[J]. J Met Soc Japan, 74(6):891-908.

DUNN L B, 1991. Evaluation of vertical motion: Past, present, and future[J]. Wea Forecasting, 3(1):65-73.

ELIASSEN A, KLEINSCHMIDT E Jr, 1957. Dynamic Meteorology[M]. Handbuch der Physik, Springer-Verlag Berlin/Heidelberg/New York:112-129. http://dx.doi.org/10.1007/978-3-642-45881-1_1.

FRITSCH J M, CHAPOELL C F, 1980. Numerical prediction of convectively driven mesoscale pressure systems. Part I: Convective parameterization[J]. J Atmos Sci, 37(8):1722-1733.

GAMACHE J F, HOUZE R A Jr, 1982. Mesoscale air motions associated with a tropical squall line

[J]. Mon Wea Rev, 110:118-135.

GAO S T, LEI T, ZHOU Y S, 2002. Moist potential vorticity anomaly with heat and mass forcings in torrential rain systems[J]. Chin Phys Lett, 19(6):878-880.

HAKIM G J, KEYSER D, 1996. The Ohio valley wave-merger cyclogenesis event of 25—26 January 1978. Part I : Diagnosis using quasigeostrophic potential vorticity inversion[J]. Mon Wea Rev, 124(10):2176-2205.

HISE E Y, 1987. MM4(Penn State/NCAR) mesoscale of model[Z]. Version 4 Documentation.

HOSKINS B J, DRAGHICI I, DAVIES H C, 1978. A new look at the ω-equation[J]. Q J Roy Meteor Soc, 104:31-38.

HOSKINS B J, MCINTYRE M E, ROBERTSON A W, 1985. On the use and significance of isentropic potential vorticity maps[J]. Q J Roy Meteor Soc, 111:877-946.

HOSKINS B J, PEDDER M, 1980. The diagnosis of middle latitude synoptic development[J]. Q J Roy Meteor Soc, 106(450):707-719.

HUO Z H, ZHANG D L, 1998. An application of potential vorticity inversion to improving the numerical predication of the March 1993 superstorm[J]. Mon Wea Rew, 126(2):426-439.

JOLY A, THORPE A J, 1990. Frontal instability generated by tropospheric potential vorticity[J]. Q J Roy Meteor Sci, 116:525-560.

JUSEM J C, ATLAS R, 1998. Diagnostic evaluation of vertical motion forcing mechanisms by using Q vector partitioning[J]. Mon Wea Rev, 126(8): 2166-2184.

KEYSER D, SCHMIDT B D, DUFFY D G, 1992. Quasigeostrophic vertical motions diagnosed from along-and cross-isentrope components of the Q vector[J]. Mon Wea Rev, 120(5):731-741.

KRISHNAMURTI T N, 1968. A diagnostic balance model for studies of weather systems of low and high latitudes, Rossby number less than 1[J]. Mon Wea Rev, 96(4): 197-207.

LEFTWICH P W Jr, 1990. On the use of helicity in assessment of severe local storm potential[C]. Preprints, 16th of Conf. on Severe Local Storm, Am Meteor Soc, 306-310.

LI Y H, SHOU S W, FAN K, 2002. Isentropic potential vorticity analysis on the mesoscale cyclone development in a torrential rain process[J]. Acta Meteorologica Sinica, 16(4):75-85.

LILLY D K, 1986. The structure, energetics and propagation of rotating convective storm. Part II : Helicity and storm stability[J]. J Atoms Sci, 43(2):126-140.

MADDOX R A, 1980. An objective technique for separating macroscale and mesoscale features in meteorological data[J]. Mon Wea Rev, 108:1108-1121.

MOLINARI J, DUDEK M, 1992. Parameterization of convective precipitation in mesoscale numerical models: a critical review[J]. Mon Wea Rev, 120(2):326-344.

MOLINARI J, SKUBIS S, VOLLARO D, et al, 1998. Potential vorticity analysis of tropical cyclone intensification[J]. J Atmos Sci, 55:2632-2644.

MOLLER J D, JONES S C, 1998. Potential vorticity inversion for tropical cyclones using the asymmetric balance theory[J]. J Atmos Sci, 55:259-282.

NORDENG T E, 1987. The effect of vertical and slantwise convection on the simulation of polar lows [J]. Tellus, 39A:354-357.

NORDENG T E, 1993. Parameterization of cumulus convection in numerical weather prediction models. The representation of cumulus convection in numerical models[J]. Meteorological Monographs, 24(46):195-202.

O'BRIEN J J, 1970. Alterative solutions to the classical vertical velocity problem[J]. J Appl Meteor, 9(2):197-203.

PLATZMAN G W, 1949. The motion of barotropic disturbances in the upper troposphere[J]. Tellus, 1(3):53-64.

QIN X, 1989. Frontal circulation in the presence of small viscous moist symmetric stability and weak forcing[J]. Q J Roy Meteor Soc, 115:1325-1352.

QIN X, 1992. Conditional symmetric of frontal rainbands and geostrophic potential vorticity anomalies [J]. J Atmos Sci, 49(8):629-648.

REED R J, SANDERS F, 1953. An investigation of the development of a mid-tropospheric frontal zone and its associated vorticity field[J]. J Meteor, 10:338-349.

ROBERT D J, BURGESS D W, 1990. Test of helicity as a tornado forecast parameter[C]. Preprints, 16th of Conf. on Severe Local Stroms, Am Meteor Soc, 588-592.

SANDERS F, HOSKINS B J, 1990. An easy method for estimation of Q vectors from weather maps [J]. Wea Forecasting, 5(2):346-353.

SCHAR C, WERNLI H, 1993. Structure and evolution of an isolated semi-geostrophic cyclone[J]. Q J Roy Meteor Soc, 119(509):57-90.

SHAPIRO R, 1970. Smoothing filtering and boundary effects[J]. Rev of Geophys Space Phys, 8: 357-387.

SHOU S W, YAN F X, ZHANG Y L, et al, 2004. Effects of dry intrusion in a rainstorm process [C]. International Conference on Storms, 5—9 July, 2004, Brisbane Australia.

SHOU S W, LIAO S S, ZHANG Y L, et al, 2004. Numerical simulation and diagnostic analysis of a rainstorm near Meiyu front occurred in eastern China[C]. International Conference on Storms, 5—9 July, 2004, Brisbane Australia.

SHOU S W, LI Y H, 1999. Study on moist potential vorticity and symmetric instability during a heavy rain event occurred in the Jiang-Huai Valleys[J]. Adv Atmos Sci, 16(2):312-321.

SHOU S W, LIU Z X, 2002. A diagnosis analysis of the inferences of dry intrusion on the development of cyclone during a heavy rain process[C]. Proceedings of Summer Workshop On Severe Storms And Torrential Rain, Chengdu, China.

SHOU Y X, LI S S, SHOU S W, et al, 2006. Application of a cloud texture analysis scheme to the cloud cluster structure recognition and rainfall estimation in a mesoscale rainstorm process[J]. Adv Atmos Sci, 23(5): 767-774.

SHOU Y X, LU F, SHOU S W, 2016. Satellite assessments of tropopause dry intrusions correlated to mid-latitude storms[J]. Atmosphere, 7(10):128. doi:10.3390/atmos7100128.

WANG W, SEAMAN N L, 1997. A comparison of convective parameterization schemes in a mesoscale model[J]. Mon Wea Rev, 125(2):252-278.

WILLIAM F M, 1983. The cumulus parameterization problem [J]. Mon Wea Rev, 111 (9): 1859-1871.

WOODALL G R, 1990. Qualitative analysis and forecasting of tornadic activity using storm-relative helicity[C]. Preprints, 16th of Conf. on Severe Local Strom, Am Meteor Soc, 311-315.

XU Q, 1992. Ageostrophic pseudovorticity and geostrophic C-vector forcing-a new look at Q vector in three dimensions[J]. J Atmos Sci, 49(12):981-990.

YANG D S, KRISHNAMURTI T N, 1981. Potential vorticity of monsoonal low level flows[J]. J Atmos Sci, 38:2676-2695.

YAO X P, YU Y B, SHOU S W, 2004. Diagnostic analyses and application of the moist ageostrophic vector Q[J]. Adv Atmos Sci, 01:96-102.

YUE C J, SHOU S W, 2008. A modified moist ageostrophic Q vector[J]. Adv Atmos Sci, 2008, 06: 1053-1061.

YUE C J, SHOU S W, Li X F, 2009. Water Vapor, Cloud, and surface rainfall budgets associated with the landfall of Typhoon Krosa(2007): A two-dimensional cloud-resolving modeling study [J]. Adv Atmos Sci, 06:1198-1208.

Chapter 9
Mesoscale Weather Forecasting

The mesoscale weather systems may cause severe weather, so that the forecasting of the mesoscale systems is very important for preventing and fighting the natural adversities. In this chapter the concepts of nowcasting and very short range forecasting (VSRF) and the forecasting methodology for heavy rain and severe convective weather etc. will be introduced.

§ 9.1 Methodology of mesoscale weather forecasting

9.1.1 Nowcasting and very short range forecasting (VSRF)

According to the suggestion of the Atmospheric Science Committee, WMO in 1981, two new kinds of forecasting, the nowcasting and very short range forecasting (VSRF) with the forecasting validity periods of 0—2 h and 0—12 h respectively, are added, beside the traditional long range forecasting, middle range forecasting and short range forecasting. Their forecasting objects are mesoscale systems especially the mesoscale convective systems and their relative weather phenomena including heavy rain and severe convective weather such as hailstorm, tornadoes, and damage windstorms etc. Nowcasting and VSRF have presented their distinguish importance in preventing the natural adversities and rising the economic benefits.

As mentioned above that the forecasting validity periods of nowcasting and VSRF are 0—2 h and 0—12 h respectively. The length of the forecasting validity periods is determined by the two basic weather forecasting methodologies, i. e. the linear extrapolation method and the nonlinear forecasting methods.

The basic idea of the linear extrapolation is that to suppose the weather system will keep its past evolution tendency and then its position and intensity in future may be derived based on the tendency. This method is only valid in a certain time period, which is called "valid time period of extrapolation". In the valid time period the error between the forecasting value and the observation is kept in an allowed scope, while the forecasting will be no longer valid beyond the scope (as shown in Fig. 9.1). In general speaking, the length of the valid time period of extrapolation is only about 1/4 of the life cycle of the weather system. For the synoptic scale systems with the life cycle longer than 3 days, their valid time period of extrapolation may be as long as 18 hours or longer. While for the mesoscale systems with the life cycle only few hours to ten hours, their valid time periods of extrapolation normally are only about 0—2 hours.

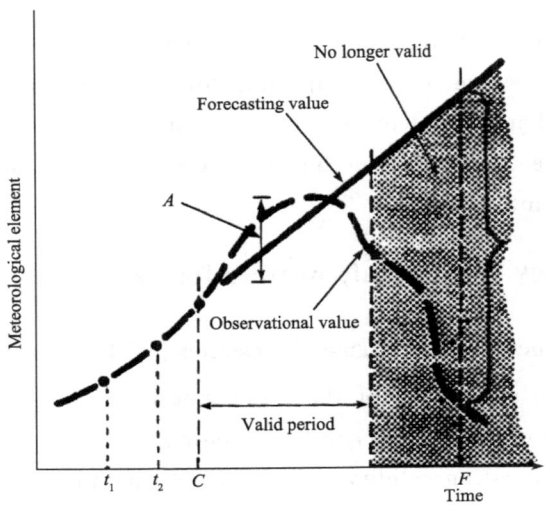

Fig. 9.1 An illustration of the conception of the valid time period of extrapolation. (The vertical axis indicates the meteorological element, horizontal axis is time, A is the scope of the allowed maximum forecasting error, C is the initial time to make forecasting, F is the time forecasted.)(Adapted from McGinley, 1986)

The forecasting ability of the non-linear forecasting represented by the numerical model forecasting is normally decreased with the decreasing of the horizontal scale of foresting objects and the shortening of the forecasting validity periods. In general speaking, the global numerical model has good accuracy for the forecasting of 12—72 h or longer. However, ability of the forecasting with the validity periods shorter than 12 hours is normally lower than the climatological forecasting.

Based on the above simple analysis about the linear extrapolation method and non-linear numerical model forecasting method we know that for the some mesoscale and small scale weather systems, we may use the linear extrapolation method to make forecasting with the validity period of 0—2 h by employing the detail observational and detecting data in their valid time period of extrapolation. This forecasting is called "nowcasting". Meanwhile, it can be seen from Fig. 9.2 that for the large scale numerical model there is the gap of the forecasting ability in the scope 0—12 h of the validity period. This needs to develop new forecasting methods to fill the mesoscale gap. Normally, the forecasting with the validity period of 0—12 h is called very short range forecasting (VSRF). There is a fold time period (0—2 h) between nowcasting and VSRF, but they are different. The later is non-linear forecasting emphasized the genesis of the system, while not only the extrapolation. There are distinct differences between VSRF and the conventional short range forecasting. A comparison of them is shown in Table 9.1.

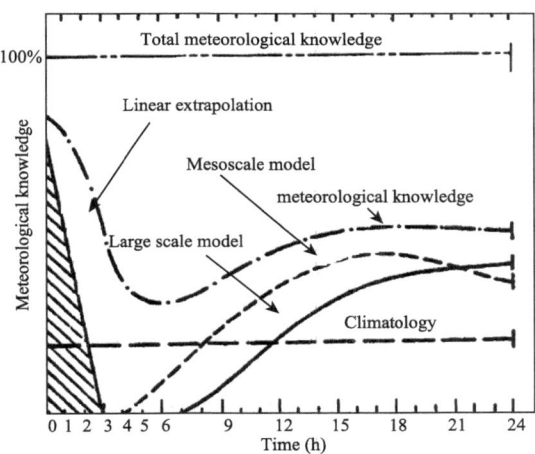

Fig. 9.2 An illustration of the forecasting ability gap of the large scale numerical model in the forecasting validity period of 0—12 h. (Adapted from Doswell, 1986)

Table 9.1 A comparison between the conventional short and mid-long ranges forecasting and the VSRF and Nowcasting. (Adapted from Tao, 1986)

Items	The conventional short and mid-long ranges forecasting	Nowcasting and VSRF
The forecasting validity period	>12 h	0—12 h
Scale concerned	Synoptic scale, planetary scale	Mesoscale
Scope concerned	Global or continental	Small region or local area
Nature of forecasting	General roughly	Confirm forecasting of location and certain meteorological elements
Observational data	Conventional surface data and sounding station network data	Surface mesoscale observational network data, satellite data, radar and other remote sensing data other surface data and upper air sounding data

Continued

Items	The conventional short and mid-long ranges forecasting	Nowcasting and VSRF
Density of the observational station network	Several hundred kilometers	$\leqslant 50$ km
Time apart of the observations	Surface obs. 3 h once, upper air obs. 12 h once.	Surface obs. 1 h, 30 min or few minutes once
Data quantity	Relatively small, about 10^6 bits $\cdot$ h^{-1}	Large quantity, about 10^8 bits $\cdot$ h^{-1}
Transport speed of data	Slowly (from several minutes to several hours)	quickly (from few seconds to few minutes)
Forecasting methods	Numerical prediction and statistics forecasting methods	Observation + extrapolation, mesoscale conceptual models, forecaster's experiences, numerical prediction and statistics forecasting
Forecast transmiting	Slow, in passivity way	Quick, in initiative and positive ways

The various forecasting including mid-long range forecasting, short range forecasting, VSRF and nowcasting are all important for weather forecasting but plays different roles. The short and mid-long range forecasting may normally provide roughly potential forecasting in future while the accurate fine weather forecasting with exact location, time and intensity will be provided by VSRF and nowcasting.

9.1.2 Basic types of the mesoscale weather forecasting methods

Many ways such as the linear extrapolation, the mesoscale numerical forecasting, the climatological methods and the conceptual models based on the mesoscale meteorological knowledge etc. may make contribution for closing the ability gap of the 0—12 h forecasting.

First, the linear extrapolation methods are usually used for forecasting the successive displacement of mesoscale features in the satellite and radar data and estimating the rainfall and variation of the weather phenomena. When we extrapolate the position of an echo or cloud mass at the moment $t+\Delta t$ based on the digital radar and satellite data, it is normally to use the method by calculating the crossing correlation coefficient. For the scattered echoes or clouds, it is normally to extrapolate their main part. The environmental flow is usually regarded as the steering flow to determine the basic moving direction and velocity of the echo and cluster and then to make some necessary revision. The valid time period of the extrapolating forecasting of the convective echoes is normally very short and the forecasting accuracy decreases nearly exponentially with the increasing of the forecasting time.

The extrapolation method may also be used to forecast many other features such as the thunderstorm lightning area detected by the lightning detecting instrument, the special echo features such as hock echo, bow echo, etc. in the radar reflectivity picture and the mesoscale cyclone, downburst shear lines etc. in the Doppler radar data. Many of the above features have very short life cycles, so that their extrapolation valid time periods are very short, sometimes maybe only few minutes even few seconds, but it is still valuable to avoid life losses.

Second, the mesoscale numerical model is the represent of the non linear forecasting. Since the conventional large scale numerical model forecasting has no valid for 0—12 h forecasting, it is important to develop the mesoscale numerical models. In recent years many mesoscale numerical models are developing quickly.

Third, the climatological forecasting is also valuable for raising the accuracy especially in some locations with special topographical effecting such as the areas near the coast, mountains, lakes etc. To know the climatological background is important for assuring certain forecasting accuracy. In the special topographic area some phenomena may frequently occur. For example, over some large lake the snowstorm easy to be formed and in some mountain area easy to form hailstorms or rainfall enhancement etc. The climatological mean fields may be introduced for correcting the forecasting results or as a predictor putting into the prediction equation when making the statistics forecasting.

Finally, the forecasting method by using the meteorological knowledge is also one of the important lings to close mesoscale forecasting gap. By using the conceptual models and empirical rules leaded from the research and operational practices, the human factors are involved in the forecasting process. However, the mesoscale meteorological knowledge of people is still insufficient so far and needs to be improved.

9.1.3 Application of the conceptual models

The conceptual models is the concentrated reflection of the knowledge about the structure, mechanism and life cycle of the observed phenomena. According to the conceptual model the observational features and the possible variation in future can be explained and expected. For example, Fujita (1978) presented a model about the formation and variation process of a bow echo as shown in Fig. 5.27. According to the model we know that the initial severe convective echo will develop into a bow echo and form cyclonic and anti-cyclonic rotations at two ends of the bow echo. The cyclonic rotation will often increase with time to form the rotating head and a comma echo. The downbursts are usually founded near the front of the bow echo and the hock echo at the rotating head. By this model we can determine the occurring possibility and location of the downburst based on the radar echo features.

Leary and Houze(1979) proposed a life cycle model of the mesoscale convective system as shown in Fig. 9.3, which shows the variation of the mixture of the convective and stratiform cloud precipitation during the developing period of the mesoscale systems in tropical and mid-latitude region. At the initial stage (t) and intensifying stage $(t+3\text{ h})$ the stronger convective cells played major role. At the mature stage $(t+6\text{ h})$ there is the mixture precipitation of the convective clouds and stratiform clouds. At the dissipating stage $(t+9\text{ h})$, it is majorly the light stratiform precipitation. In the satellite picture, the feature at dissipating stage is the broad high clouds and possibly the wide distributed lightning. For this situation the forecasters, who is not familiar the model, may possibly make mistake for thinking the frequent lightning phenomena as the successive outbursts of the disaster weather.

Except above examples, many other conceptual models mentioned in previous such as the structure models of the isolated convective systems, the squall lines, the mesoscale rain bands near the frontal cyclones and typhoons, and the MCC etc. may be applied for making nowcasting and VSRF.

9.1.4 Application of the empirical rules

Although the objective forecasting methods have played more and more important roles in the mesoscale weather forecasting nowadays, the empirical rules are still needed to be applied. Following are some useful empirical rules applied in the convective weather forecasting by the forecasters.

The formation of the common convective weather normally needs three conditions, i.e. the water vapor, convective instability and initiating mechanisms. The pre-convection conditions may be summarized as following: ① there is a moist layer near the surface; ② the atmosphere layer is unstable with the higher equivalent potential temperature θ_e and lower θ_e in the middle layer; ③ there is a stable layer or a capping temperature inversion layer playing the effect to keep the instability energy; and ④ there is a mechanism for triggering the convection.

The low level water vapor condition may be determined based on the analyses of

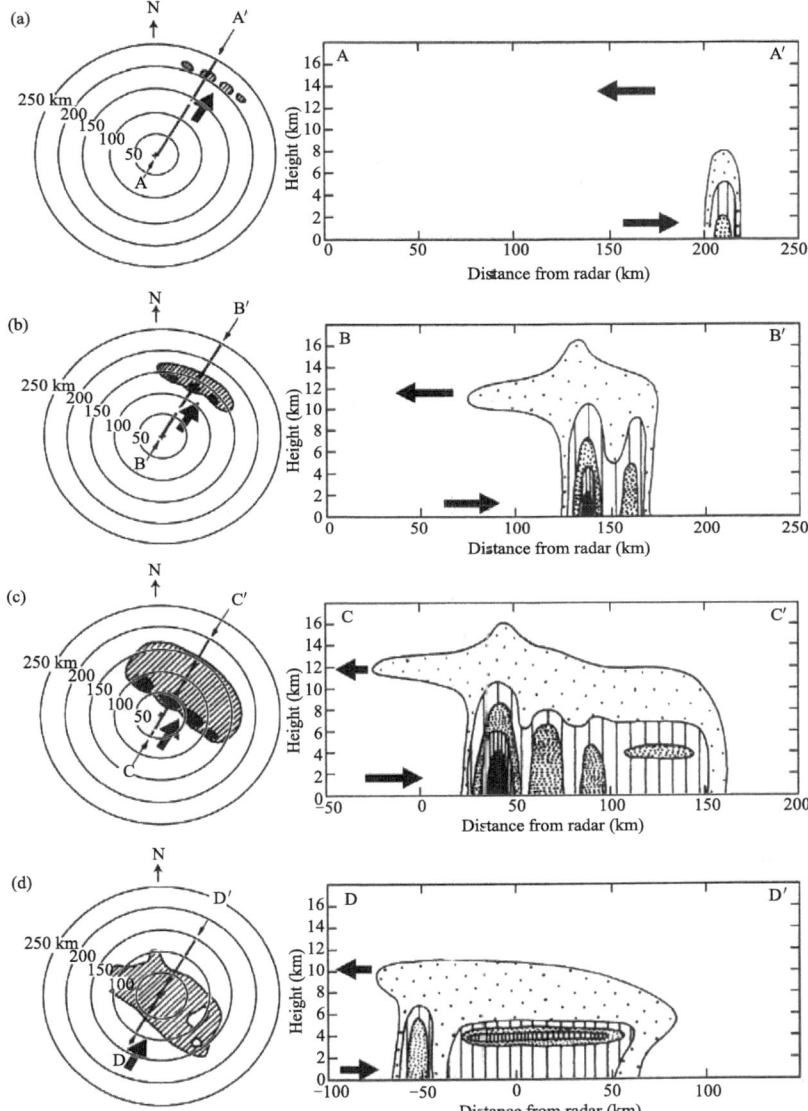

Fig. 9.3 The conceptual model of the horizontal and vertical sections of the radar reflectivity at the different stages of the mesoscale convective systems' life cycle. (a) initial stage; (b) intensifying stage; (c) mature stage; and (d) dissipating stage. (the convective heavy rain core indicated by shading area, and the stratiform precipitation area indicates by the characteristics with horizontal uniform bright belt)(Adapted from Leary and Houze, 1979)

the visible cloud picture or the infrared images for making the short and very short range forecasting. The instability may be determined by sounding data. Since two soundings normally have a 12 hours interval, so that it is necessary to analyze the variation of the instability during the time period by using all the data obtained possibly, for instance, the hourly surface observational data of temperature, cloud base, cloud top and other meteorological parameters as well as the radar and aircraft reports etc. The variation of the surface temperature and moisture may reflect the variation of the vertical structure of atmosphere. For example, when the surface temperature and moisture increase may signify the exist of the capping inversion layer. The visibility decreasing may indicate the moisture increasing and the capping inversion intensifying. While the surface heating rate and humidity decreasing possibly means the capping inversion layer has been eliminated. The shrinking or extruding at the mid part of the cumulus cloud body may indicate the cumulus has penetrated the inversion layer (as shown in Fig. 9.4b).

Fig. 9.4 The speculated atmospheric vertical structure based on the surface observation. (a) The surface temperature decreasing indicates the capping inversion layer has possibly been removed; (b) the shrinking or extruding may indicate the cumulus has penetrated the inversion layer; (c) A is the sounding outside the storm, the cloud bottom corresponding to the CCL (convective condensation level), B is the sounding under the storm, the cloud bottom corresponding to the lifted condensation level (LCL). (Adapted from McGinley, 1986)

The initiating mechanisms may be various discontinuity lines such as the fronts, convergence line, shear line etc. The observational surface wind field analysis should be paid much attention based on which the locations of the convergence area, outflow boundary and low level jet flow etc. may be determined. The convections are usually closely related with these wind field features. The wind vector variation diagram is a useful tool for forecasting. The convective weather is often found in the convergence area in the wind vector variation. The initiating mechanisms consist of two scales' effects. One of them is the effect for removing stability. This effect is given by the mesoscale or sub synoptic scale motions within the 3—12 h time period. It will cause the original stable atmosphere to be lifted and become unstable gradually. Another effect is the triggering mechanism, which is normally the thermal bubbles with 1 m·s^{-1} order of vertical motion and to cause the weakening capping inversion layer destroyed so that the convection breaks out. When the mesoscale system is strong, the situation can not only play the role to remove stabile layer, but also play the role to cause the convection directly. While under the situation that the mesoscale system is weaker, the triggering effect of the local thermal bubbles plays important role.

When the forecasters considering the initiating conditions, they should pay attention to the variation of the capping inversion layer at first. If the surface temperature increased quickly, it means the existing of the capping inversion and its strongest area is located near the area with highest temperature. Then they should pay attention to the mechanisms for removing the stability. The front, dry line, old outflow boundary, mesoscale convergence line or convergence zone, shear line, warm advection area, explosive moisture enhancement area etc. are all possibly to be the stability removing mechanisms.

General speaking, forecasters should carefully analyze the morning sounding to decide the intensity of the capping inversion layer and determine the possible stability removing mechanisms by using various data.

The pre-storm conditions for the severe convective storms are similar to the common convective storms as mentioned above, but there are two special conditions, i. e. ① the dry air over the low level moist layer; and ② the strong vertical wind shear, may play important role for the formation of the severe storms. These two conditions plus the strong instability and the strong capping inversion layer may act as filter effect to cause the severe convective be filtered from the common convective storms.

§ 9.2 Diagnosis and forecasting of heavy rain

9.2.1 Definition and types of heavy rain

Heavy rain is generally referred to the event with strong precipitation rate and greater rainfall. In operational work the heavy rain is normally defined by 24 hour rainfall (R_{24}). The standard of the heavy rain may be different. In the Yangtze River area of China the standard of heavy rain is normally set as $R_{24} \geqslant 50$ mm and is furtherly divided into four classes: 50.0—99.9 mm, 100.0—199.9 mm, 200.0—399.9 mm, and >400.0 mm. The standard for heavy rain may be different in different areas, but some people suggests to take the R_{24} value greater than the 1/15 of the mean annual rainfall as the unified standard of the heavy rain for different locations. In general speaking, when the value of R_{24} is big, then correspondingly the 1 hour rainfall is also big. For example, in the Yangtze River area for the R_{24} of 50.0—99.9 mm and 100.0—199.9 mm, the corresponding one hour rainfall is about 10.0—20.0 mm and 20.1—40.0 mm respectively. But such strong rainfall is normally concentrated only in a few hours. In a heavy rain process the temporal and spatial distributions of rainfall are not uniformly. The heavy rain processes may be classified into different types such as the local or large scope precipitation, short lasting or successive precipitation, and fast moving or stationary precipitation etc. according to their temporal and spatial scales and movement. The heavy rain processes may be classified into different types such as typhoon rainstorm, frontal rainstorm etc. according to their influence systems.

9.2.2 Conditions of the heavy rain formation and their diagnoses

Set the sign "I" as the precipitation rate or precipitation intensity, then "I" can be expressed as following:

$$I = -\frac{1}{g}\int_0^{p_0} \frac{dq_s}{dt} dp \tag{9.2.1}$$

where q_s is the saturated specific humidity. And W, the total rainfall in the time period t_1-t_2, may be expressed as

$$W = -\frac{1}{g}\int_{t_1}^{t_2}\int_0^{p_0} \frac{dq_s}{dt} dp dt \tag{9.2.2}$$

Since

$$\frac{dq_s}{dt} = \delta F \omega \tag{9.2.3}$$

where ω is vertical velocity, $\omega<0$ indicates rising. When $q \geqslant q_s$, and $\omega<0$, $\delta=1$; when $q>q_s$, or $\omega>0$, $\delta=0$. F is the condensation function, which can be expressed as following:

$$F = \frac{q_s T}{p}\left(\frac{LR - C_p R_w T}{C_p R_w T^2 + q_s L^2}\right) \tag{9.2.4}$$

where the parameters q_s, T, p, R, R_w, L, C_p represent the saturated specific humidity, temperature, pressure, gas constant, gas constant of water vapor, evaporation (condensation) latent heat and specific heating at constant pressure respectively. F is the function of humidity. By substituting equation (9.2.3) into equation (9.2.2), we can get:

$$W = -\frac{1}{g}\int_{t_1}^{t_2}\int_0^{p_0} \delta F\omega \, dp \, dt \qquad (9.2.5)$$

According to the equation (9.2.5) we know the total rainfall is the function of humidity, vertical velocity and the time duration ($\Delta t = t_2 - t_1$) of the precipitation from time t_1 to t_2, i.e.:

$$W \propto (q_s, -\omega, \Delta t) \qquad (9.2.6)$$

Hence, the basic conditions for the formation of heavy rain include the abundant water vapor, strong vertical and convective motion, which implying the higher atmospheric instability, as well as the favorable situation to keep the precipitation stationary for longer time period.

The water vapor condition may be indicated by using the parameters such as specific humidity (q), dew point temperature (T_d), the difference between temperature (T) and dew point temperature (T_d), $D = T - T_d$, etc. Besides, "the precipitable water vapor (R_p)" is expressed as

$$R_p = \int_0^\infty \rho q \, dz \qquad \text{or} \qquad R_p = \frac{1}{g}\int_0^{p_0} q \, dp \qquad (9.2.7)$$

which may be used for indicating the water vapor contained in whole atmosphere layer; "the water vapor flux ($\mathbf{F}$)" and "the water vapor flux divergence (D)" are expressed respectively as

$$\mathbf{F} = \int_0^\infty \rho q \mathbf{V} \, dz \qquad \text{or} \qquad \mathbf{F} = \frac{1}{g}\int_0^\infty q \mathbf{V} \, dp \qquad (9.2.8)$$

And

$$-D = -\frac{1}{g}\int_0^{p_0} \nabla \cdot (q\mathbf{V}) \, dp \qquad (9.2.9)$$

which may be used for indicating the water vapor transportation and concentration respectively.

The instabilities closely related with the heavy rain are expressed by the instability energy parameters such as the moist static energy, the convective available energy (CAPE), and MCAPE, NCAPE, CIN, DCAPE, TCAPE, SCAPE, LI, SI, SSI, TT, A index, K index and so on, which are mentioned in the previous chapters, may all be used for the heavy rain analysis and diagnosis.

The diagnoses of divergence, vorticity and vertical velocity play the important roles in heavy rain analysis and forecasting.

Normally we can use the Bellamy's three points method, the three points flux method, the square mesh method and the longitude-latitude grid method etc. to calculate the divergence and vorticity. When using the Bellamy's three points method, the formulas for calculating divergence (D) and vorticity (ζ) may be expressed as follows respectively:

$$D = \frac{V_A}{H_A}\cos(\gamma_A - \alpha_A) + \frac{V_B}{H_B}\cos(\gamma_B - \alpha_B) + \frac{V_C}{H_C}\cos(\gamma_C - \alpha_C) \quad (9.2.10)$$

$$\zeta = \frac{V_A}{H_A}\sin(\gamma_A - \alpha_A) + \frac{V_B}{H_B}\sin(\gamma_B - \alpha_B) + \frac{V_C}{H_C}\sin(\gamma_C - \alpha_C) \quad (9.2.11)$$

where A, B, C are the vertex points of a triangle; H_A, H_B, H_C are the heights from the vertex points to their opposite sides respectively; γ_A, γ_B, γ_C are the direction angles of the heights; α_A, α_B, α_A are the wind direction angles at the vertex points of the triangle; V is the total wind speed.

By using the square mesh method, the formulas for calculating divergence and vorticity may be expressed as follows respectively:

$$D_0 = \frac{\Delta u}{\Delta x} + \frac{\Delta v}{\Delta y} = \frac{1}{2d}(u_1 - u_3 + v_2 - v_4) \quad (9.2.12)$$

$$\zeta_0 = \frac{\Delta v}{\Delta x} - \frac{\Delta u}{\Delta y} = \frac{1}{2d}(v_1 - v_3 - u_2 + u_4) \quad (9.2.13)$$

where the subscripts 0, 1, 2, 3, 4 are represents of the grid points at the center, east, north, west and south respectively, and d is the grid interval.

Upper level divergence and low level convergence will be favorable to cause ascending motion and favorable to form precipitation. The difference of the divergence is called "relative divergence". The following equation represented the relationship between the relative divergence ($\Delta D = D_2 - D_1$) and the rainfall (R):

$$R \approx -\frac{p_0}{g}\frac{dq_1}{dt} + \frac{p_0 q_1}{g}(D_2 - D_1) \quad (9.2.14)$$

where p_0 is the surface pressure; g is the gravity acceleration; q_1 is the mean specific humidity at low level; D_2 and D_1 are the divergence at upper and lower levels respectively.

The vertical velocity is normally calculated by using the continuity equation as follows:

$$\omega_p = \omega_0 + \int_p^{p_0} D\,dp \quad (9.2.15)$$

where ω_p, ω_0, D are the vertical velocities at constant pressure (p) surface and at the ground surface and the divergence respectively. As to the vertical velocity caused by the topographic raising the calculating formula may be expressed as following:

$$w_l = \mathbf{V}_l \cdot \nabla h = u_l \frac{\partial h}{\partial x} + v_l \frac{\partial h}{\partial y}$$

or

$$\omega_l = -\rho_l g \left(u_l \frac{\partial h}{\partial x} + v_l \frac{\partial h}{\partial y} \right) + \rho_l g \zeta_l \sqrt{\frac{K}{2f}} \sin 2\mu \quad (9.2.16)$$

The vertical velocity may be calculated by using ω equation, Q vector divergence as mentioned in previous chapters.

9.2.3 Forecasting methods of heavy rain

In the operational service of meteorology there are many methods used for forecasting heavy rain processes. Some of them are introduced as follows.

(1) Comprehensive superposition method

The "comprehensive superposition" method is used for determining the location or the surrounding areas of the heavy precipitation by putting the favorable factors for forming heavy rain together in a same synoptic map. For example, the area comprehensively with the negative value of Showalter index (SI), the moist tongue at 850 hPa with the specific humidity greater than 15 g $\cdot$ kg^{-1}, or the water vapor convergence area, the area with negative vorticity at 300 hPa and positive vorticity at 850 hPa may be regarded as the most favorable area to generate the heavy rain process.

(2) The grouped index method

Heavy rain is a comprehensive result effected by various favorable factors. We may combine the factors together to make a comprehensive parameter, which is called the grouped index, to be used for heavy rain forecasting. As an example, the grouped index Q is defined as following:

$$Q = h\zeta - hD + J_{sw} - \frac{3SI}{1 + (T - T_d)_{850}} \quad (9.2.17)$$

where

$$h\zeta = V_{AS} + V_{BS} + V_{CS}$$
$$hD = V_{AN} + V_{BN} + V_{C_nN}$$

where A, B, C are the vertex points of the triangle $\triangle ABC$ and represent Nanjing, Changsha and Fuzhou stations respectively as shown in Fig. 9.5. V_{AS}, V_{BS}, V_{CS} represent respectively the wind components of the three stations along the direction S, i.e. the direction parallel to the opposite side of the triangle; V_{AN}, V_{BN}, V_{C_nN} represent respectively the wind components of the three stations along the direction N, i.e. the direction perpendicular to the opposite side of the triangle; J_{Sw} represents the projection of the wind at Changsha station in the direction to the southwest-northeast direction. $J_{Sw} > 0$

means the projection pointed to the northeast direction, which is the favorable situation for precipitation; if $J_{Sw}<0$ (i. e. the projection pointed to southwest direction), then to set J_{Sw} as 0; SI is the Showalter index; $(T-T_d)_{850}$ is the difference of temperature and dew-point temperature at 850 hPa constant pressure surface. According to the experiments of forecasting, $Q=12$ can be regarded as a critical value. When $Q\geqslant 12$, it is greatly possible to occur heavy rain in Zhejiang Province in 24 hours.

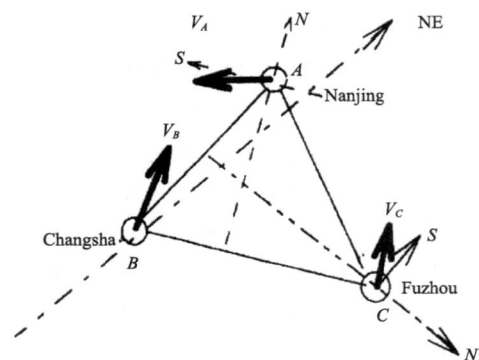

Fig. 9.5 An illustration of the regular triangle $\triangle ABC$ formed by Nanjing, Changsha and Fuzhou stations.

(3) The expert system for heavy rain forecasting

The expert system for heavy rain forecasting is a kind of computer's software system with artificial intelligence. It consists of knowledge base, data base, inference engine, man-machine dialog and learning system etc. The knowledge base contains amount of the expert experiences about heavy rain forecasting. Each of them are expressed by the computer's language in form IF E THEN H(CF). Where E is precondition, H is Conclusion or hypothesis, CF is reliability. If the precondition is not singular, for example there are E1,E2,E3 etc., then it can be expressed as AND or OR, and therefore it may be written as the following forms

    IF E1 AND E2   THEN   H(CF)
    IF E1 AND(E2 OR E3) THEN   H(CF)

These languages may be also expressed by the following logical implication forms:

$$E \xrightarrow{CF} H$$

$$E1 \wedge E2 \xrightarrow{CF} H$$

$$E1 \wedge (E2 \vee E3) \xrightarrow{CF} H$$

Finally link up all the individual expressions in the semantic network form to make them as an entirety.

(4) MOS

MOS(model output statistics) is one of the interpretation methods of numerical prediction products. At first to select the predictor vectors $\hat{x}_t$ from the historical data base of the numerical prediction products and then to derive the statistics relation between the predictand vectors $\hat{y}_t$ and the simultaneous predictor vectors $\hat{x}_t$ as following:

$$\hat{y}_t = f(\hat{x}_t) \tag{9.2.18}$$

Input the numerical prediction results of the predictor vectors $\hat{x}_t$ into the equation the predictand $\hat{y}_t$ can be derived. MOS method has been widely applied for precipitation and other meteorological elements predictions.

(5) MD, MDC or MDCE predict methodology

Where M indicates the output of the numerical prediction model, D indicates the diagnosis analysis, C indicates the satellite cloud data, E is the experiences of experts. MD means the method combined M and D; MDC means the method combined M, D and C; MDCE means the method combined M, D, C and E.

(6) Ingredients based Methodology

Doswell (1996), John and Doswell (1992) suggested a forecasting ideal based on the available model output, which is called the ingredients based methodology and used for making the forecasting of heavy rain and severe convective weather. According to this ideal, we have to understand the physical mechanisms of the heavy rain and severe convective weather formation at first. The major factors and physical quantities to cause the severe weather is called components or ingredients. Secondly it is necessary to analyze the synoptic scale and mesoscale environment fields at the time previous the formation of the severe weather and to determine the danger area that the severe weather may occur. And then to trace the variation of the ingredients and the danger area based on the numerical prediction products. Hence, the ingredients based methodology is actually a kind of interpretation methods of the numerical prediction products.

§ 9.3 Analysis and forecasting of the severe convective weather

9.3.1 Definition of the severe convective weather

The thunderstorms can be divided into common thunderstorm and severe thunderstorm according to their intensities of the relative weather. The typical weather phenomena of the common thunderstorm are thunder, lightning, gust and shower or sometime small hails etc., while the severe thunderstorms are normally companied with strong electronic activity, high precipitation rate, bigger hailstones, damage wind, and sometimes tornadoes etc. The severe thunderstorm is usually called severe convective weather.

The standard of the severe convective weather is not unique. In different country maybe the standard is different. According the National Weather Service (NWS) the severe convective weather normally implies the weather accompanying with the straight wind velocity over 26 m $\cdot$ s^{-1}, the hailstone bigger than 1.9 cm in diameter, and tornadoes etc.

9.3.2 The favorable synoptic situation of the severe convective weather

The severe convective weather is usually formed under certain favorable synoptic situation. In general speaking, the typical situation favorable to the development of severe convective weather is characterized by the upper level and low level jet streaks as shown in Fig. 9.6. There is a warm and moist tongue stretched at low level. On the middle level a dry tongue overlays over the low level moist tongue. The instability of stratification is very strong.

Similarly, based on some case studies, Hamill et al. (2005) summarized the favorable situation for the severe convective weather as shown in Fig. 5.40 of Chapter 5. The prime convective weather area indicated by the triangle in the figure, which is formed by the sides nearly parallel to the surface dry line, the surface warm front and the upper level jet stream axis respectively and with the gravity center located near the point of intersection of the upper jet axis and the lower level jet axis.

Polston (1996) summarized the synoptic situations favorable to form larger hailstones, with the largest diameter $\geqslant$10.16 cm, into two types. The first type (Type A) is characterized by high and low level jet streams, which are crossed up and down, strong wind shear near the surface warm front (discontinuity line) and unstable stratification. The supercells and large hails occurred normally near the area with strongest

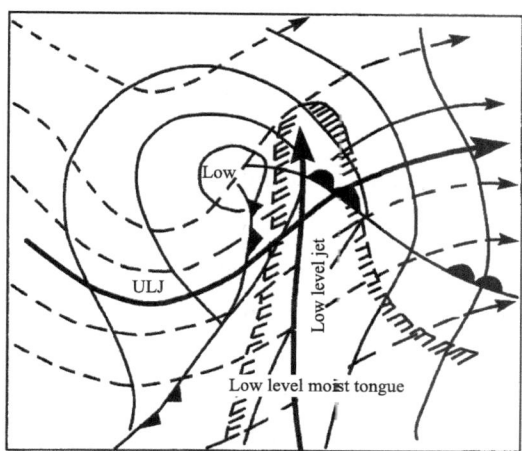

Fig. 9.6 The situation favorable to develop the severe convective weather (the thin solid lines are isobars on the sea level surface; the dashed lines are the stream lines on the high level; the shaded area is the low level moist tongue. ULJ=upper level jet stream). (Adapted from Kessler, 1987)

warm advection in the north side of the discontinuity line. The second type (Type B) is characterized by upper level dry and low level moist, very high convective instability, high level of free convection (LFC), which is normally at 680—720 hPa. The hails occur in the area with greater gradient of moisture and the area coupling upper level divergence and low level convergence, i.e. the area along the dry line or nearby. According to central meteorological station of China, in China the synoptic situation favorable to hailstorm may be divided into four types. They are the upper level cold trough (including forward tilted or backward tilted troughs), the upper level cold vortex, the north-west flow at the rear part of the upper level trough and the south branch trough. Their basic features about the upper level and low level situation are similar to that in the United States mentioned above.

9.3.3 Some indexes used for forecasting of the severe convective weather

For forecasting the severe convective weather such as thunderstorms, tornadoes, hailstorms, and thunderstorm gales etc. it is very often to use various forecasting indexes, some of which are introduced in the following.

(1) The intensity of the dry and warm lid (L_s)

The development of the severe convection needs strong instability energy, which is normally accumulated under some favorable situations, one of which is the inversion layer in the low troposphere and called as the dry and warm lid.

The intensity of the dry and warm lid can be expressed by the index L_s, which may be defined as:

$$L_s = (\theta_w)_{max} - \bar{\theta}_w \qquad (9.3.1)$$

where $(\theta_w)_{max}$ indicates the saturated wet bulb potential temperature at the top of the inversion layer; $\bar{\theta}_w$ is the mean value of the wet bulb potential temperature of the atmospheric layer from surface to 500 hPa constant pressure surface. The big L_s indicates the strong dry and warm lid, which implies that the convective energy stored in the atmosphere is strong.

(2) The storm strength index (SSI)

Strong vertical wind shear is also an important condition favorable to cause severe convection. In the previous text we have mentioned the relationship between Bulk Richardson Number (BRN) and the intensity of convection. BRN reflected the synthetic action of CAPE and vertical wind shear. According to the statistics the small BRN corresponding to supercell, while the middle and big numbers of BRN are normally corresponding to multi-cell and common cell storms respectively.

Turcotte et al. (1987) proposed the storm strength index (SSI), which reflected the synthetic action of the buoyancy energy and vertical wind shear on the storm intensity. The definition of SSI is as following:

$$SSI = 100 \times [2 + 0.276 \ln(Shr) + 2.011 \times 10^{-4} \times Eh] \qquad (9.3.2)$$

where Eh is the convective available energy (CAPE), Shr is the weighted average value of the vertical wind shear in the atmospheric layer from surface to the height H and may be calculated by the following equation:

$$Shr = \left[\int_0^H \rho(Z) |V(Z)| dz / \int_0^H \rho(Z) dz \right] - 0.5 [|V(0) - V(0.5 \text{ km})|] \qquad (9.3.3)$$

According to the situation in the United States and Canada, $SSI = 100$ or 120 can be regarded as the critical value between severe convective storm and non-severe convective storm. The bigger SSI implies the higher possibility to occur severe convective storms.

(3) The deep convection index (DCI_{NS})

Nitta et al. (1994) proposed the deep convection index (DCI_{NS}), which may be expressed as following:

$$DCI_{NS} = \begin{cases} 250 - T_{BB} & \text{when} \quad T_{BB} < 250 \text{ K} \\ 0 & \text{when} \quad T_{BB} \geq 250 \text{ K} \end{cases} \quad (9.3.4)$$

Since 250 K is at about the height of 400 hPa, the bigger number of DCI_{NS} means the bigger height of the convective cloud top exceed the height of 400 hPa. Murakami et al. proposed another deep convection index (DCI_M), which is written as the following:

$$DCI_M = 10 \times [(T_{BB} - T_{400})/(T_{tr} - T_{400})] \quad (9.3.5)$$

where T_{400} and T_{tr} represents the temperature on the height of 400 hPa and on the tropopause respectively. Bigger number of DCI_M means the higher convective cloud top. Barlow (1993) proposed a new deep convection index (NDCI), which may be expressed as following:

$$NDCI = (T_{850} + T_{d850}) - LI \quad (9.3.6)$$

where T_{850}, T_{d850}, and LI are the temperature, dew point temperature at 850 hPa and the lifted index respectively.

(4) Severe weather threat index (SWEAT)

The severe weather threat index (SWEAT) is used as an index for the tornado forecasting very often in the United States. The index also reflected the synthetic action of the instability energy and vertical shear of wind velocity and wind direction on the strength of the convective storm. The index may be expressed as following:

$$SWEAT = 12D + 20(TT - 49) + 2f_8 + f_5 + 125(S + 0.2) \quad (9.3.7)$$

where $D = T_{d850}$ (℃) (if D is negative, then this term equals zero);

f_8 = the wind velocity on the 850 hPa constant pressure surface(n mile · h^{-1} *);

f_5 = the wind velocity on the 500 hPa constant pressure surface(n mile · h^{-1});

S = sin(the wind direction on 500 hPa-the wind direction on 850 hPa);

$TT = T_{850} + T_{d850} - 2T_{500}$, if $TT < 49$, then the term $(TT - 49)$ equals zero;

The term 125 $(S + 0.2)$ is the shear term, which will be set as zero when one of the following conditions is not provided with

The wind direction on 850 hPa is between 130°—250°;

The wind direction on 500 hPa is between 210°—310°;

The difference of 500 hPa and 850 hPa is positive;

---

* 1 n mile · h^{-1} = 1 kt = 0.514 m · s^{-1}.

The wind speed on 850 hPa or 500 hPa at least equals 15 n mile · h^{-1}*。

Note: there is no any term equals zero in the equation (9.3.7)

Higher *SWEAT* value implies higher possibility to form tornadoes or other severe storms.

(5) Tornado index

The tornadoes' intensity may be indicated by Fujita scale F as mentioned before. Colquhoun (1996) proposed a progress equation of the tornadoes' intensity F as following:

$$F = -0.145(LI) + 0.136(S6) - 1.5 \qquad (9.3.8)$$

where LI is the lifted index, $S6$ is the vertical wind shear between surface and 600 hPa.

(6) Total potential instability index (*LAPOT*)

The total potential instability index (*LAPOT*) is also an index may be used for indicating the potential possibility of the tornadoes and severe convective storms, which is defined as:

$$LAPOT = \frac{1}{\bar{\theta}_E \Delta P}\left(\Delta\theta_E + \frac{\bar{\theta}_E}{\bar{\theta}_{SEE}}\Delta\theta_{SEE}\right) \qquad (9.3.9)$$

or
$$LAPOT = \frac{-2[\theta_E(P_2) + \theta_{SE}(P_2) - 2\theta_E(P_1)]}{[\theta_E(P_2) + \theta_E(P_1)](P_2 - P_1)} \qquad (9.3.10)$$

where $\bar{\theta}_E$, $\bar{\theta}_{SEE}$ are the mean equivalent potential temperature and mean pseudo-equivalent potential temperature on the P level respectively, $\Delta P = P_2 - P_1$, ($P_1 > P_2$). $\bar{\theta}_{SEE} = [\theta_{SE}(P_2) + \theta_E(P_1)]/2$ are the average value of the pseudo-equivalent potential temperature at the top of the atmospheric layer (P_2) and the equivalent temperature at the bottom of the atmospheric layer (P_1); $\Delta\theta_E$ and $\Delta\theta_{SE}$ are the differences of θ_E and θ_{SE} of the atmospheric layer respectively; $\Delta\theta_{SEE} = [\theta_{SE}(P_2) - \theta_E(P_1)]$ is the difference between θ_{SE} at the top of the atmospheric layer and θ_E at the bottom of the air layer. Following is a simple way to calculate θ_E:

$$\theta_E = \frac{E_T}{C_p} = \frac{1}{C_p}\left(C_p T + gz + L_0 r + \frac{V^2}{2}\right)$$
$$\approx T(K) + 2.5r + 9.8z \qquad (9.3.11)$$

where T is temperature, Z is the height above sea surface of the air parcel (in unit: geopotential height), L_0 is evaporation latent heat, r is mixing ratio (in unit: g · kg^{-1}), V is the vector of air velocity.

Following is another simple way to calculate:

$$\theta_E = \theta + B_0 \frac{P_0}{PE_w(T_d)}$$

or
$$\theta_E = \theta_w + B_0 E_w(\theta_w) \qquad (9.3.12)$$

where θ is the potential temperature at the height with pressure P, $P_0 = 1000$ hPa, $E_w(T_d)$ is the water vapor pressure under the dew point temperature T_d, $B_0 = 0.622 L_0/C_p P_0 = 1.555$ K·hPa^{-1}. θ_w is the wet bulb potential temperature at pressure P, $E_w(\theta_w)$ is the water vapor pressure at θ_w.

In the tornado break process, the features of the distribution of the $LAPOT$ index are that to the north of the tornado area the $LAPOT$ index in the 700—850 hPa atmospheric layer has bigger negative value (about -50×10^{-5} hPa^{-1}), while in the layer of 500—700 hPa the index is a positive value (about $+25 \times 10^{-5}$ hPa^{-1}).

(7) Wind index ($WINDEX$)

McCann (1994) introduced an empirical index called $WINDEX$ or WI, which may be used for forecasting the potential of the thunderstorm gust wind and the downburst, the index may be written as following:

$$WI = 5[H_M R_Q (\Gamma^2 - 30 + Q_L - 2Q_M)]^{0.5} \qquad (9.3.13)$$

where H_M is the melting height above ground level (AGL), (in unit: km); $R_Q = Q_L/12$, but may not greater than 1, in unit: g·kg^{-1}; Γ is the vertical temperature lapse rate between the surface and the melting level, in unit: ℃·km^{-1}; Q_L is the mixing ratio in the layer 1 km above the surface, in unit: g·kg^{-1}; Q_M is the mixing ratio at the melting level, in unit: g·kg^{-1}. WI is a non-dimensional number, but it may be regard as a dimensional number with the unit n mile·h^{-1}, or about 2 m·s^{-1} when it is used for operational wind storm forecasting. The numerical result calculated from equation (9.3.13) may be regarded as the possible wind speed of the thunderstorm gale. According to the equation (9.3.13), we know that the higher melting level, bigger Q_L (low level moist), smaller Q_M (middle level dry) will be favorable to form strong thunderstorm winds.

(8) Parameters used for hailstorm forecasting (WBZ, R_{max} and W_{max} etc.)

According to the cloud physics analysis of the hailstorm formation, we know that the hailstorm formation requires some necessary conditions including the suitable height of the wet bulb temperature of 0 ℃ (WBZ) (which is at about 600 hPa or 4 km), the suitable height of the -20 ℃ level (which is at about 400 hPa or 7.5 km), and the suitable ratio between the thicknesses of the cold cloud layer and the warm cloud layer etc., moreover the hailstorm formation also requires stronger vertical velocity. According to the statistics data from Russia, there is following relationship between the maximum diameter of the hailstone (R_{max}) and the maximum vertical velocity inner cloud (W_{max}):

$$R_{max} \approx \frac{W_{max}^2}{\beta^2} \qquad (9.3.14)$$

where $\beta \approx 2.2 \times 10^3$ cm$^{1/2}$ · s^{-1} (or $\beta \approx 2.6 \times 10^3$ cm$^{1/2}$ · s^{-1} in some references). Hence, we can use WBZ and W_{max} to estimate the diameter of the hailstone. In general speaking, under the situation with certain WBZ the diameter of the hailstone increases with W_{max}. Normally the bigger hail events are all occurred under distinct background situation, for instance according to the average data of the seven bigger hail events in the United States, in which the maximum diameter of the hail exceed 10.16 cm, the convective available energy ($CAPE$) equals 3352, the storm relative helicity index (SRH) at 2 km ($SRH_{2\,km}$) equals 130, at 3 km ($SRH_{3\,km}$) is 153, the energy helicity index (EHI) is 2.75. All the amounts of the above indexes are in standard units.

9.3.4 Forecasting of the severe convective weather

The above indexes can be used to forecast the severe convective weather, but it has to be noted that these indexes just indicated the potential of the severe weather. The formation of the severe weather still needs other important conditions, for instance the vertical motion, including the synoptic scale and mesoscale ascending motions. The synoptic scale vertical motions normally are provided by the mid to high level trough, the jet streak core, warm advection, cyclone, fronts, and shear lines etc. The secondary circulation near the front-jet system plays important role in the developing process of the severe convective weather. The severe convective weather as the mesoscale weather phenomena is more directly related with the mesoscale vertical motions, which are usually related with the mesoscale ascending zones caused by the discontinuity interface, inhomogeneous heating, interaction between airflow and topography such as the mesoscale cyclones, mesoscale fronts and shear lines or convergence line, the thunderstorm outflows. sea-land breeze and mountain-valley breeze circulations etc. To carefully analyze and determine the near-surface mesoscale lifting source is the key to forecast the location and time of the severe convective weather developing.

After the convective storms occurring it should estimate their moving direction and speed correctly for making the forecasting and warning of the storm for the downstream region. The forecasting of the storm moving is mainly based on the satellite data, radar echo data and lightning detecting network data etc. by using extrapolation method. The storm motion consists of two parts, i, e. the translation of the cells along the mean environmental wind and the propagation caused by the metabolism of the storm itself. Therefore the storm motion is normally different from the mean environmental wind direction and speed. The motion speed is normally less than the mean environmental wind speed and the motion direction may be defected to the right or left sides of the mean wind direction and most of them are defected to the right side. According to the investigation given by Maddox (1976), the storm moving speed is about 75% of the mean wind speed of the environment and the moving direction of the storm defected about 30° to the right side of the mean environmental wind direction. Davies et al. (1993, 1998) amended the conclusions of Maddox (1976) and suggested that the environmental winds should be divided into different scales. Under the different scale of the environmental wind the degree of the storm moving speed and direction defecting the wind may be different. For instance, when the environmental mean wind is less than 15 m $\cdot$ s^{-1}, the storm moving speed is about 75% of the environmental mean wind speed and the moving direction is defected about 30° to the environmental wind direction at the height 0—6 km; while for the situation that the environment wind is greater than 15 m $\cdot$ s^{-1}, the storm moving speed will be about 85% of the environmental wind speed and defected about 20° to the environmental wind direction at the height 0—6 km.

Mills and Colquhoun (1998) proposed a decision tree combining many indexes, criteria together. By using the decision tree the different type of weather phenomena such as heavy rain and severe convective storm, severe thunderstorm and common thunderstorm, tornadic storm and non-tornadic storm, fast moving storm and quasi stationary storm etc. may be separated and distinguished. The decision tree may provide empirical rules for forecasters and the knowledge base for object forecasting system to make forecasting by computer expert system automatically or by man-machine combination.

9.3.5 The estimation of the forecasting accuracy rate

It is necessary to set some objective criteria to estimate the accuracy of the severe convective weather forecasting. Donaldson et al. (1975) proposed a criteria success index (*CSI*) as an accuracy rate of the forecasting to estimate objectively the forecasting technique level. The index *CSI* may be expressed as following:

$$CSI=\frac{x}{(x+y+z)} \qquad (9.3.15)$$

where x is the success forecast reports, y is the missing forecast reports, z is empty forecast reports. The definitions of the success, missing and empty forecast reports are as illustrated in Table 9.2. The amount of the CSI can be 0—1, 0 means the forecasts are totally failed, 1 means the forecasts are totally correct.

Table 9.2 An illustration of the success, missing and empty forecast reports (ST means severe thunderstorm).

Forecast \ Observation	Yes (ST)	No (ST)
Yes (ST)	X	Z
No (ST)	Y	

Other indexes such as Skill index etc. may be also used for estimating the score of forecasting. The Skill index may be defined by the ratio of the difference between the correct prediction times and the correct blind prediction times and the difference between the total prediction times and the total blind prediction times. Obviously, the higher the value of CSI or Skill indexes means the better the effect of the prediction. Since the amount of CSI is inversely proportional to the empty forecast times. Therefore it is the way to increase the amount of CSI by reducing the empty forecast frequency.

§ 9.4 Data and tools applied in nowcasting and VSRF

Except the conventional and enhancement surface and upper air observational data, the nowcasting are mainly based on the data from the remote-sensing data including the automatic meteorological station network data, lightning detecting data, satellite and radar data etc., which provide the successive information about the variation of the storm motion, storm intensity and weather phenomena.

9.4.1 Application of the automatic meteorological network data

The automatic meteorological station can automatically measure the meteorological elements such as the pressure, temperature, moisture, rainfall, wind direction and wind speed, sunshine and radiation etc. Some of them can also observe the visibility, cloud (height, shape, amount), weather phenomena, lightning and thunders, surface temperature and soil temperature etc. All the observed data from the automatic meteorological station data will be concentrated to the analysis and forecasting center and treated. By using the successive detecting data the various mesoscale systems and their variation can be correctly presented. These information will be very important for making the nowcasting.

9.4.2 Application of the lightning detecting data

When the electromagnetic waves produced by thunderstorm propagate in the sky, the phase difference between the electronic signal and the magnetic signal will be varied with the propagating distance. The variation is distinctly in the scope of 50—300 km. By using this character the position of the thunder-streak can be measured. The single station lightning tester receives the electronic signals by using a whip antenna and the magnetic signals by using a pair of orthogonal ring antennas. After the antenna received the electric and magnetic signals forming the electric potential difference and after the composite and integral treatments the positions of the thunder-lightning activities will be then displayed on the fluorescent screen of the cathode-ray tube (CRT). The single station lightning tester has bigger errors. For reducing the errors we may use three lightning testers to form a triangle and then to measure the thunder and lightning at the same location from three different directions to form more accurate distribution map. The thunder-lightning distribution map may combine with the radar echo picture for determining the thunderstorm activities.

9.4.3 Application of the satellite remote sensing data

The satellite remote sensing data includes the satellite visible cloud imagery, infrared cloud image and the satellite remote sensing profiles of the atmospheric temperature, humidity as well as the digital cloud images etc. The satellite visible cloud imagery and infrared cloud images may provide important information about the cloud features of synoptic and mesoscale systems, which are useful for short range and very short range forecasting. The parameters such as temperature, humidity and instability etc. inversed from the satellite data can provide important base for the very short range forecasting of severe convective weather. For instance, based on the functional relationship between the lightning frequency with the height of cloud top (H) and $CAPE$, Prince and Rind (1992) estimated the maximum vertical velocity W_{max} and the lightning frequency on land by using the cloud top height H derived from the high resolution satellite imagery. They get the empirical equations as following:

$$H = 0.47(CAPE)^{0.44} \tag{9.4.1}$$

$$W_{max} = 0.27 H^{1.73} \tag{9.4.2}$$

$$F = 1.439 \times 10^{-8} H^{7.86} \tag{9.4.3}$$

The observation shows that the severe convective storms are usually related with deep convection, the depth of which may be indicated by satellite T_{BB} (the infrared equivalent black body temperature on the cloud top). Nitta, Murakami and Barlow et al. proposed the indexes indicated the convection depth such as DCI_{NS}, DCI_M, and DCI_B, which may be used as the supplementary tools of diagnose and forecasting of the severe storm systems such as MCC, squall line and supercells etc. The satellite data will be also helpful for understanding the mesoscale atmospheric dynamic processes. For instance there is good corresponding relations between the potential vorticity (PV) feature and the feature of the satellite water vapor pictures. Demirtas et al. found that the high and low value areas of potential vorticity on 315 K isentropic surface are corresponding to the dark and bright areas on the satellite water vapor image respectively. NOAA/ERL and NESDIS of United States proposed the idea of "Satellite forecasting funnel" (Fig. 9.7) for making the 0—48 h forecasting of the convective weather, which is a multi-scale concatenation event. So called the "Satellite forecasting funnel" is a concatenation of the various scale weather systems and weather processes by using the satellite water vapor image, infrared and visible light images data compositely. This method combined the satellite data analysis and mesoscale dynamic diagnosis makes a dynamic interpretation of the satellite data.

9.4.4 Application of radar data

Weather radar, especially the Doppler radar is one of the most important equipment in the modern meteorological observational network. Doppler radar can measure not only the reflectivity like an operational radar, but also the radial velocity etc. and also may measure the tree dimensional distribution of the air flow in cloud based on 2—3 Doppler radar's cooperate detecting for describing the structure of the storm. The weather radar may also be used for detecting the mesoscale systems such as mesoscale front, cyclone, shear line, low level jet (LLJ), cold high pressure etc. and to recognize hailstorm, tornado. downburst. The radar data may also be retrieved as wind velocity fields etc. so that it will be helpful to improve numerical modeling,

9.4.5 VSRF and nowcasting system

VSRF (very short range forecasting) and nowcasting are very different from the

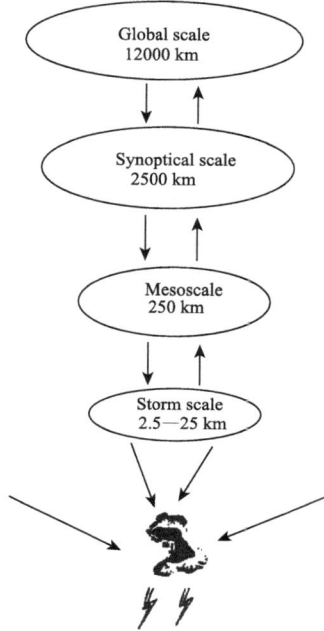

Fig. 9.7　Forecasting funnel.

traditional mid-short range forecasting. The forecasting objects of VSRF and nowcasting are mainly the mesoscale and small scale systems. The data needed for mesoscale analysis should have much higher temporal and spatial resolutions and much greater amount. The data collection, transmission, processing as well as the forecasting and warning making and distributing should be much quickly. All these difficult tasks to be done must rely on the high speed communication and a big power computer system as well as a real time effective VSRF and nowcasting system, which should include the subsystems such as observation, communication, analysis, forecasting and services systems etc. VSRF and nowcasting are the products of such a modern meteorological prediction operational system.

Recent years many countries are developing the VSRF and nowcasting systems. Take China as an example, in 2008, China Meteorological Administration (CMA) has launched a campaign on the development of its first version of integrated nowcasting system SWAN (Severe Weather Analysis and Nowcasting system). This currently-developed nowcasting system aims at providing an integrated, state-of-the-art and swift severe weather nowcasting platform that is suitable to operational applications.

SWAN ingests data from China's new generation radars, automatic weather station, satellite, and mesoscale numerical weather prediction model. It offers a tool for severe weather monitoring, analysis, warning and prediction. The current version of SWAN system provides a software package that integrates a series of nowcast algorithms and functions, including 3D radar mosaic, storm identification, tracking and prediction, quantitative precipitation estimate, short-range quantitative precipitation forecast, real time forecast verification, severe weather warning, and interactive forecast preparedness. The system has deployed in China early Spring in 2009. CMA plans to upgrade SWAN with more algorithms and functions.

References

ANTHES R A, 1986. The General Question of Predictability. Mesoscale Meteorology and Forecasting[M]. Am Meteor Soc.

ANTHES R A, CALSON T N, 1982. Conceptual and numerical models of evolution of the environment of severe local storms[C]. Joint U S China Workshop on Mountain Meteorology, Beijing, China.

ATLAS D R, SRIVASTAVA C, SEKHON R S, 1973. Doppler radar characteristics of precipitation at vertical incidence[J]. Rev Geophys Space Phys, 11:1-35.

BARLOW W R, 1993. A new index for the prediction of deep convection[C]. Preprints 17th Conf on Severe Local Storms, St Louis, Am Meteor Soc, 129-132.

BELLON A, AUSTIN G L, 1978. The evolution of two years of real time operation of a short-term precipitation forecasting (SHARP)[J]. J Appl Meteor, 17:1778-1787.

BLANCHARD D O, 1998. Assessing the vertical distribution of convective available potential energy [J]. Wea Forecasting, 13:870-877.

BROWNING K A, 1982. Nowcasting[M]. London:Academic Press.

BROWNING K A, 1989. The mesoscale data base and its use in mesoscale forecasting[J]. Q J Roy Meteor Soc, 115:717-762.

CHARBA J P, 1979. Two to six hour severe local storm probabilities: an operational forecasting system[J]. Mon Wea Rev, 107:268-282.

CHISHOLM A J, ENGLISH M, 1973. Alberta Hailstorms[M]. AMS Met Monographs, 14.

COLQUHOUN J R, 1987. A decision tree method of forecasting thunderstorms, severe thunderstorms and tornadoes[J]. Wea Forecasting, (2):337-345.

COLQUHOUN J R, 1996. Relationship between tornado intensity and various wind and thermodynamic variables[J]. Wea Forecasting, 11:126-136.

COOK T M C, SHIREY M S, 1998. Verification and analysis of the 48 km eta model best CAPE and best LI forecast[C]. Preprints, 16th Conf on Weather Analysis and Forecasting, Phoenix City, Am Meteor Soc, 173-175.

DESAUTELS G, VERRET R, 1996. Canadian Meteorological Centre summer severe weather package (storm relative helicity)[C]. Preprints, 18th Conf on Severe Local Storms, San Francisco, CA, Amer Meteor Soc, 689-692.

DONALDSON R J Jr, DYER R M, KRAUSS M J, 1975. An objective evaluator of techniques for predicting severe weather events[C]. Preprints, 9th Conf Severe Local Storms, Norman, Oklahoma, Am Meteor Soc, 321-326.

DOSWELL C A III, BROOKS H E, MADDOX R A, 1996. Flash flood forecasting an ingredients based methodology[J]. Wea Forecasting, 11:560-581.

DOSWELL C A III, WEISS S J, JOHNS R H, 1993. Tornado forecasting: A review[J]. Amer Geophys Union, 557-571.

DOSWELL C A III, 1986. Short Range Forecasting. Mesoscale Meteorology and Forecasting[M]. Am Meteor Soc, 689-719.

EMANUEL K A, 1994. Atmospheric Convection[M]. New York: Oxford Univ Press: 168-173.

FUJITA T T, 1978. Manual of downburst identification for project NIMROD[R]. SMRP Research Paper, Chicago University.

GILMORE M S, WINKER L J, 1996. The Influence of DCAPE on supercell dynamics[C]. Preprints, 18th Conf on Severe Local Storms San Francisco, CA, Am Meteor Soc, 723-727.

GLAHN H R, LOWRY D A, 1972. The use of model output statistics (MOS) in objective weather forecasting[J]. J Appl Meteor, 11:1203-1211.

HARRY R G, 1984. Surface wind forecasts from the Local AFOS MOS Program (LAMP)[C]. 10th Conference on Weather Forecasting and Analysis, 78-86.

HOUZE R A, BETTS A K, 1981. Convection in GATE[J]. Rev Geophy Space Phys, 19(4):541-576.

JOHNS R H, DOSWELL C A III, 1992. Severe local storm forecasting[J]. Wea Forecastng, 7:588-612.

KUO Y-H, CHEN G T J, 1992. The international conference on mesoscale meteorology and TAMEX 3—6 December 1991, Taipei, Taiwan[J]. Bull Am Meteor Soc, 73:1611-1622.

LIOU Y C, GAL-CHEN T, LILLY D K, 1991. Retrieval of winds, temperature, and pressure from single Doppler radar and a numerical model[C]. Preprints, 25th Int Conf on Radar Meteorology, Paris, France, Am Meteor Soc, 151-154.

MAGLARAS G J, LAPENTA K D, 1997. Development of a forecast equation to predict the severity of thunderstorm events in New York State[J]. Nat Wea Dig, June 3—9.

MCGINLEY J, 1986. Nowcasting mesoscale phenomena. Mesoscale Meteorology and Forecasting [M]. Ray P S (ed.). Am Meteor Soc:657-688.

MCNULTY R P, 1995. Severe and convective weather: A central region forecasting challenge[J]. Wea Forecasting, 10:187-201.

MILLER R G, 1981. GEM: A statistical weather forecasting procedure[R]. NOAA Technical Report, NWS, (28).

MILLS G A, COLQUHOUN J R, 1998. Objective prediction of severe thunderstorm environments: Preliminary results linking a decision tree with an operational regional NWP model[J]. Wea Forecasting, 13:1078-1092.

NITTA B T, SEKINE S, 1994. Diurnal variation of convective activity over the tropical Western Pacific[J]. J Met Soc Japan, 72:627-641.

OSTBY F P, 1992. Operations of the National Severe Storms Forecast Center[J]. Wea Forecasting, 7:546-563.

REIFF J, BLAAUBOER D, DE BRUIN H A R, et al, 1984. An air mass transformation model for short range forecasting[J]. Mon Wea Rev, 112(3):393-412.

SHOU S W, 1989. Aeronautical meteorology[C]. WMO Regional Seminar for National Instructors of RA II and RA V, Jakata, Indonisia, Sept.

SHOU S W, ZHANG S Y, 1991. An objective forecasting system of severe convective weather[C]. 4th International Aeronautical Meteorological Meteorology Conference, June.

SMITH R K, 1997. Thermodynamics of Moist and Cloudy Air. The Physics and Parameterization of Moist Atmospheric Convection[M]. Netherlands:Kluwer Academic Publishers:29-58.

THOMPSON R L, 1998. Eta model storm relative winds associated with tornadic and nontornadic supercells[J]. Wea Forecasting, 13:125-137.

TURCOTTE V, VIGOUX D, 1987. Severe thunderstorms and hail forecasting using derived parameters from standard RACBS data[C]. Preprints, Second workshop on Operational Meteorology. Halifax. NS Canada Atmospheric Environment Service/Canadian Meteor And Oceanogr Soc, 142-153.

WAKIMOTO R M, WILSON J W, 1989. Non-supercell tornadoes [J]. Mon Wea Rev, 117: 1113-1140.

WALDVOGEL A, FEDERER B, GRIMM P, 1979. Criteria for the detection of hail cells[J]. J Appl Meteor, 18:1521-1525.

WU B J, ZHAO X Y, XU C H, et al, 1999. A multi indictor superposition method for hailfall forecast [C]. The WMO Scientific conf on Wea Modif Chiang Mai, Thailand, February 12—17, 486-489.

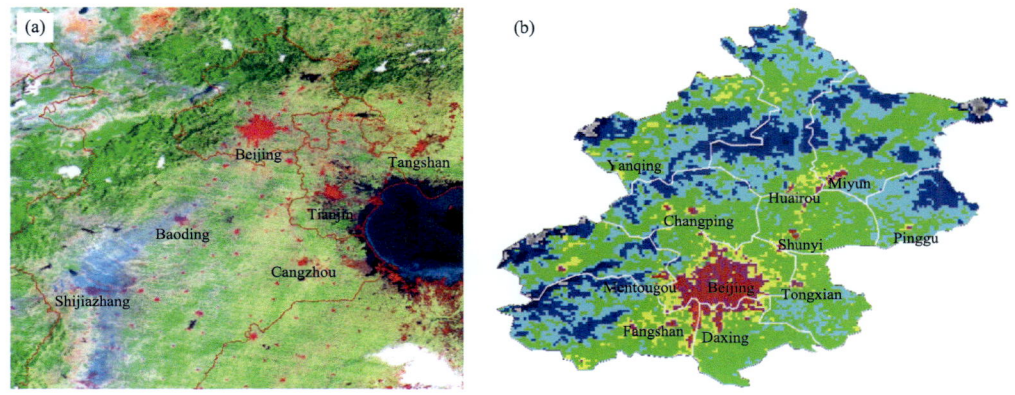

Fig. 2.19 The urban island detecting maps of satellite FY-1C of Hebei China. (a) at 0831 BJT Sept. 1, 2000; (b) at 1332 BJT Sept. 1, 2001, Beijing

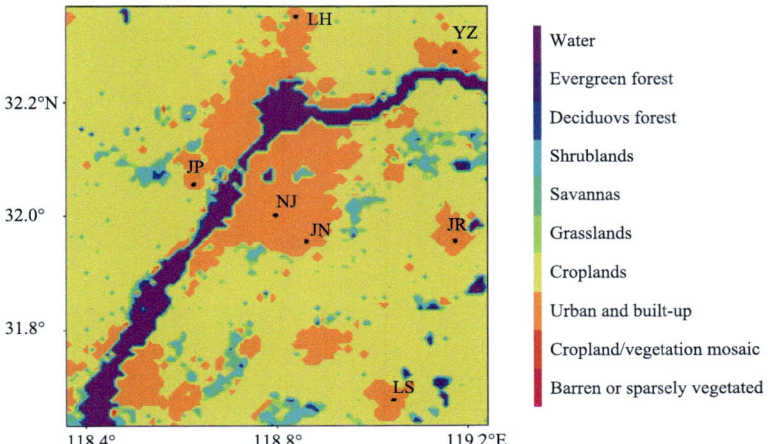

Fig. 2.20 The urban heat island effect near the area of Nanjing City based on the MODIS data. The horizontal and vertical axises of the diagram are the latitude and longitude respectively. (After Wang et al., 2008)

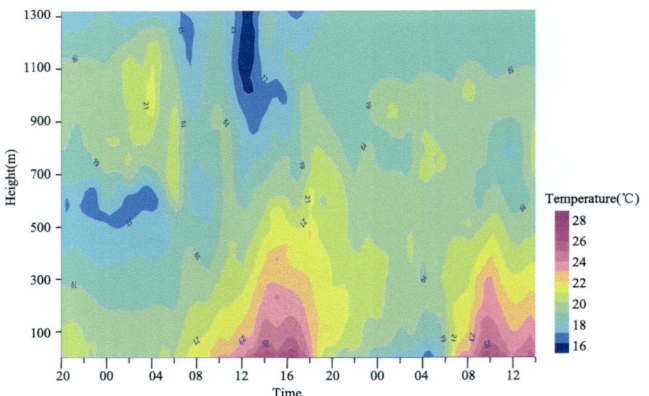

Fig. 2.22 The pictures of the vertical distribution of temperature near the coast of Beilun Port, Ningbo, China observed by kite balloon. (After Song et al., 2008)

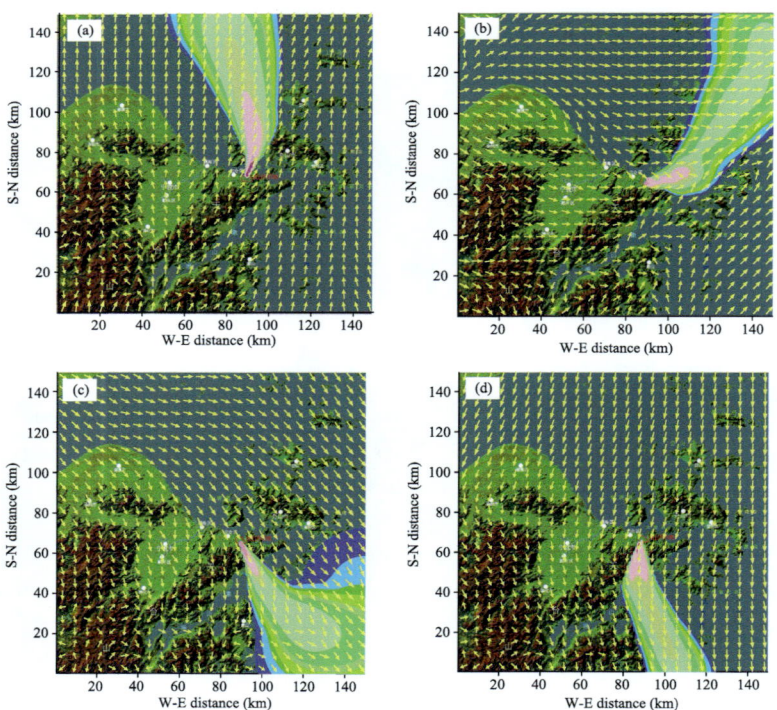

Fig. 2.24 Variation of the shifting direction of the pollutant tracers and the wind fields on (a) 1700 BJT 21 Oct.; (b) 2300 BJT 21 Oct. (c) 0500 BJT 22 Oct.; (d) 1100 BJT 22 Oct., 2006 near Beilun Port, Ningbo, China based on Mesoscale numerical simulation. (Adapted from Song et al., 2008)

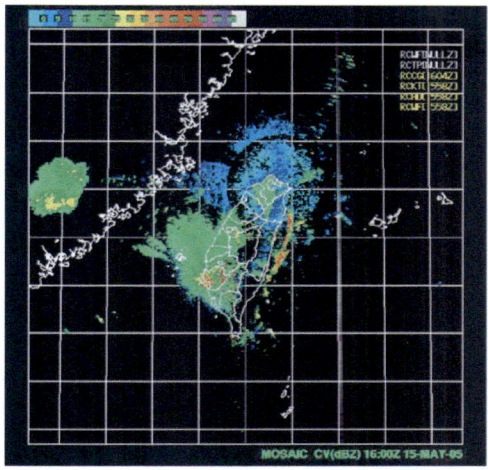

Fig. 2.26 Weather radar image acquired at 1600 UTC 15 May, 2005 (0000 LST 16 May), that is, 10 h, 24 min before the ERS-2 SAR data acquisition. (After Werner et al., 2007)

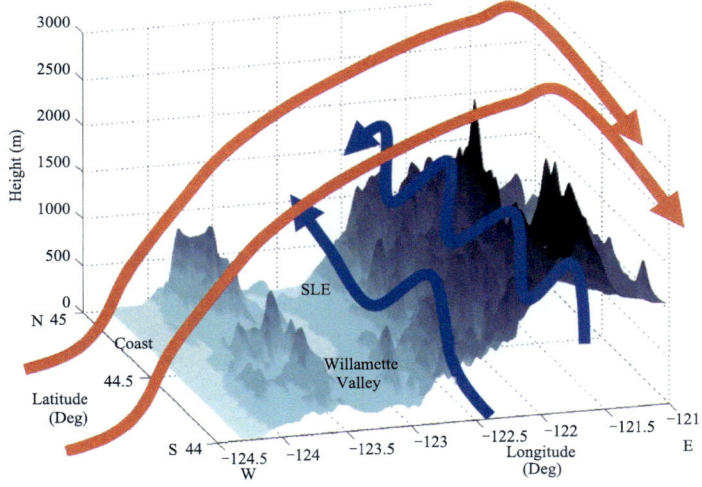

Fig. 2.32 Three-dimensional idealized schematic of topography and wind flow over the study area from 2300 to 0100 UTC 13—14 Dec., 2001. Blue arrows show strong southerly low θ_e airflow at low levels along the windward (west facing) slopes of the Cascade range which was subsequently involved in wave generation over multiple small-scale east-west-oriented ridges-valleys within the Cascade foothills. Red arrows show the high θ_e cross barrier flow that surmounted the low θ_e air and exhibited a vertically propagating mountain wave structure anchored to the mean north-south Cascade crest. (After Garvert et al., 2007)

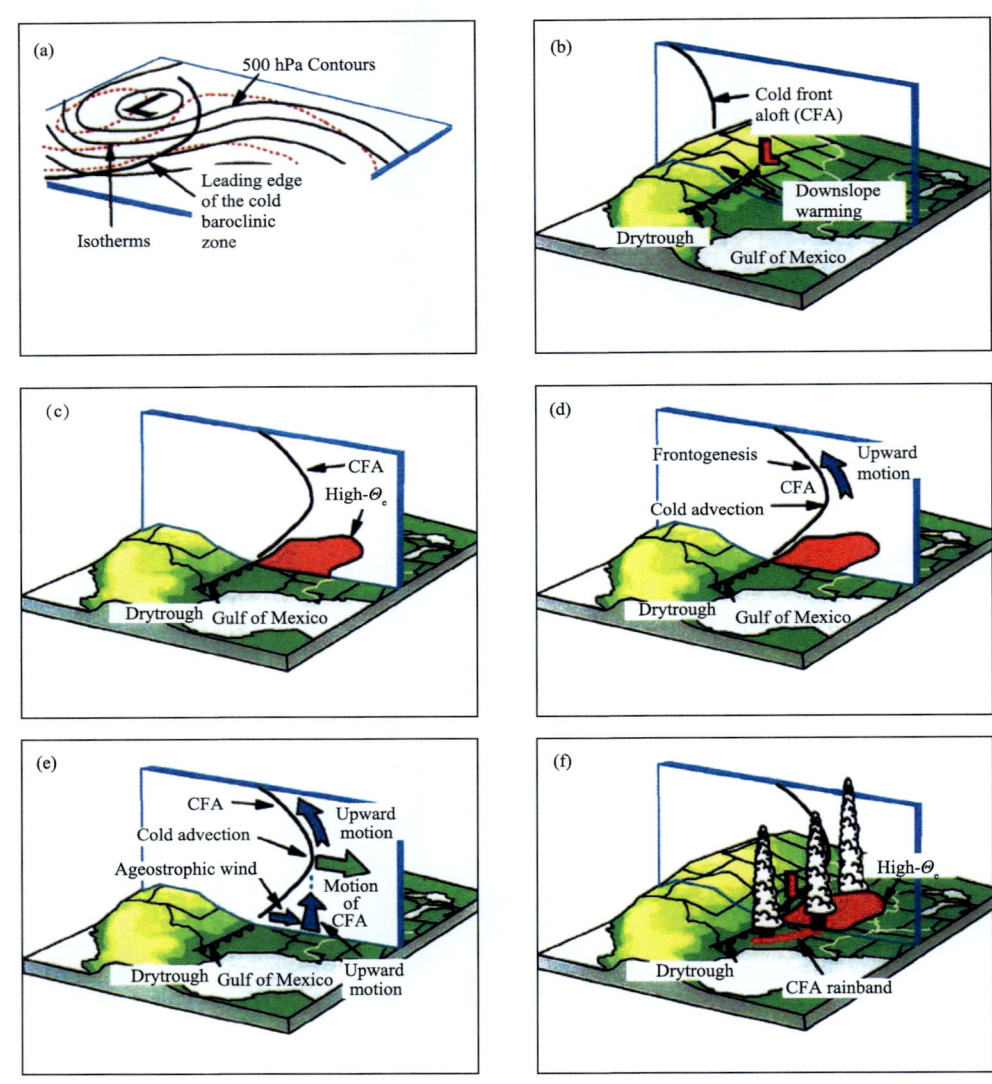

Fig. 2.33 Schematics illustrating the approach of the cold front aloft (CFA) and the formation of the CFA rain band. (After Hobbs, 1996)

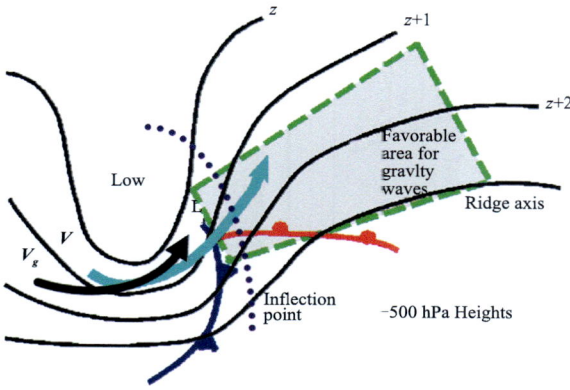

Fig. 3.11 Schematic of the environment conducive for mesoscale inertia-gravity wave generation by geostrophic adjustment. Depicted are the 300 hPa jet, 500 hPa height field (solid contours), and surface synoptic features. Positions of all atmospheric features shown on the schematic are approximate means during first half of wave episode, when wave generation mechanisms are assumed to operate most efficiently. The shaded region represents the area favorable for gravity wave activity during entire wave episode. Jet streak positions representing core of maximum wind speeds within the 300 hPa jet stream before (V_g) and after (V) the wave generation are denoted by thick arrows. (Adapted after Uccellini and Koch, 1987)

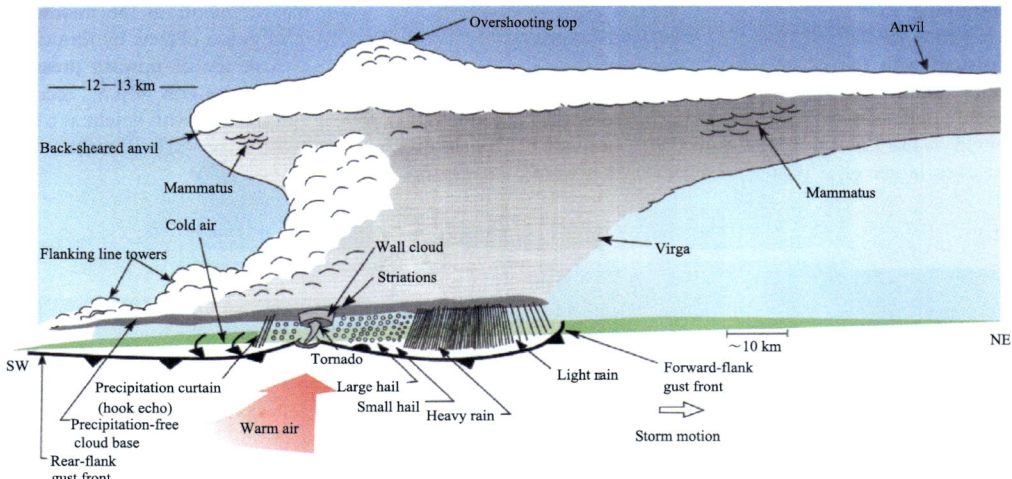

Fig. 5.10 The appearance of a supercell storm accompanied with tornado. Motion of the warm air is relative to the ground. (After Houze, 1977, 2014; Wallace, 2006)

· 5 ·

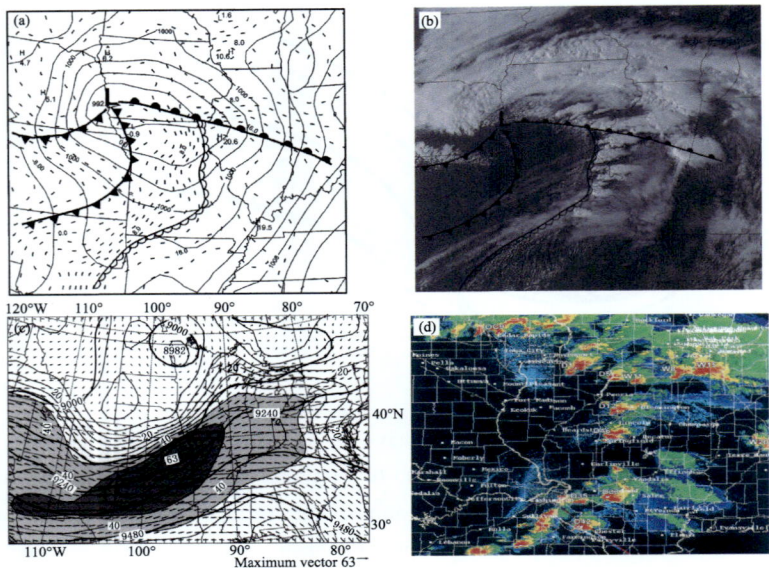

Fig. 5.38 (a) Objectively analyzed synoptic surface pressure and dew-point analysis and subjective frontal analysis for 2100 UTC 19 Apr., 1996. Isobars (isodrosotherms) are shown every 2 hPa (2 ℃); (b) GOES-8 visible imagery at 2145 UTC 19 Apr., 1996 with superposed frontal positions and surface low pressure center at 2100 UTC; (c) Geopotential height field (m, thick lines), isotachs (m·s^{-1}, thin lines), and vector winds at 300 hPa at 0000 UTC 20 Apr., 1996 (from MM5 RAWINS analysis). The light and dark gray shaded areas represent wind speeds greater than 35 and 45 m·s^{-1}, respectively. The maximum vector wind scale is shown in the lower right; (d) Radar reflectivity (0.5°) composite from KILX, KLSX, and KDVN at approximately 0000 UTC. (After Lee et al., 2006)

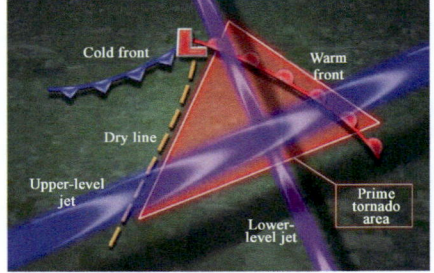

Fig. 5.40 Conceptual model of synoptic conditions typically associated with a large tornado outbreak. The triangle area indicates region of expected tornadoes. (After Hamill et al., 2005)

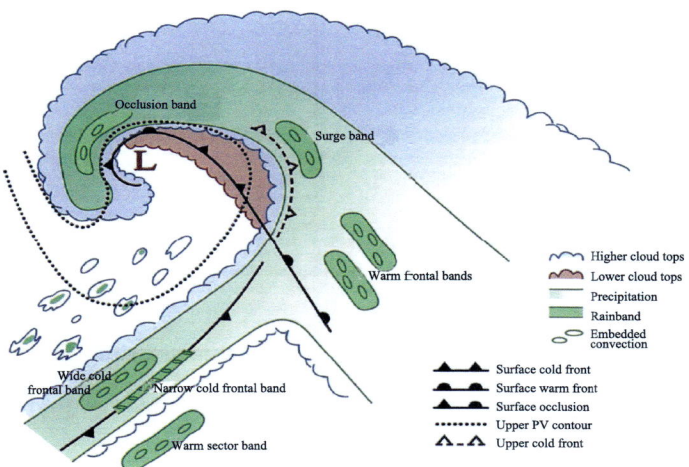

Fig. 5.53 Idealization of the cloud and precipitation pattern associated with a mature extratropical cyclone. (After Houze, 2014)

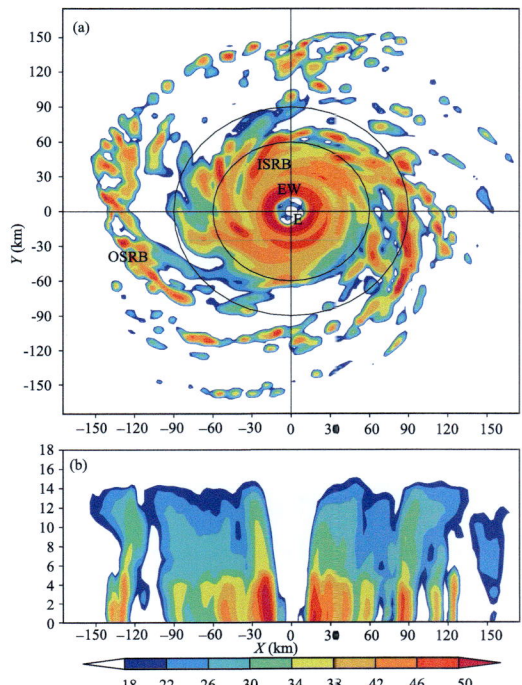

Fig. 5.69 plan view (a) and cross section (b) of radar reflectivity in a typhoon. (E=eye, EW=eye wall, ISRB=inner spiral rainband, OSRB=outer spiral rainband)

· 7 ·

Fig. 5.73 Infrared satellite image of a mesoscale convective system over Missouri. Courtesy of J. Moore, St. Louis University, St. Louis, Missouri. (After Houze, 2014)

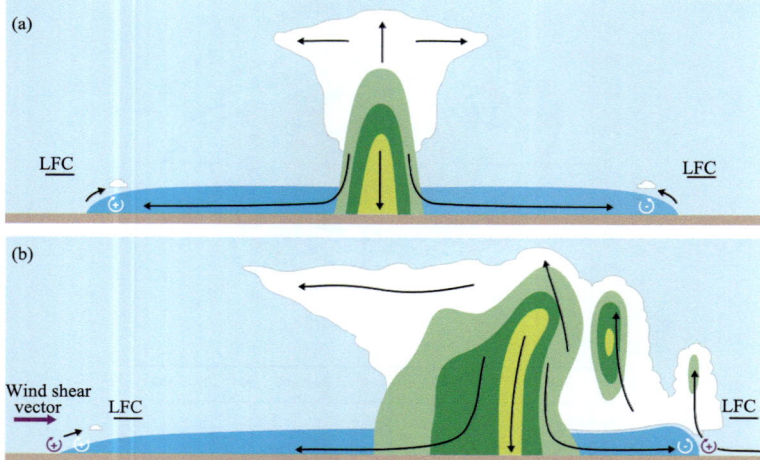

Fig. 7.4 An illustration shown the effect of the environmental vertical wind shear on the propagation of the storm. When the environmental vertical wind shear is weak the propagation of the storm has no priority direction (a); while when the environmental vertical wind shear is strong, the propagation of the storm has priority direction (b). (After Markowski, 2011)

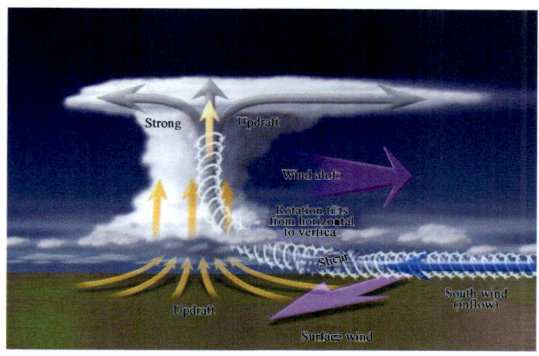

Fig. 8.12 Illustration of the storm relative helicity.

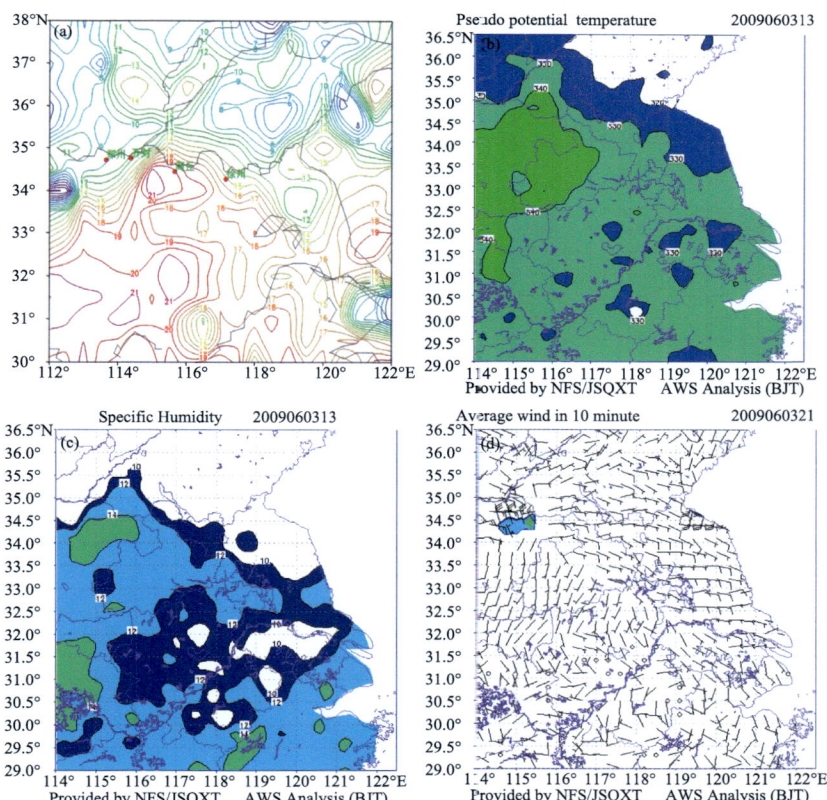

Fig. 8.20 Automatic meteorological station network data analyses on Jun. 3, 2009. (a) dew point temperature. The red point indicates the position of Shangqiu City; (b) pseudo potential temperature; (c) specific humidity; (d) wind field.

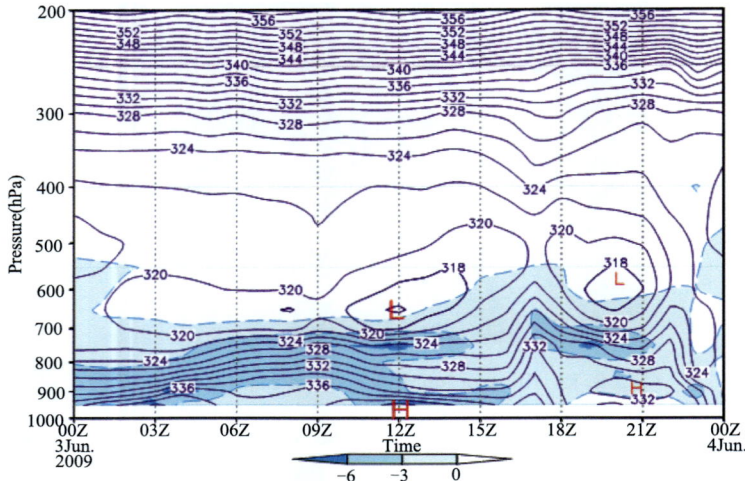

Fig. 8.21 The time-height (pressure) section of pseudo potential temperature Θ_{se} (solid lines, in unit: K) during 00Z Jun. 3—00Z Jun. 4, 2009 over the area near Shangqiu. Letters L and H indicate the low and high value of Θ_{se}, the shadow indicates the instability ($\Delta\theta/\Delta z$).